BIOCONTROL SYSTEMS AND PLANT PHYSIOLOGY IN MODERN AGRICULTURE

Processes, Strategies, Innovations

BIOCONTROL SYSTEMS AND PLANT PHYSIOLOGY IN MODERN AGRICULTURE

Processes, Strategies, Innovations

Edited by
Romeo Rojas, PhD
Guillermo Cristian Guadalupe Martínez Ávila, PhD
Juan Antonio Vidales Contreras, PhD
Cristóbal Noé Aguilar, PhD

AAP | APPLE ACADEMIC PRESS

First edition published 2023

Apple Academic Press Inc.
1265 Goldenrod Circle, NE,
Palm Bay, FL 32905 USA

760 Laurentian Drive, Unit 19,
Burlington, ON L7N 0A4, CANADA

CRC Press
6000 Broken Sound Parkway NW,
Suite 300, Boca Raton, FL 33487-2742 USA

4 Park Square, Milton Park,
Abingdon, Oxon, OX14 4RN UK

Library and Archives Canada Cataloguing in Publication

Title: Biocontrol systems and plant physiology in modern agriculture : processes, strategies, innovations / edited by Romeo Rojas, PhD, Guillermo Cristian Guadalupe Martínez Ávila, PhD, Juan Antonio Vidales Contreras, PhD, Cristóbal Noé Aguilar, PhD.
Names: Rojas, Romeo, editor. | Martínez Ávila, Guillermo Cristian Guadalupe, editor. | Vidales-Contreras, Juan Antonio, editor. | Aguilar, Cristóbal Noé, editor.
Description: First edition. | Includes bibliographical references and index.
Identifiers: Canadiana (print) 20220195226 | Canadiana (ebook) 20220195277 | ISBN 9781774639788 (hardcover) | ISBN 9781774639795 (softcover) | ISBN 9781003277118 (ebook)
Subjects: LCSH: Agricultural innovations. | LCSH: Plant physiology. | LCSH: Pests—Biological control.
Classification: LCC S494.5.I5 B56 2023 | DDC 338.1/6—dc23

Library of Congress Cataloging-in-Publication Data

....

CIP data on file with US Library of Congress

....

ISBN: 978-1-77463-978-8 (hbk)
ISBN: 978-1-77463-979-5 (pbk)
ISBN: 978-1-00327-711-8 (ebk)

About the Editors

Romeo Rojas, PhD

Coordinator of the Research Center and Development for Food Industries. Scientist, Universidad Autonoma de Nuevo Leon, School of Agronomy, Nuevo León, México.

Romeo Rojas, PhD, is Professor at the School of Agronomy, Autonomous University of Nuevo León, Mexico, as well as Coordinator of the Research Center and Development for Food Industries. His research and teaching are in food, biotechnology, and agricultural sciences. Dr. Rojas is currently working on several research and collaboration projects funded by the Agricultural Secretariat of Mexico, National Forestry Commission, and the National Mexican Research Council. He has published over 20 original research papers in indexed journals and 16 book chapters and has participated in and contributed to over 50 scientific meetings. He is a member of the Mexican Society of Biotechnology and Bioengineering (SMBB), the National System of Researchers (SNI) (since 2015 [level I]), and other organizations. He directed one PhD thesis, two MSc theses, and four BSc theses. Dr. Rojas became a Food Science and Technology Engineer (2007) from the Autonomous Agrarian University Antonio Narro, with specialization in agricultural sciences. His PhD degree in Food Science and Technology was awarded in 2013 by the Autonomous University of Coahuila, Mexico. He has had academic research stays in Portugal (Univeridade do Minho) and in Queretaro, México (Universidad Nacional Autónoma de Mexico).

Guillermo Cristian Guadalupe Martínez Ávila, PhD

Scientist, Universidad Autonoma de Nuevo Leon, School of Nuevo León, México.

Guillermo Cristian Guadalupe Martínez Ávila, PhD, is Professor at the School of Agronomy, Autonomous University of Nuevo León, Mexico, specializing in research on foods, biotechnology, and agricultural sciences. Dr. Ávila has published 24 original research papers in indexed journals and 10 book chapters and has participated in and contributed to over 50 scientific meetings. He was named as an outstanding reviewer for journals such as Food Research International, Heliyon, LWT-Food Science and Technology, etc. He is a member of the National System of Researchers (since 2012 [level 1]), Mexican Society of Biotechnology and Bioengineering, and Mexican Association of Food Chemistry. He has been a thesis director of two MSc theses and three BSc theses. Dr. Martínez has collaborated and worked on several research projects funded by the Agricultural Secretariat of Mexico, National Forestry Commission and the National Mexican Research Council (SAGARPA-CONACYT; CONAFOR-CONACYT). He became a Chemical Pharmacobiologist (2005) at the Autonomous University of Coahuila, Mexico, and holds a specialization in food sciences. His PhD degree in Biotechnology was awarded in 2011 by the Autonomous University of Coahuila. He has been in two academic stays in Ireland (University College Cork) and in Fortaleza, Brazil (Universidade de Fortaleza).

Juan Antonio Vidales Contreras, PhD

Scientist, Universidad Autonoma de Nuevo Leon,
School of Agronomy, Nuevo León, México.

Juan Antonio Vidales Contreras, PhD, has been a Professor since 1985 at the Agronomy School, Autonomous University of Nuevo Leon, Mexico, specializing in the field of environmental science and engineering. Currently, he is head of the graduate and research office. Dr. Vidales Contreras currently has developed more than 10 research projects funded by diverse local and national institutions. He has published over 58 original research papers on wastewater treatment and agricultural systems in indexed journals and three book chapters and has participated in and contributed to over 20 scientific meetings. He is a member of the CONACYT National System of Researchers and has desirable PRODEP profile desirable from the Mexican Education Agency. Dr. Vidales Contreras has worked on more than 10 research projects funded by diverse local and national institutions. He has been a thesis director of one PhD dissertation and four and five MSc and BSc theses, respectively. He became an Agronomy Engineer (1984) in the Autonomous University of Nuevo Leon (UANL), with a specialization in groundwater hydrology. He was awarded a PhD in Soil Water and Environmental Science from the Autonomous University of Nuevo Leon and University of Arizona.

Cristóbal Noé Aguilar, PhD

Scientist, Department of Food Research, School of Chemistry,
Universidad Autónoma de Coahuila, Saltillo, Coahuila, Mexico.

Cristóbal Noé Aguilar, PhD, is Director of Research and Postgraduate Programs at the Universidad Autonoma de Coahuila, Mexico. He is also a member of the Bioprocesses and Bioproducts Research Group and Professor in the Food Research Department in the School of Chemistry at the Universidad Autónoma de Coahuila, Saltillo, Mexico. Dr. Aguilar is Associate Editor of *Heliyon* (Microbiology) and *Frontiers in Sustainable Food Systems* (Food Processing) and has published more than 330 papers in indexed journals, more than 40 articles in Mexican journals, and 250 contributions in scientific meetings. He has also published many book chapters, several Mexican books, four editions of international books, and more. He has been awarded several prizes and awards, the most important of which are the National Prize of Research 2010 from the Mexican Academy of Sciences; the Prize "Carlos Casas Campillo 2008" from the Mexican Society of Biotechnology and Bioengineering; National Prize AgroBio–2005; and the Mexican Prize in Food Science and Technology. Dr. Aguilar is a member of the Mexican Academy of Science, the International Bioprocessing Association, Mexican Academy of Sciences, Mexican Society for Biotechnology and Bioengineering, and the Mexican Association for Food Science and Biotechnology. He has developed more than 21 research projects, including six international exchange projects.

Contents

Contributors

Victor H. Avendaño Abarca
Universidad Autonoma de Nuevo Leon, School of Agronomy. 66054, General Escobedo, Nuevo León, México

Pedro Aguilar-Zárate
Departamento de Ingenierías. Tecnológico Nacional de México, Campus Ciudad Valles. Carretera al Ingenio Plan de Ayala Km. 2, Colonia Vista Hermosa, Ciudad Valles, San Luis Potosí. C. P. 79010

Naomi Gabriela Álvarez-Díaz
Departamento de Ingenierías. Tecnológico Nacional de México, Campus Ciudad Valles. Carretera al Ingenio Plan de Ayala Km. 2, Colonia Vista Hermosa, Ciudad Valles, San Luis Potosí. C. P. 79010

Fernanda Andrade-Damián
Departamento de Ingenierías. Tecnológico Nacional de México, Campus Ciudad Valles. Carretera al Ingenio Plan de Ayala Km. 2, Colonia Vista Hermosa, Ciudad Valles, San Luis Potosí. C. P. 79010

Juana Aranda-Ruíz
Universidad Autonoma de Nuevo Leon, School of Agronomy. 66054, General Escobedo, Nuevo León, México

Cesar de Jesus Ayala-Meza
Universidad Autonoma de Nuevo Leon, School of Agronomy. 66054, General Escobedo, Nuevo León, México

Israel Bautista-Hernández
Universidad Autonoma de Nuevo Leon, School of Agronomy. 66054, General Escobedo, Nuevo León, México

Rafael Germán Campos-Montiel
Autonomous University of the State of Hidalgo. Institute of Agricultural Sciences. College Ranch. Av. Universidad Km 1 Ex-Hacienda de Aquetzalpa A. P. 32. C. P. 43600. Tulancingo, Hidalgo

María Luisa Carrillo-Inungaray
Autonomous University of San Luis Potosi. Multidisciplinary Academic Unit, Huasteca Zone. Romualdo del Campo No. 501. Fracc. Rafael Curiel. C. P. 79060. Ciudad Valles, S. L. P., Mexico

Cecilia Castro-López
Laboratorio de Química y Biotecnología de Productos Lácteos, Centro de Investigación en Alimentación y Desarrollo A. C. (CIAD, A. C.), Gustavo Enrique Astiazarán Rosas 46, Hermosillo, Sonora 83304, Mexico

José Sandoval Cortés
Universidad Autónoma de Coahuila, School of Chemistry, Department of Food Science and Technology, Blvd. V. Carranza, 25280, Saltillo, Coahuila, México

M. Mousumi Das
Centre for Bio-innovation and Product Development, Department of Biotechnology and Microbiology, Kannur University, Dr. Janaki Ammal Campus, Thalassery, Kannur 670661, Kerala, India

Marisol Galicia-Juárez
Universidad Autonoma de Nuevo Leon, School of Agronomy. 66054, General Escobedo,
Nuevo León, México

Nateily Gallo de la Paz
Universidad Autonoma de Nuevo Leon, School of Agronomy. 66054, General Escobedo,
Nuevo León, México

Celestino García-Gómez
Universidad Autonoma de Nuevo Leon, School of Agronomy. 66054, General Escobedo,
Nuevo León, México

Donaji Josefina González-Mille
Universidad Autonoma de Nuevo Leon, School of Agronomy. 66054, General Escobedo,
Nuevo León, México

Dulce Concepción González-Sandoval
Universidad Autonoma de Nuevo Leon, School of Agronomy. 66054, General Escobedo,
Nuevo León, México

M. Haridas
Centre for Bio-innovation and Product Development, Department of Biotechnology and Microbiology,
Kannur University, Dr. Janaki Ammal Campus, Thalassery, Kannur 670661, Kerala, India

Juan Pablo Hernández-Rodríguez
UANL, Facultad de Ciencias Biológicas, Av. Pedro de Alba S/N, Ciudad Universitaria,
San Nicolás de los Garza, N. L. México

Raul Rodriguez Herrera
Facultad de Ciencias Químicas, Universidad Autónoma de Coahuila, Blvd. V. Carranza y Jose,
Cardenas Valdez s/n, Col. Republica Ote. 25280 Saltillo Coahuila, Mexico

U. Aranda Lara
INIFAP-Campo Experimental Río Bravo, 88900. Río Bravo, Tamaulipas, México

Luis Enrique Ordóñez López
Universidad Autonoma de Nuevo Leon, School of Agronomy. 66054, General Escobedo,
Nuevo León, México

Araceli Loredo-Treviño
Universidad Autónoma de Coahuila, School of Chemistry, Department of Food Science and Technology,
Blvd. V. Carranza, 25280, Saltillo, Coahuila, México

Alejandro Isabel Luna-Maldonado
Universidad Autonoma de Nuevo Leon, School of Agronomy. 66054, General Escobedo,
Nuevo León, México

Brenda Luna-Sosa
Universidad Autonoma de Nuevo Leon, School of Agronomy. 66054, General Escobedo,
Nuevo León, México

Julia Mariana Márquez-Reyes
Universidad Autonoma de Nuevo Leon, School of Agronomy. 66054, General Escobedo,
Nuevo León, México

C. Martínez-Ávila
Universidad Autonoma de Nuevo Leon, School of Agronomy. 66054, General Escobedo,
Nuevo León, México

Guillermo Cristian Guadalupe Martínez-Ávila
Universidad Autonoma de Nuevo Leon, School of Agronomy. 66054, General Escobedo,
Nuevo León, México

C. G. Guillermo Martínez-Ávila
Universidad Autonoma de Nuevo Leon, Research Center and Development for Food Industries,
School of Agronomy, 66050 General Escobedo, Nuevo León, México

Mariela R. Michel
Departamento de Ingenierías. Tecnológico Nacional de México, Campus Ciudad Valles. Carretera al
Ingenio Plan de Ayala Km. 2, Colonia Vista Hermosa, Ciudad Valles, San Luis Potosí. C. P. 79010

Diana Beatríz Muñiz-Márquez
Departamento de Ingenierías. Tecnológico Nacional de México, Campus Ciudad Valles. Carretera al
Ingenio Plan de Ayala Km. 2, Colonia Vista Hermosa, Ciudad Valles, San Luis Potosí. C. P. 79010

Juan Napoles-Armenta
Universidad Autonoma de Nuevo Leon, School of Agronomy. 66054, General Escobedo,
Nuevo León, México

Antonio Flores Naveda
Centro de Capacitación y Desarrollo en Tecnología de Semillas, Universidad Autónoma Agraria
Antonio Narro, 25315, Buenavista, Saltillo, Coahuila, México

Guillermo Niño-Medina
Universidad Autonoma de Nuevo Leon, School of Agronomy. 66054, General Escobedo,
Nuevo León, México

Orquídea Pérez-González
UANL, Facultad de Ciencias Biológicas, Av. Pedro de Alba S/N, Ciudad Universitaria,
San Nicolás de los Garza, N. L. México

Diana Jaqueline Pimentel-González
Autonomous University of the State of Hidalgo. Institute of Agricultural Sciences. College Ranch. Av.
Universidad Km 1 Ex-Hacienda de Aquetzalpa A. P. 32. C. P. 43600. Tulancingo, Hidalgo

Abigail Reyes-Munguía
Autonomous University of San Luis Potosi. Multidisciplinary Academic Unit, Huasteca Zone.
Romualdo del Campo No. 501. Fracc. Rafael Curiel. C. P. 79060. Ciudad Valles, S. L. P., Mexico

B. Rincón-López
Programa de Maestría en Ciencias en Fitomejoramiento,
Universidad Autónoma Agraria Antonio Narro, 25315, Buenavista, Saltillo, Coahuila, México

Humberto Rodríguez-Fuentes
Universidad Autonoma de Nuevo Leon, School of Agronomy. 66054, General Escobedo,
Nuevo León, México

Romeo Rojas
Universidad Autonoma de Nuevo Leon, School of Agronomy. 66054, General Escobedo,
Nuevo León, México

Hector Arturo Ruiz Leza
Biorefinery Group, Food Research Department, Faculty of Chemistry Sciences,
Autonomous University of Coahuila. Saltillo, Coahuila. Mexico

A. Sabu
Centre for Bio-innovation and Product Development, Department of Biotechnology and Microbiology,
Kannur University, Dr. Janaki Ammal Campus, Thalassery, Kannur 670661, Kerala, India

Saúl Saucedo-Pompa
Universidad Autónoma de Coahuila, School of Chemistry, Department of Food Science and Technology,
Blvd. V. Carranza, 25280, Saltillo, Coahuila, México

Julio César Tafolla-Arellano
Departamento de Ciencias Básicas, Laboratorio de Biotecnología y Biología Molecular,
Universidad Autónoma Agraria Antonio Narro, 25315, Buenavista, Saltillo, Coahuila, México

Martin E. Tiznado Hernández
Research Center in Food and Development A. C. 83304, Hermosillo, Sonora, México

Milton Torres-Cerón
Department of Ecology and Conservation Biology, Texas A&M AgriLife, Texas A&M University,
534 John Kimbrough Blvd., Room 214, Building 1537, 2258 TAMU, College Station,
Texas 77843-2258

Mayra Treviño-Garza
Universidad Autónoma de Nuevo León, Research Center and Development for Biological Sciences,
School of Biological Sciences. Pedro de Alba, Niños Héroes, Ciudad Universitaria,
San Nicolás de los Garza, N. L., México

Mireya Vázquez-Aguilar
Universidad Autonoma de Nuevo Leon, School of Agronomy. 66054, General Escobedo,
Nuevo León, México

Fabiola Veana-Hernández
Departamento de Ingenierías. Tecnológico Nacional de México, Campus Ciudad Valles.
Carretera al Ingenio Plan de Ayala Km. 2, Colonia Vista Hermosa, Ciudad Valles,
San Luis Potosí. C. P. 79010

Janeth Margarita Ventura-Sobrevilla
Universidad Autónoma de Coahuila, School of Chemistry, Department of Food Science and Technology,
Blvd. V. Carranza, 25280, Saltillo, Coahuila, México

Juan Antonio Vidales-Contreras
Universidad Autonoma de Nuevo Leon, School of Agronomy. 66054, General Escobedo,
Nuevo León, México

Jorge Enrique Wong-Paz
Departamento de Ingenierías. Tecnológico Nacional de México, Campus Ciudad Valles.
Carretera al Ingenio Plan de Ayala Km. 2, Colonia Vista Hermosa, Ciudad Valles,
San Luis Potosí. C. P. 79010

Francisco Zavala-García
Universidad Autonoma de Nuevo Leon, School of Agronomy. 66054, General Escobedo,
Nuevo León, México

Abbreviations

BCAs	bio-control agents
BHA	butylated hydroxyanisole
BHT	butylated hydroxytoluene
CCD	central composite design
DLI	daily light integral
EC	electrical conductivity
EW	epicuticular waxes
ES	effective salinity
FYM	farm yard manure
GC-MS	gas chromatography-mass spectrometry
GLS	glucosinolates
HD	hydrodistillation
HPβCD	hydroxypropyl-β-cyclodextrin
HPLC	high-performance liquid chromatography
HPS	high pressure sodium
HRT	hydraulic residence time
IPM	integrated pest management
LC-ESI-MS/MS	liquid chromatography–electrospray–tandem mass spectrometry
LED	light-emitting diodes
LID	daily integral light
MCM	modified czapec medium
MRSA	meticillin-resistant Staphylococcus aureus
NBS-LRR	nucleotide binding site and leucine rich repeats
NFT	nutrient film technique
NS	nutrient solution
PDA	potato dextrose agar
PF	plant factory
PFAL	plant factory with artificial lighting systems
PFPS	plant factory production system
PS	potential salinity
PSII	photosystem II
RH	relative humidity
RKN	root-knot nematode

RLNs	root-lesion nematodes
ROS	reactive oxygen species
RSC	residual sodium carbonate
RSM	response surface methodology
SAR	sodium adsorption ratio
SCFE	Extracción Mediante Fluidos Supercríticos
SFME	solvent-free microwave extraction
SmF	submerged fermentation
SSF	solid-state fermentation
SSR	simple sequence repeats
WTR	waste rubber tire
UAAAN	Universidad Autonoma Agraria Antonio Narro
UV	ultraviolet radiation
VMHD	vacuum microwave hydrodistillation

Preface

In recent years, interest in the consumption of natural, organic, pesticide-free, and healthy foods has increased. For this, there are production alternatives that do not include pesticides, herbicides, and chemicals for primary food production, specifically, the use of biological control, controlled systems of production, and the study of the physiology of plants to know their resistance to different environments in modern agriculture. This can contribute to organic agriculture and green and sustainable technologies. The focus of biocontrol systems is to reduce or eliminate the use of agrochemicals by controlling plant diseases by minimizing environmental damage through the use of antagonistic organisms. In addition, plant genes and/or products are also used as metabolites to reduce the negative effects of plant pathogens and promote positive response in plants. The International Biocontrol Manufacturers' Association defines biocontrol as the use of agents or products that naturally affect pathogens by limiting their spread. The main agents used are microorganisms, chemical mediators, and natural chemicals. It is also environment friendly.

New strategies of cultivation that maximize production are optimizing light, temperature, humidity, nutrients, and humidity in a controlled environment. This allows for highly optimized environmental control and more precise production accuracy through active monitoring than traditional greenhouses. In summary, the new processes and innovations on agricultural technology identify the optimal growth points in closed areas using light sources such as LEDs and artificially control the growth environment that allows crops throughout the year, eliminating limitations such as climate, pollution, or geographical association. Both processes maximize the benefits.

The physiology of plants for adaptation to different intensive production systems is a necessity due to the high incidence and resistance of crops to pests and diseases. By knowing their physiology, it is possible to know, understand, and establish the climatic conditions in an open and closed field for optimal growth. With this, knowledge of the development and/or the appearance of secondary metabolites as a defense mechanism for plants is essential for understanding their physiology, as well as for identifying the adaptation of crops to different areas.

All the topics included in this book are part of the consolidated work of 40 years of postgraduate experience in agricultural sciences. It includes the most recent advances innovations in the use of agricultural technologies that are a viable alternative for the control of pests, food production, environment friendly by which costs are reduced and that improve the quality of products.

PART I

Biocontrol Systems: Processes and Strategies in Modern Agriculture

CHAPTER 1

Botanical Compounds as Adjuvants: An Alternative to Reduce the Pesticide Use

SAÚL SAUCEDO-POMPA[1],
GUILLERMO CRISTIAN GUADALUPE MARTÍNEZ-ÁVILA[2],
JOSÉ SANDOVAL CORTÉS[1], ARACELI LOREDO-TREVIÑO[1], and
JANETH MARGARITA VENTURA-SOBREVILLA[1*]

[1]*Department of Food Science and Technology, Universidad Autónoma de Coahuila, School of Chemistry, Blvd. V. Carranza, 25280, Saltillo, Coahuila, México*

[2]*Universidad Autónoma de Nuevo León, Facultad de Agronomía, Francisco Villa s/n. Col. Ex Hacienda El Canadá, 66050, General Escobedo, Nuevo León, Mexico*

Corresponding author. E-mail: janethventura@uadec.edu.mx

ABSTRACT

In the formulation of agrochemicals, there are various types of components: active and inert ingredients, additives, and adjuvants. The adjuvants are intended to improve the activity of the agrochemical or facilitate its application. There are also products that are only adjuvants, which are mixed with pesticides to potentiate their action: conditioning the irrigation water to avoid its degradation when mixed, protecting the active ingredient from the climatic environment, or improving the wetting of the solution to be applied when coming into contact with the leaf surface of the plant. At present, most of the agricultural adjuvants used are of synthetic origin; however, there are materials of natural origin that can be used as agricultural adjuvants, with the advantage of being biodegradable materials and that do not cause damage to the biological environment where these treatments are applied. There are from biopolymers, waxes, and oils of natural origin that have a great potential for its use as an agricultural adjuvant agent.

1.1 INTRODUCTION

The trends of the agrochemical market present a growing demand for increasingly selective, less toxic, and biodegradable products; there are organic products with a low environmental impact that are made from raw materials of natural origin and low impact on the environment and human health; these organic agrochemicals are intended for the replacement of synthetic chemicals. There is another type of product that are called adjuvants, and this family of products is intended not to replace synthetic chemicals but rather to potentiate their effect and therefore reduce application dose. The adjuvants are tools that can improve the effectiveness of agrochemical products, improve and facilitate their characteristics, maximizing the effect of the products to be used. The properties of an adjuvant determine its functionality and these in turn are determined by the design and characteristics of the formulation. The use of adjuvants offers economic benefits and the environment due to the possibility of potential action of the active principles to be applied. The type of adjuvants that may be needed depends on the product to be used, the crop, the pest, and the environmental conditions. In response to this question, we need to consider the characteristics of the target pest, the chemical and biological nature of the agrochemical, and the properties of the adjuvant.

These products may or may not be present in the original formulation of the agrochemical to be applied and even when they are present, the combination and proportion in which they are found cannot contemplate all situations of application. This usually means that the spray solution must be supplemented with specific additives to optimize each particular situation.

1.1.1 ADJUVANTS

Adjuvants are chemical ingredients that are added or mixed with pesticides to improve their effectiveness, increase their stability, and facilitate the application process (Mullin et al., 2015), They are added to herbicides, insecticides, fungicides, and other pesticide formulations (including bactericides), and they are classified as humectants, emulsifiers, dispersants, adherents, buffers, antifoam, drying retarder, etc. The adjuvants are much lower costs than pesticides; however, when used properly it is possible to reduce the application rate of conventional pesticides up to 10 times, that is to say they manage to reduce the dose of application and therefore the cost of pest control (Stark and Walthall, 2003).

Generally, adjuvants are classified as inert inputs in agricultural applications and are generally not included in the risk assessments required to register pesticides (Markets and Market, 2018). However, some conventional adjuvants have certain levels of toxicity, and some even become more harmful than the active ingredients (Li et al., 2017). Faced with this problem, there are adjuvants of natural origin, which can also reduce the dose of application of pesticides and on the other hand are much less toxic and harmful to the environment due to their chemical nature and origin (Zhu et al., 2014). They also have the advantage that due to their chemical nature they tend to fulfill their function and begin a process of degradation in the face of the environmental conditions of agricultural crops (Koul et al., 2008), so they do not present the problem of other adjuvants agents that tend to accumulate in the cultivation soils and generate havoc in the long term (Oancea et al., 2017).

1.2 TYPES OF ADJUVANTS

The adjuvants are chemical products that have one or more surfactant properties, adherents, humectants (surfactants), pH correctors, sequestrants, antiderivatives, etc. Some of them simultaneously fulfill several of these functions and are compatible with the different types of pesticides.

1.2.1 EMUSIFIERS

The emulsifiers stabilize the application mixtures of incompatible compounds in water, helping to form an emulsion and stabilizing the product to be applied. They have a molecule with a highly polar part that attracts water and a nonpolar that attracts oil, grease, or wax (Rosen, 2012). This type of material is mainly used in insecticide products that are sold in oil insoluble in water, which is necessary to emulsify. They can also be used to improve oil-based agrochemical products that require improving their emulsifying properties (Hernández-Tenorio and Orozco-Sánchez, 2020).

1.2.2 ADHERENTS

They adhere drops of water that are sprinkled on the foliar surface of the plants or insects so that the treatment that is applied remains adhered more time, and avoid that they are washed by the rain or water of irrigation. There are different types of materials that provide adherent capacity, among the most used are fatty acids, which are hydrolyzed on the leaves forming a layer that acts as a protector against washing the product. They also have an adhesive effect on certain insect pests which increases the efficacy of the insecticide treatments to be applied (Barrenechea et al., 2004; Kala et al., 2020). Other materials used are emulsifiable polyethylene that forms films that are resistant to water and do not affect the transpiration of the leaves; some colloidal aluminum and magnesium silicates that form a layer on the leaves and when dried retain the product. Other common adherent substances are albumins, caseinates, gums, mucilages, and starches that encapsulate and adhere to the treatments to be applied (Wang et al., 2020a).

1.2.3 PENETRATING

They favor the absorption or penetration of certain products in plant tissues; these substances can disturb the waxy cuticle of the leaf, so the entry of the agrochemical occurs faster, also by decreasing the surface tension, the sprinkled drops more easily reach the stomata and the capillary movement of the liquid through these tissues is greater. Certain substances can act as a mobilizer of active plant ingredients; this action being independent of the moisturizing effect (Peroukidis et al., 2020).

1.2.4 MOISTURIZERS

Its main mode of action is to reduce the surface tension of the spray drops. The surfactants are classified by the ionic charge of the active part of the molecule as ionic, nonionic, and ambivalent or amphoteric. Nonionic surfactants are the most commonly used in foliar applications and the most commonly used in mixtures with pesticides are organo-siliconized ones (Kemmitt et al., 2006). Its use is recommended for phytosanitary products with contact action; they can be of natural or synthetic origin and are designed to delay the evaporation of water in the air or in the leaf, that is, these compounds reduce the drying process of the solution and retard the crystallization of the pesticide, and the active ingredient to remain longer in liquid form on the sheet greater penetration opportunity will have. They are useful adjuvants in crops established in arid or semi-arid zones (Rosen, 2012).

1.2.5 ACIDIFIERS AND BUFFERS OR BUFFER

pH adjusters are adjuvants that maintain or prolong the stability of a pesticide by modifying the final pH of the spray solution. Most pH adjusters are acidifiers that reduce pH and adjuvants that increase pH are less common. The buffer and the acidifier are often used interchangeably, but they are very different. Water has little or no buffering capacity, so a small amount of a pH adjuster can significantly alter the pH of the spray solution. An acidifier will lower the pH, but the pH of the resulting solution is still sensitive to any other compound that can be added to it. To some extent, a buffer can maintain the pH of the solution at a constant level even when more products are added; these are useful when using pH-sensitive pesticides. For most pesticides (herbicides, insecticides, and fungicides), the optimum pH range in water before mixing should be 5.5–6.5, while at a pH above 7.0 the molecules are at risk of degradation and thus the loss of their biological effectiveness (Bogdanov, 2004).

1.3 MARKET OF AGRICULTURAL ADJUVANTS

The global adjuvant market grows periodically, for 2018 it was estimated at 3.13 billion dollars, by 2023 it is projected to reach a value of 4.04 billion with an annual growth rate of 5.24% during this period (https://www.epa.gov/pesticide-registration/conventional-reduced-risk-pesticide-program#howapply, 2018). This market is driven by the growing demand by government regulatory agencies for the reduction of pesticide products because it is expected to decrease the growth of resistance of pests, the impact on human health, the damage to nontarget organisms (birds, fish, plants, and beneficial insects), groundwater contamination, and amount of pesticide residues in food (EPA, United States Environmental Protection Agency, 2018), in addition to the growing adoption of integrated pest management (https://www.epa.gov/pesticide-registration/conventional-reduced-risk-pesticide-program#howapply, 2018).

Generally, the adjuvant manufacturing companies do not reveal their composition; this is due to the need to protect their information and prevent their products from being copied without any compensation; generally, they are only mentioned in the formulas as conditioning additives.

1.3.1 MARKET OF AGRICULTURAL ADJUVANTS BY APPLICATION

According to its application, the market of agricultural adjuvants is segmented into herbicides, insecticides, fungicides, and others. Since herbicides are the largest area of application, occupying more than 50% of the market of agricultural adjuvants (Global Market Insights, 2018), this segment is of great interest because if it is not controlled properly the growth of weeds can reduce the quality and yield of the crop agricultural objective. The weeds are covered by a membrane called the cuticle, which is a barrier that prevents an herbicide from having an effective effect, and the composition of this cuticle is very diverse since it varies according to the species of plants (Curran et al., 1999). It is generally composed of waxes, cutin, and pectins (Bourgault et al., 2020). The type of adjuvants most used in combination with herbicides are surfactants, waxes, and salts; these adjuvants help to penetrate the cuticle of plants and potentiate the effect of herbicides (Curran et al., 1999).

The second segment of the market of agricultural adjuvants are insecticides, since insects are one of the main pests that affect crops because they feed directly from crops or transmit various forms of bacterial infection to the crop, and adjuvants modify the properties and ability of insecticides to penetrate and attack the target pest (Global Market Insights, 2018).

1.3.2   *MARKET OF AGRICULTURAL ADJUVANTS, BY REGION*

The Asia Pacific region consumes most of the global market of agricultural adjuvants (Global Market Insights, 2018). This is due to the increase of the population in these regions which increases the demand for food crops. Another factor is the limited availability of arable land of which its use must be optimized (FAO, Food and Agriculture Organization, 2018). In this region, the market consumed around 950 million dollars in 2017. The second market of agricultural adjuvants is Europe, presenting a growing demand in countries like the United Kingdom, Germany, Spain, and France; the high demand for this type of product in these regions is due to the growing need for sustainable agricultural techniques (Global Market Insights, 2018).

1.4   USE OF MATERIALS OF NATURAL ORIGIN AS AGRICULTURAL ADJUVANTS

At the moment, the most of the adjuvant products that are in the market are of the synthetic base, but in nature different materials that have a great potential to replace these compounds exist (Table 1.1). In all areas, gums and waxes of natural origin that can be used as adherent agents, fatty acids of natural oils that can act as penetrants, botanical surfactants that act as a surfactant or

TABLE 1.1 Classification and Characteristics of Natural Materials with Potential as Agricultural Adjuvants.

Classification	Material	Characteristics	References
Natural Waxes	Beeswax	It will be based on saturated and unsaturated monoesters, saturated and unsaturated hydrocarbons, and hydroxypolyesters. Beeswax is a protector of solar radiation. It has great benefits as an ingredient encapsulant and protector, adherent, moisture barrier, and sunscreen.	Jiménez et al. (2004), Vandenburg and Wilder (1970), Fabra et al. (2009), and Milanovic et al. (2010)
	Carnauba wax	It is composed of diesters of hydroxycarboxylic acids and fatty acid alcohols, and its characteristic component is methoxycinnamic acid. It is used in formulations as a carrier and volume agent, as well as its compatibility in the mixtures, it is an excellent encapsulating agent and emulsion stabilizer.	Chen et al. (2020), Kowalczyk and Baraniak (2014), and Mellema et al. (2006)
	Candelilla wax	It is composed of esters of fatty acids with alcohols of high molecular weight. It is used like coating or substance transport agent, has a high potential as an emulsifier and formula stabilizer. It has properties as encapsulating agent, adherent, and moisture barrier.	Cabello Alvarado et al. (2013), Tadesse et al. (2020), and Yadav et al. (2007)
Biopolymers	Arabic gum	It is composed of D-galactose, L-arabinose, L-rhamnose, D-glucuronic, and 4-O-methyl-D-glucuronic acid mainly. It is widely used for its ability to form stable emulsions at low temperatures. It is an excellent encapsulating agent.	Niu et al. (2018), Garcia-Ochoa et al. (2000), and Vidhyalakshmi et al. (2009)
	Xanthan gum	It is composed of pentasaccharide repeat units, comprising glucose, mannose, and glucuronic acid. It is used as emulsion stabilizer is very compatible with other biopolymers. It is a stabilizing, adherent, and encapsulating material.	Mudgil et al. (2014) and Kumbar et al. (2001)
	Guar gum	It is an exo-polysaccharide composed of the sugars galactose and mannose. It produces very viscous solutions, is a stabilizer in suspensions and foams, encapsulation in adherent.	Sajilata et al. (2006) and Esfanjani et al. (2015)
	Starch	It is a mixture of two very similar polysaccharides, amylose, and amylopectin, has properties like adhesive, binder, clouding agent, film former, foam stabilizer, anti-aging bread agent, gelling agent, glazing agent, humectant, stabilizer, texturizer, and thickener.	Wang et al. (2020b) and Esfanjani et al. (2015)

TABLE 1.1 *(Continued)*

Classification	Material	Characteristics	References
	Pectins	Pectins have three main domains: homogalacturonans, rhamnogalacturonan I, and rhamnogalacturonan II. It has excellent properties as an encapsulant and ingredient hauling agent. It has excellent properties as an encapsulating agent as it is an agent with a high level of controlled release of ingredients.	Novoa and Castillo (1998) and Nesterenko et al. (2013)
Proteins	Vegetable proteins	It has properties as ingredient encapsulator; they are an option for the substitution of proteins of animal origin and petroleum derivatives as encapsulating agents.	Chambi and Grosso (2006) and Jafari et al. (2008b)
	Proteins of animal origin	Casein and collagen are the main proteins of animal origin used as adjuvants; they are very effective encapsulating agents.	Guamán et al. (1996)
Natural oils	Soy oil	Soy oil is abundant in polyunsaturated fatty acids, such as Omega 3 and Omega 6. It has a high potential as an encapsulating agent and also has an effect as an insecticide, so it can potentially potentiate agricultural insecticide products.	Ogunniyi (2006) and Kaur and Bhaskar (2020)
	Castor oil	About 90% of this oil is composed of ricinoleic acid. It also contains a huge amount of vitamin E, proteins, minerals, and has insecticide, antifungal, anti-inflammatory, and antibacterial properties.	Rubio-Rodríguez et al. (2012), Jafari et al. (2008a), and Ozyurt and Ötles (2014)
	Fish Oil	Fish oils are oils obtained from the tissues of some fish species. They are mainly composed of fatty acid methyl ester, saturated fatty acid, monounsaturated fatty acid, polyunsaturated fatty acid, and polyunsaturated fatty acid. They are widely used for their encapsulating and adherent capacity.	Shasha and McGuire (1999) and Haidar et al. (2020)
Saponins	Saponins	Chemically, they are glycosides with a polycyclic aglycone, and they are chemical compounds with detergent properties. Due to their nature, they can be used to improve the applications of agrochemicals improving applications in terms of surface function.	Augustin et al. (2011) and Mao et al. (2012)

humectant, in addition to acidifying materials extracted from plants, which can regulate the pH of irrigation waters. In this review, we mention the main natural adjuvants with potential to be exploited.

1.4.1 NATURAL WAXES

There are three types of natural waxes that dominate the market due to their functionality, beeswax, carnauba wax, and candelilla wax. These materials have the capacity as an adherent and emulsifying agent since they can encapsulate the active ingredient to be applied and adhere it on the foliar surface of the plants and protect it from the environment; besides these waxes also have properties as emulsifiers because they manage to stabilize and improve an agricultural emulsion. Here, we describe the three main ones.

1.4.1.1 BEESWAX

The beeswax is the material of which the bee panels are composed. Bees secrete a liquid through their cherry glands that harden in contact with air. A small flake forms at the bottom of the bee that will be deposited to form the bee panels. One million of these flakes form approximately 1 kg of beeswax, bees make the hexagonal alveoli from this wax. These hardened alveoli form the panels of bees and will protect honey and pollen for bees. The queen bee deposits eggs in the wax panel, and the young grow inside. There are different types of beeswax according to the type of bee that creates the wax (Bonaduce and Colombini, 2004). Beeswax is a substance that is composed of 284 different components. It will be based on saturated and unsaturated monoesters, saturated and unsaturated hydrocarbons, and hydroxypolyesters (Jiménez et al., 2004).

Beeswax has multiple applications in the industry, being its main applications in the pharmaceutical, cosmetology, and food industries, due to its chemical composition, insolubility in water, and effect as a protector of solar radiation, it has great benefits as an ingredient encapsulant and protector (Vandenburg and Wilder, 1970; Fabra et al., 2009). As an adjuvant it is very appreciated because it is easy to disperse in irrigation waters, has a high capacity as emulsifier, and represents a barrier to moisture which ensures the integrity and adherence of the pesticides to be protected (Milanovic et al., 2010).

1.4.1.2   CARNAUBA WAX

Carnauba wax is a material that is extracted from the *Copernica cerifera* palm (or also *Copernicia prunifera*). It is used in the food industry as a shining agent for products as well as a carrier and volume agent. It is composed of diesters of hydroxycarboxylic acids and fatty acid alcohols; one of its characteristic components is methoxycinnamic acid (Chen et al., 2020). This wax is obtained from the leaves of the *Cerifera copernicia*, a palm tree that grows in different parts of the world, but that only produces wax in five or six states of the semiarid northeast part of Brazil. Its high melting point, its hardness, its ability to give a luster of great quality and duration, as well as its compatibility in the mixtures make the carnauba the most important wax of the vegetable group (Kowalczyk and Baraniak, 2014).

Carnauba wax is classified by its degree of refining since this gives added value to the product is a classification that has been adopted in the wax industries. It is a good encapsulating agent and has excellent adhesion properties (Mellema et al., 2006).

1.4.1.3   CANDELILLA WAX

Candelilla is an herb that belongs to the Euphorbiaceae family, wild in the dry regions of northern Mexico and southern Texas. Candelilla wax is chemically composed of esters of fatty acids with alcohols of high molecular weight; molecules that are obtained by a chemical reaction between an acid and an alcohol. In its raw form, the candelilla wax is brown and changes to a yellow color; once refined, it is of undefined structure and its hardness is of an intermediate degree between that of carnauba wax and that of bees. The proportion of its components determines the hardness, impermeability to water, brightness, and other characteristic characteristics of the wax as its melting temperature (Cabello Alvarado et al., 2013).

Candelilla wax is one of the most used natural products in the industry and is an ingredient used for its characteristics as a shining agent, coating agent, or substance transport agent. There are other applications that include coatings for cardboard and paper, crayons, paints, wax candles, lubricants, adhesives, anticorrosives, drugs, inks, and more (Tadesse et al., 2020). As an adjuvant agent, it represents great benefits as it is a very stable encapsulating agent, in addition to preserving properties as an emulsifying and stabilizing agent (Yadav et al., 2007).

1.4.2 BIOPOLYMERS

There is a wide range of biopolymers of natural origin that can be used as encapsulating agents protective of the active ingredients, also that has great potential as an adherent agricultural agent, due to its characteristics and chemical composition. The sources of these materials can vary, since their origin can come from both botanical and animal sources, the main natural biopolymers with high potential as agricultural support agents are described below.

1.4.2.1 ARABIC GUM

The Arabic gum is considered as a polysaccharide of natural origin, which is produced as a resinous secretion that is known as "gummosis"; it appears on lesions and cracks in the bark of trees such as *Acacia senegal* and *Acacia seyal*. The Arabic gum appears in the trees with the sole purpose of protecting them against the occurrence of diseases and the spread of pests. In this way, if the bark of a tree is cut, the gum will be produced to close the holes (Niu et al., 2018).

This resin known as Arabic gum is generally amber colored and is normally collected by hand once dried. Arabic gum is composed of different polysaccharides and acids, for example, D-galactose, L-arabinose, L-rhamnose, D-glucuronic, or 4-O-methyl-D-glucuronic acid. Arabic Gum is easily soluble in water at room temperature but is insoluble in alcohol (Garcia-Ochoa et al., 2000). Due to its stability, it is an excellent encapsulating and adherent agent (Vidhyalakshmi et al., 2009).

1.4.2.2 XANTHAN GUM

It is a relatively recent product and used only since 1969. It was developed in the United States as part of a program to look for new applications of corn, since it is produced by fermentation of sugar, which can be obtained previously from corn starch, by the *Xanthomonas campestris* bacteria. Xanthan gum is stable in a wide range of acidity; it is soluble in cold and hot and it resists very well the processes of freezing and thawing. It is used in emulsions and mixed with other polysaccharides, especially with locust bean gum; it is capable of forming gels. It is practically not metabolized in the digestive tract, being eliminated in the feces. No adverse effect is known and has a behavior similar to that of fiber naturally present in food (Mudgil et al., 2014). It is very compatible with other materials to achieve formulas of agricultural treatments and enhance their encapsulating and protective effect (Kumbar et al., 2001).

1.4.2.3   GUAR GUM

It is obtained from a vegetable native to India (*Cyamopsis tetragonolobus*), which is also grown in the United States. For hundreds of years, the plant has been used for human and animal food. Gum is used as a food additive only since the 1950s. It produces very viscous solutions, is able to hydrate in cold water, and is not affected by the presence of salts. It is used as a stabilizer in ice cream, in products that must be subjected to sterilization treatments at high temperatures and in other dairy products, also as a stabilizer in suspensions and foams. No adverse effects are known in its use as an additive (Sajilata et al., 2006). It is a highly effective encapsulating agent for pesticides because it is very compatible with a wide range of pesticide agents, encapsulating, and stabilizing them, thus enhancing its effect (Esfanjani et al., 2015).

1.4.2.4   STARCH

Chemically, it is a mixture of two very similar polysaccharides, amylose, and amylopectin; they contain crystalline and noncrystalline regions in alternating layers. Starch is a food reserve polysaccharide predominant in plants. Commercial starches are obtained from cereal seeds, particularly maize (*Zea mays*), wheat (*Triticum spp.*), several types of rice *(Oryza sativa)*, and some roots and tubers, particularly potato (*Solanum tuberosum*), sweet potato (*Ipomoea batatas*), and cassava (*Manihot esculenta*). Both starches and modified starches have a huge number of possible applications in foods, including the following: adhesive, binder, clouding agent, film former, foam stabilizer, anti-aging bread agent, gelling agent, glazing agent, humectant, stabilizer, texturizer, and thickener. Starch differs from all other carbohy-drates in that, in nature, it appears as complex discrete particles (granules). The starch granules are relatively dense, insoluble, and are very poorly hydrated in cold water. They can be dispersed in water, giving rise to the formation of low viscosity suspensions that can be easily mixed and pumped, even at concentrations higher than 35% (Wang et al., 2020b). Starches are very compatible with other materials to enhance the encapsulation of pesticide agents, besides being very compatible with pesticide agents of different chemical nature (Esfanjani et al., 2015).

1.4.2.5   PECTINS

Pectins are a mixture of acidic and neutral branched polymers. They constitute 30% of the dry weight of the primary cell wall of plant cells. In the presence of water, they form gels. They determine the porosity of the wall, and therefore the degree of availability of the substrates of the enzymes involved in the modifications of the same. Pectins also provide charged surfaces that regulate pH and ion balance. Pectins have three main domains: homogalacturonans, rhamnogalacturonan I, and rhamnogalacturonan II. Pectin is a natural polysaccharide, one of the major constituents of plant cell walls, and is obtained from the remains of the orange and lemon juice manufacturing industry and from the manufacture of cider (Novoa and Castillo, 1998). It has excellent properties as an encapsulating agent as it is an agent with a high level of controlled release of entrants (Nesterenko et al., 2013).

1.4.2.6   PROTEINS

Natural polymers, such as proteins, offer a great opportunity to be used as raw material in the processing of agricultural adjuvants due to their biodegradability and the fact that they can protect the active ingredients of pesticides.

1.4.2.6.1   Vegetable Proteins

Fruits and vegetables generally have a low protein content, polysaccharides being the major component and which have been used as raw material for obtaining encapsulating compounds. On the other hand, cereals and legumes have a higher content of proteins in their composition (8–15% and 17–45%, respectively), being isolated and studied to study their effect as protectors of ingredients (Chambi and Grosso, 2006). They are an option for the substitution of proteins of animal origin and petroleum derivatives as encapsulating agents. Proteins extracted from soybean, pea, and wheat have already been studied as carrier materials for microparticles. Some other plant proteins, such as rice, oat, or sunflower, with interesting functional properties could be investigated as potential matrices for microencapsulation (Jafari et al., 2008b; Haidar et al., 2020; Chatterjee and Bhattacharjee, 2013).

1.4.2.6.2   *Proteins of Animal Origin*

Milk-derived proteins, such as whey and casein, have been studied extensively due to their high nutritional value.

Casein is used to encapsulate ingredients from aqueous solutions due to its ability to form intermolecular bonds (hydrogen, electrostatic, and hydrophobic), which increase the cohesion of the polymer. Another protein of animal origin is collagen, it is classified as fibrous, and it is present in the skin and connective tissue. It has intra and intermolecular bridges due to the formation of covalent bonds (ester, amide, and peptide bonds) that are formed in the polymer matrix, although disulfide bonds also play an important role in the structure; however, these are not very numerous due to the low content of cysteine present in this type of protein (Guamán et al., 1996).

1.4.3   NATURAL OILS

Oils of natural origin can be used as adherent, wetting, and penetrating agents in agricultural applications, being the oils with the greatest potential for agricultural application: vegetable oils such as soybean and castor oil in addition to fish oil. These materials can be used to fulfill the function of protecting the active ingredient as well as to disperse and adhere it on the foliar surface of the plants. The main materials with the greatest potential are described below.

1.4.3.1   SOY OIL

Soy oil, also called soya oil, is the most consumed food oil in the world. It is unctuous, light, and yellowish in color. It is obtained from the pressing of soy and is abundant in polyunsaturated fatty acids, such as Omega 3 and Omega 6. It is used mostly in gastronomy and can be found in salad dressings and oils for frying food. In recent years, and as a result of the flourishing of the biodiesel industry, there are also several published studies evaluating its effectiveness as an insecticide agent (Geranpour et al., 2020; Ogunniyi, 2006). Soybean oil is used in emulsions to encapsulate active ingredients in other areas of the industry; therefore, it has a high potential as an encapsulator in agricultural applications to protect pesticide ingredients (Geranpour et al., 2020).

1.4.3.2 CASTOR OIL

Castor oil is extracted from the seeds of the castor plant, also known as *Ricinus communis*, a plant native to Africa and India. About 90% of this oil is composed of ricinoleic acid. This unique fatty acid is found in lower concentrations in some other seeds and oils (0.27% in cottonseed oil and 0.03% in soybean oil) and is believed to be responsible for the unique properties of castor oil. It also contains a huge amount of vitamin E, proteins, minerals, and has antifungal and antibacterial properties (Jafari et al., 2008; Rubio-Rodríguez et al., 2012).

1.4.3.3 FISH OIL

Fish oils are oils obtained from the tissues of some fish species. For human food, they can be obtained either by eating fish or by taking supplements. Fish that are especially rich in oils that are beneficial to the body and are known by the name of omega-3 fatty acids include mackerel, tuna, salmon, sturgeon, mullet, anchovies, sardines, herring, trout, and menhaden. They provide around 1 g of omega-3 fatty acids in about 3.5 ounces (100 g) of fish (Shasha and McGuire, 1999). They are mainly composed of fatty acid methyl ester, saturated fatty acid, monounsaturated fatty acid, polyunsaturated fatty acid, and polyunsaturated fatty acid. They are widely used for their encapsulating and adherent capacity (Ozyurt and Ötles, 2014).

1.4.4 SAPONINS

The term "saponin" is derived from the Latin word "toad," which means "soap," which reflects its disposition to form stable foams similar to soap in aqueous solutions. Saponins are compounds found in many plants (Augustin et al., 2011). They owe their name to the distinctive characteristic of foaming. Its name probably comes from the plant Saponaria, whose roots have been used historically to form soap (from the Latin toad = soap) (Qasim et al., 2020). Chemically, they are glycosides with a polycyclic aglycone (a free portion of glycoside) that can occur in the form of a steroid or triterpenoid choline bonded through C3 carbon via an etheric bond to a side chain of sugars. It also commonly refers to aglycone as sapogenin, while it is also commonly referred to the subset of steroidal saponins as sapogenin. Saponins are amphipathic due to their liposoluble aglycone function and their saccharide chain which, in turn, is water soluble. This characteristic is the basis of its ability to form a foam (Mao et al., 2012).

1.5   CONCLUSIONS

Currently the agrochemical market is constantly growing. This is due to the increase in population at the global level that in turn increases the demand of agricultural products, this because in previous decades have been used doses of pesticides very high, what has led to the pests species that affect agricultural fields to generate resistance. This has had as a response from producers to increase the pesticides doses to critical levels. This affects the greetings of the staff in charge of applying these pesticides, contaminating not only the soil of culture but also affects beneficial organisms for crops, as it is the case of fungi and bacteria antagonists and pollinating insects. As a first option is to change conventional agricultural treatments for pesticides of natural origin with a lower environmental impact. Another option is to promote conventional pesticides with adjuvants, which are products that are intended to make the application of conventional pesticides more efficient, increasing their effectiveness and therefore reduce the application doses for the control of pests; however, synthetic adjuvants, even if classified as inert, cause havoc in the environment, before this problem there is the possibility of using adjuvants of natural origin and a lower environmental impact, which, like conventional synthetic adjuvants, make the use of agricultural pesticides more efficient. This opens up the possibility of using natural products, which can be considered organic and used in conventional as well as organic agriculture, coupled with the benefits of In this type of products. There is a constant increase in the sector of adjuvant products in the global market, which increases the demand for products of a higher quality and efficiency. This in turn gives rise to new and more innovative research in this area.

ACKNOWLEDGMENTS

Thanks to the doctorate program in food of the Autonomous University of Coahuila

CONFLICTS OF INTEREST

The authors declare no conflict of interest.

KEYWORDS

- **natural adjuvants**
- **agrochemicals**
- **pesticide**

REFERENCES

Augustin, J. M.; Kusina, V.; Anderson, S. B.; Bak, S. Molecular Activities, Biosynthesis and Evolution of Triterpenoid Saponins. *Phytochemistry* **2011,** *72,* 435–457.

Barrenechea, I. V.; López, B. M.; Hermida, M. L. Eficacia de aceites vegetales, minerales y de pescado frente a Frankliniella occidentalis (Pergande). *Boletín de sanidad vegetal. Plagas* **2004,** *30* (1), 177–184.

Bogdanov, S. Quality and Standards of Pollen and Beeswax. *Apiacta* **2004,** *38,* 334–341.

Bonaduce, I.; Colombini, M. P. Characterisation of Beeswax in Works of Art by Gas Chromatography-Mass Spectrometry and Pyrolysis-Gas Chromatography-Mass Spectrometry Procedures. *Journal of Chromatography A* **2004,** *1028* (2), 297–306.

Bourgault, R.; Matschi, S.; Vasquez, M.; Qiao, P.; Sonntag, A.; Charlebois, C.; Molina, I. Constructing Functional Cuticles: Analysis of Relationships between Cuticle Lipid Composition, Ultrastructure and Water Barrier Function in Developing Adult Maize Leaves. *Ann. Bot.* **2020,** *125* (1), 79–91.

Cabello Alvarado, C. J.; Sáenz Galindo, A.; Barajas Bermúdez, L.; Pérez Berumen, C.; Ávila Orta, C.; Valdés Garza, J. A. Cera de Candelilla y sus aplicaciones. Avances en Quimica **2013,** *8.*

Chambi, H.; Grosso, C. Edible Films Produced with Gelatin and Casein Cross-Linked with Transglutaminase. *Food Res. Int.* **2006,** *39* (4), 458–466.

Chatterjee, D.; Bhattacharjee, P. Comparative Evaluation of the Antioxidant Efficacy of Encapsulated and Un-Encapsulated Eugenol-Rich Clove Extracts in Soybean Oil: Shelf-Life and Frying Stability of Soybean Oil. *J. Food Eng* **2013,** *117* (4), 545–550.

Chen, C.; Chen, J.; Zhang, S.; Cao, J.; Wang, W. Forming Textured Hydrophobic Surface Coatings via Mixed Wax Emulsion Impregnation and Drying of Poplar Wood. *Wood Sci. Technol* 2020, 1–19.

Curran, W. S.; McGlamery, M. D.; Liebi, R. A.; Lingenfelter, D. D. Adjuvants for Enhancing Herbicide Performance. 1999.

EPA, United States Environmental Protection Agency. Conventional Reduced Risk Pesticide Program. 2018. https://www.epa.gov

Esfanjani, A. F.; Jafari, S. M.; Assadpoor, E.; Mohammadi, A. Nano-Encapsulation of Saffron Extract through Double-Layered Multiple Emulsions of Pectin and Whey Protein Concentrate. *J. Food Eng.* **2015**, *165*, 149–155.

Fabra, M. J.; Hambleton, A.; Talens, P.; Debeaufort, F.; Chiralt, A.; Voilley, A. Influence of Interactions on Water and Aroma Permeabilities of ι-Carrageenan–Oleic Acid–Beeswax Films Used for Flavour Encapsulation. *Carbohydr. Polym* **2009**, *76* (2), 325–332.

FAO, Food and Agriculture Organization. Asia y el Pacífico. 2018. http://www.fao.org

Fomuso, L. B.; Corredig, M.; Akoh, C. C. Effect of Emulsifier on Oxidation Properties of Fish Oil-Based Structured Lipid Emulsions. *J. Agric. Food Chem.* **2002**, *50* (10), 2957–2961.

Garcia-Ochoa, F.; Santos, V. E.; Casas, J. A.; Gomez, E. Xanthan Gum: Production, Recovery, And Properties. *Biotechnol. Adv.* **2000**, *18* (7), 549–579.

Geranpour, M.; Assadpour, E.; Jafari, S. M. Recent Advances in the Spray Drying Encapsulation of Essential Fatty Acids and Functional Oils. *Trends Food Sci. Technol*. **2020**.

Global Market Insights. Agricultural Adjuvants Market Size By Product (Activator, Utility), by Chemical Group (Alkoxylates, Sulfonates, Organosilicone), by Crop Type (Cereals & Grains, Oilseeds & Pulses, Fruits & Vegetables), By Application (Herbicides, Insecticides, Fungicides), Industry Analysis Report, Regional Outlook (U.S., Canada, Germany, UK, France, Spain, Italy, Russia, China, India, Japan, Australia, Indonesia, Malaysia, South Korea, Brazil, Mexico, Argentina, South Africa, Saudi Arabia, UAE), Growth Potential, Price Trends, Competitive Market Share & Forecast, 2018–2025. 2018. https://www.gminsights.com

Guamán, J.; Andrade, V.; Peralta Salinas, L.; Triviño Gilces, C.; Espinoza Mendoza, A.; Arias de López, M.; Manzano Gavilánez, B. Manual del cultivo de soya, 1996.

Haidar, I.; Harding, I. H.; Bowater, I. C.; McDowall, A. W. (2020). Physical Characterisation of Drug Encapsulated Soybean Oil Nano-Emulsions. *J. Drug Deliv. Sci. Technol.* **2020**, *55*, 101382.

Hernández-Tenorio, F.; Orozco-Sánchez, F. Nanoformulaciones De Bioinsecticidas Botánicos Para El Control De Plagas Agricolas. *Revista de la Facultad de Ciencias* **2020**, *9* (1), 72–91.

https://www.epa.gov/pesticide-registration/conventional-reduced-risk-pesticide-program#howapply, 2018.

Jafari, S. M.; Assadpoor, E.; Bhandari, B.; He, Y. Nano-Particle Encapsulation of Fish Oil by Spray Drying. *Food Res. Int.* **2008a**, *41* (2), 172–183.

Jafari, S. M.; Assadpoor, E.; He, Y.; Bhandari, B. Encapsulation Efficiency of Food Flavours and Oils During Spray Drying. *Dry. Technol.* **2008b**, *26* (7), 816–835.

Jiménez, J. J.; Bernal, J. L.; Aumente, S.; del Nozal, M. J.; Mart??n, M. T.; Bernal Jr, J. Quality Assurance of Commercial Beeswax: Part I. Gas Chromatography-Electron Impact Ionization Mass Spectrometry of Hydrocarbons and Monoesters. *J. Chromatogr. A* **2004**, *1024* (1–2), 147–154.

Kala, S.; Sogan, N.; Agarwal, A.; Naik, S. N.; Patanjali, P. K.; Kumar, J. Biopesticides: Formulations and Delivery Techniques. In *Natural Remedies for Pest, Disease and Weed Control*; Academic Press, 2020; pp 209–220.

Kaur, R.; Bhaskar, T. Potential of Castor Plant (*Ricinus communis*) for Production of Biofuels, Chemicals, and Value-Added Products. In *Waste Biorefinery*; Elsevier, 2020; pp 269–310.

Kemmitt, S. J.; Wright, D.; Goulding, K. W.; Jones, D. L. pH Regulation of Carbon and Nitrogen Dynamics in Two Agricultural Soils. *Soil Biol. Biochem.* **2006**, *38* (5), 898–911.

Koul, O.; Walia, S.; Dhaliwal, G. S. Essential Oils as Green Pesticides: Potential and Constraints. *Biopestic Int.* **2008**, *4* (1), 63–84.

Kowalczyk, D.; Baraniak, B. (2014). Effect of Candelilla Wax on Functional Properties of Biopolymer Emulsion Films-a Comparative Study. *Food Hydrocolloids* **2014**, *41*, 195–209.

Kumbar, S. G.; Kulkarni, A. R.; Dave, A. M.; Aminabhavi, T. M. Encapsulation Efficiency and Release Kinetics of Solid and Liquid Pesticides through Urea Formaldehyde Crosslinked Starch, Guar Gum, and Starch+ Guar Gum Matrices. *J. Appl. Polym. Sci.* **2001,** *82* (11), 2863–2866.

Li, H.; Jiang, Z.; Cao, X.; Su, H.; Shao, H.; Jin, F.; Wang, J. Simultaneous Determination of Three Pesticide Adjuvant Residues in Plant-Derived Agro-Products Using Liquid Chromatography-Tandem Mass Spectrometry. *J. Chromatogr. A* **2017,** *1528,* 53–60.

Mao, Z.; Zheng, X. F.; Zhang, Y. Q.; Tao, X. X.; Li, Y.; Wang, W. Occurrence and Biodegradation of Nonylphenol in the Environment. *Int. J. Mol. Sci.* **2012,** *13* (1), 491–505.

Markets and Market. Agricultural Adjuvants Market by Function (Activator and Utility), Chemical Group (Alkoxylates, Organosilicones, and Sulfonates), Application (Herbicides, Fungicides, and Insecticides), Formulation, Crop Type, and Region - Global Forecast to 2023. 2018. www.marketsandmarkets.com

Mellema, M.; Van Benthum, W. A. J.; Boer, B.; Von Harras, J.; Visser, A. Wax Encapsulation of Water-Soluble Compounds for Application in Foods. *J. Microencapsulation* **2006,** *23* (7), 729–740.

Milanovic, J.; Manojlovic, V.; Levic, S.; Rajic, N.; Nedovic, V.; Bugarski, B. Microencapsulation of Flavors in Carnauba Wax. *Sensors* **2010,** *10* (1), 901–912.

Mudgil, D.; Barak, S.; Khatkar, B. S. (2014). Guar Gum: Processing, Properties and Food Applications-A Review. *J. Food Sci. Technol.* **2014,** *51* (3), 409–418.

Mullin, C. A.; Chen, J.; Fine, J. D.; Frazier, M. T.; Frazier, J. L. The Formulation Makes the Honey Bee Poison. *Pesticide Biochem. Physiol.* **2015,** *120,* 27–35.

Nesterenko, A.; Alric, I.; Silvestre, F.; Durrieu, V. Vegetable Proteins in Microencapsulation: A Review of Recent Interventions and Their Effectiveness. *Indust. Crops Prod.* **2013,** *42,* 469–479.

Niu, F.; Kou, M.; Fan, J.; Pan, W.; Feng, Z. J.; Su, Y.; Zhou, W. Structural Characteristics and Rheological Properties of Ovalbumin-Gum Arabic Complex Coacervates. *Food Chem.* **2018,** *260,* 1–6.

Novoa, M. A. O.; Castillo, L. O. Potencialidad del Uso de las Leguminosas como Fuente Proteica en Alimentos para Peces, 1998.

Oancea, A. M.; Turturică, M.; Bahrim, G.; Râpeanu, G.; Stănciuc, N. Phytochemicals and Antioxidant Activity Degradation Kinetics during Thermal Treatments of Sour Cherry Extract. *LWT-Food Science and Technology* **2017,** *82,* 139–146.

Ogunniyi, D. S. Castor Oil: A Vital Industrial Raw Material. *Bioresour. Technol.* **2006,** *97* (9), 1086–1091.

Ozyurt, V. H.; Ötles, S. Properties of Probiotics and Encapsulated Probiotics in Food. *Acta Scientiarum Polonorum. Technologia Alimentaria* **2014,** *13* (4).

Peroukidis, S. D.; Tsalikis, D. G.; Noro, M. G.; Stott, I. P.; Mavrantzas, V. G. Quantitative Prediction of the Structure and Viscosity of Aqueous Micellar Solutions of Ionic Surfactants: A Combined Approach Based on Coarse-Grained MARTINI Simulations Followed by Reverse-Mapped All-Atom Molecular Dynamics Simulations. *J. Chem. Theory Comput.* **2020,** *16* (5), 3363–3372.

Qasim, M.; Islam, W.; Ashraf, H. J.; Ali, I.; Wang, L. Saponins in Insect Pest Control. *Co-Evolution of Secondary Metabolites* **2020,** 897–924.

Ranjha, N. M.; Khan, H.; Naseem, S. Encapsulation and Characterization of Controlled Release Flurbiprofen Loaded Microspheres Using Beeswax as an Encapsulating Agent. *J. Mater. Sci. Mater. Med* **2010,** *21* (5), 1621–1630.

Rosen, M. J. *Surfactants and Interfacial Phenomena*, 4th ed.; Wiley University of Waikato, 2012; All Rights Reserved. Published 11 January 2012.

Rubio-Rodríguez, N.; Sara, M.; Beltrán, S.; Jaime, I.; Sanz, M. T.; Rovira, J. Supercritical Fluid Extraction of Fish Oil from Fish By-Products: A Comparison with Other Extraction Methods. *J. Food Eng.* **2012,** *109* (2), 238–248.

Sajilata, M. G.; Singhal, R. S.; Kulkarni, P. R. Resistant Starch-a Review. *Comprehen. Rev. Food Sci. Food Safety* **2006,** *5* (1), 1–17.

Shasha, B. S.; McGuire, M. R. U.S. Patent No. 5,997,945; Washington, DC: U.S. Patent and Trademark Office, 1999.

Sparg, S. G.; Light, M. E.; Van Staden, J. Biological Activities and Distribution of Plant Saponins. *J. Ethnopharmacol.* **2004,** *94,* 219–243.

Stark, J. D.; Walthall, W. K. Agricultural Adjuvants: Acute Mortality and Effects on Population Growth Rate of Daphnia Pulex after Chronic Exposure. *Environ. Toxicol. Chem.* **2003,** *22* (12), 3056–3061.

Tadesse, W.; Dejene, T.; Zeleke, G.; Desalegn, G. Underutilized Natural Gum and Resin Resources in Ethiopia for Future Directions and Commercial Utilization. *World* **2020,** *8* (2), 32–38.

Vandenburg, L. E.; Wilder, E. A. The Structural Constituents of Carnauba Wax. *J. Am. Oil Chem. Soc.* **1970,** *47* (12), 514–518.

Vidhyalakshmi, R.; Bhakyaraj, R.; Subhasree, R. S. Encapsulation "the Future of Probiotics"-a Review. *Adv. Biol. Res.* **2009,** *3* (3–4), 96–103.

Wang, Z.; Lan, L.; He, X.; Herbst, A. Dynamic Evaporation of Droplet with Adjuvants under Different Environment Conditions. *Int. J. Agric. Biol. Eng* **2020a,** *13* (2), 1–6.

Wang, Z.; Xu, B.; Luo, H.; Meng, K.; Wang, Y.; Liu, M.; Tu, T. Production Pectin Oligosaccharides Using Humicola Insolens Y1-Derived Unusual Pectate Lyase. *J. Biosci. Bioeng* **2020b,** *129* (1), 16–22.

Yadav, M. P.; Igartuburu, J. M.; Yan, Y.; Nothnagel, E. A. Chemical Investigation of the Structural Basis of the Emulsifying Activity of Gum Arabic. *Food Hydrocoll.* **2007,** *21* (2), 297–308.

Zhu, W.; Schmehl, D. R.; Mullin, C. A.; Frazier, J. L. Four Common Pesticides, Their Mixtures and a Formulation Solvent in the Hive Environment Have High Oral Toxicity to Honey Bee Larvae. *PloS One* **2014,** *9* (1), e77547.

CHAPTER 2

Customization of Bioreactor Technology for On-Site Production of Biocontrol Agents

M. MOUSUMI DAS[1], M. HARIDAS[1], HECTOR ARTURO RUIZ LEZA[2], and A. SABU[1*]

[1]Department of Biotechnology and Microbiology, Centre for Bio-innovation and Product Development, Kannur University, Dr. Janaki Ammal Campus, Thalassery, Kannur 670661, Kerala, India

[2]Food Research Department, Biorefinery Group, Faculty of Chemistry Sciences, Autonomous University of Coahuila, Saltillo, Coahuila, Mexico

*Corresponding author. E-mail: drsabu@gmail.com

ABSTRACT

Bioreactors are used for the production of cell metabolites and biomass. Biomass, including bio-control agents (BCAs), is used to control pests, pathogens, and diseases in various crops. However, the BCAs are expensive and rural farming communities cultivating economically important crops are unable to apply quality BCAs to their crops. On-site production of BCAs using customized bioreactors with locally available substrates under solid-state fermentation (SSF) is one approach to overcome this hurdle. This chapter throws lights into the innovative approach adopted for the development of an economically feasible tray fermenter designed by appropriating bioreactor technology. The designed fermenter was suitable for on-site production with minimum energy requirement. Porous trays and a central shaft in the fermenter allows proper aeration through substrate bed and easy recovery of biocontrol agents. A series of cotton wicks immersed in water provides adequate humidity within the fermenter. Provision for water circulation allows easy spore wash for collecting viable spore for aerial applications. This novel SSF apparatus is a promising device for mass multiplication of BCAs in SSF. Imparting training to the rural farming community for operating the fermenter is another important aspect while going for on-site production of BCAs. Training of the rural farmers for operating the fermenter for production of BCAs and its application in the crop field helped the farmers in changing their farming activities to a sustainable way with multiple economic and environmental benefits.

2.1 INTRODUCTION

The increasing demand for chemical-free food has led to the development of effective and sustainable *pest-management* strategies for crop protection. Biological control of phytopathogens by microorganisms appears to be an excellent alternative to the existing chemical treatment methods (Das et al., 2019; Méndez-González et al., 2018). Microbial biocontrol agents have different mechanisms of action for dealing with pests and pathogens. Among various bio-control methods, fungal BCAs are promising and there is a lot of scope for research in the development of novel fungal agents. In recent decades, there is a renewed interest in the production and use of fungal BCAs to control the pests and soil-borne phytopathogens, mainly due to the need to reduce the chemical pesticide application and the consequent benefits for the environment and for public health (Das and Sabu, 2020). However, commercially available bio-control agents (BCAs) are expensive and impoverished farmers cultivating economically important crops are unable to apply quality BCAs with high-viable biomass content to their field. The lack of knowledge and unavailability of inexpensive BCAs for these farmers are the limiting factors for adopting better practices of cultivation. To evaluate the efficacy of fungal BCAs in field, it is necessary to produce the inoculum in sufficient quantity. Different types of agricultural byproducts are used as solid support in solid-state fermentation (SSF) for large-scale production of fungal BCAs (Das and Sabu, 2020; Moo Young et al., 1983).

The recent advances in modern microbial biotechnology involve new methods and initiatives to improve the growth and biomass production of fungal BCAs by SSF. Isolation and characterization of native strains is the best possible way to ensure the successful use of these microbial antagonists as they are then adapted to the environment in which they will be used (Tranier et al., 2014). Also, the success of the SSF process depends on the bioreactor system and the nature of the solid substrate used. In the SSF process, the bioreactor system provides a controlled environment for the microbial growth and biological activity (Krishna, 2005). Since the last few decades, SSF bioreactors are being used for the production of antimicrobial metabolites and biomass including BCAs to control pests and pathogens in an agricultural crop (Lizardi-Jiménez and Hernández-Martínez, 2017; Webb and Manan, 2017). Researchers have shown a renewed interest in SSF and numerous studies have been performed dealing with the design and development of low-cost bioreactors for enhancing biomass production of BCAs by SSF. Several types of bioreactors have been used for SSF technologies in the production of bio-fertilizers and bio-pesticides. SSF bioreactors can be conveniently classified into four types according to the type of aeration (with or without forced aeration) or the mixing/agitation system employed. These can be classified as follows: tray bioreactor (unforced aeration, without mixing or agitation); packed-bed bioreactor (forced aeration, without mixing or agitation); rotating-drum and stirred-drum bioreactors (unforced aeration, with continuous or intermittent mixing); fluidized-bed, rocking-drum, and stirred-aerated bioreactors (forced aeration, with continuous mixing or agitation) (Singhania et al., 2009; Webb and Manan, 2017). SSF has emerged as a promising technology and is reliable for the development of several bioprocesses and value added products like antibiotics, enzymes, and organic acids. It offers many advantages for cost-effective production of fungal spores or conidia at a large scale. This chapter first presents the importance of BCAs for sustainable agriculture and then on on-site production of BCAs in a newly designed bioreactor/ tray fermenter under SSF.

2.1.1   SCOPE OF BCAs

BCAs are eco-friendly and will definitely become a suitable substitute for the chemical pesticides and insecticides, which are now facing serious resistance from the farmers and common public. They reduce the hazards associated with the use of toxic chemicals in agriculture as well as the agrochemical resistance of the target organisms (Das and Sabu, 2020). They play a significant role in controlling plant pests like parasitic nematodes, insects, weeds, and mites and also helps in maintaining and balancing the plant species along with their natural enemies. BCAs include fungi, bacteria, oomycetes, viruses, and protozoa that control different kinds of pests and pathogens infecting agricultural crops, although each organism has a narrow target range and highly specific mode of action for its target pest (Usta, 2013). Over the past decade, there is an increasing demand for bacterial and fungal BCAs to control microbial pathogens, pests, and plant-parasitic nematodes of economically important crops. Compared with other BCAs such as bacteria and virus, there are some distinct advantages in utilizing fungi. They have the ability to infect their target host by direct adhesion to the surface and penetration through the cuticle. The attractiveness of myco-biocontrol agents, in particular, is due to their high-reproductive capacity, host specificity, short-generation time, persistence in the field, dispersal efficiency, and their ease of cultivation and maintenance in the laboratory (Jyoti and Singh, 2016; Das and Sabu, 2020). Emergence of fungal antagonists has made it a promising biological-control strategy to control the pests and plant diseases. Certain rhizosphere microbes have gained substantial research attention as promising fungal bio-control agents; the best studied of these are *Beauveria bassiana, Verticillium lecanii, Metarhizium anisopliae, Purpureocillium lilacinum, Trichoderma harzianum, Trichoderma viride, Trichoderma asperellum, Trichoderma virens,* etc. (Das and Sabu, 2020; Sandhu et al., 2012; Petrot et al., 2015; Prakash et al., 1999; Li et al., 2016).

2.1.2   FUNGAL BCAs IN SUSTAINABLE AGRICULTURE

Fungal BCAs with extensive host specificity and environmental safety are an attractive alternative for hazardous agrochemicals. They may be used either as the sole control agent or in combination with others, as in integrated pest-management programs. The most-commonly used BCAs are living organisms, which are pathogenic for the pests, insects, weeds, and parasitic nematodes. These include bio-fungicides (*Trichoderma harzianum, Trichoderma viride*), bio-herbicides (*Phytopthora*), bio-insecticides (*Beauveria bassiana, Metarhizium anisopliae*), and bio-nematicide (*Paecilomyces lilacinus*) (Gupta and Dikshit, 2010). Unlike other BCAs such as viruses and bacteria, fungal BCAs have received extensive attention, because of their ability to thrive on inexpensive agro-industrial residues, diverse bio-control mechanisms, and rapid production of biomass. *Trichoderma* is the most widely used and effective bio-fungicide with high reported success rate compared to other BCAs. *Trichoderma* species are free-living endophytic fungi spread in all soils and plant root ecosystems. *Trichoderma viride* and *Trichoderma harzianum* have carved out a niche in Indian market as potential and most important fungal BCAs for the management of various crop diseases. *Trichoderma*-based cost-effective formulations are prepared using different organic and inorganic carriers either through solid- or liquid-fermentation technologies. They are delivered through either soil application, foliar spray, seed treatment, seedling lip, or bio-priming (Kumar et al., 2014).

The most important factor to be considered while selecting a BCA for commercial application is the availability of an environmentally sustainable and cost-effective production and stabilization technology for the development of an effective formulation of viable micro-organisms. Mass production of fungal antagonists by SSF has emerged as a sustainable production method, which is preferred over submerged fermentation (SmF) as it is eco-friendly, cost-effective, and given the advantage of using agro-processing residues as solid substrate (Das and Sabu, 2020). Most fungi sporulate well on solid substrates, mainly agro-processing wastes such as wheat bran, beer waste, coffee husk, sugarcane bagasse, etc. (Ooijkaas, 2000; Sala et al., 2019; Das et al., 2020). Various researchers have claimed that the SSF has multiple benefits over conventional SmF including lower capital and operational cost, lower space and energy requirements, simpler equipment, inexpensive substrates or media, and less-expensive downstream processing (Weiland, 1988; Pandey, 1991). Besides, BCAs produced by SSF are cost-effective and present good stability, better quality, and high-density propagules with higher conidia content than liquid fermentation. SSF bioreactors recycles the agro-industrial wastes without environmental issues for various applications in biological processes such as, bioremediation and biodegradation of toxic chemicals, biological detoxification of agro-processing residues, nutritional enrichment, bio pulping, and production of biomolecules such as enzyme, antibiotics, biomass, spores, organic acids, aroma compounds, pigments, bio-pesticides or bi-control agents, mushrooms, xanthan gum, alkaloids, and plant growth factors (Soccol and Vandenberghe, 2003).

2.2 BIOREACTORS FOR THE PRODUCTION OF BCAs UNDER SSF

SSF is defined as the microbial fermentation process occurring in the absence or near-absence of free water under controlled conditions. Also, SSF provides a solid medium that resembles the natural habitat of several micro-organisms. In SSF, the growth and metabolism of micro-organism occurs on natural raw material with low moisture contents and has been identified as a promising methodology and technique in biotechnology (Webb and Manan, 2017). SSF has been reported to have many biotechnological advantages over SmF such as higher yield and productivity, extended stability of products, lower catabolic repression, less effort in downstream processing, reduced energy and cost requirements, and resembles the natural habitat of several micro-organisms (Couto and Sanroman, 2006; Singhania et al., 2009). In addition, sterilization of the substrate or solid medium is not necessary because of the low water-activity requirements used in SSF (Singhania et al., 2009). This process facilitates the possibilities for the recycling of agro-processing residues without economic fate for numerous applications in bioprocesses such as nutritional enrichment, bio-remediation, bio-degradation of hazardous materials, biological detoxification, production of bio-molecules, or value-added products such as enzymes, organic acids, bio-fertilizers, bio-pesticides, pigments, food aroma compounds, xanthan gum, mushrooms, and vegetable hormones (Soccol and Vandenberghe, 2003).

Bioreactors are central part of the biotech process and present great importance in biochemical transformation during the industrial fermentation processes. In SSF, treated materials promotes biotransformation by the action of enzymes or microorganisms and mammalian or plant cell systems (Pandey et al., 2008). The most important aspect to be considered during the development of SSF process is the design, operation, and scale-up of bioreactors. Various bioreactor designs have been developed and operated, in which a chemical process is carried out where microbes are used for the manufacturing of biologically active substances and enzymes. Bioreactors are enclosed vessels or tanks that maintain optimal conditions for the growth and development of a selected microorganism used in fermentation process. The microorganism undergoes fermentation to produce large quantities of desired metabolites and value-added products, such as enzymes, antibiotics, organic acids, alcohols, etc. (Spier et al., 2011; Gaikwad et al., 2018). Bioreactor is the heart of the fermentation process and has a wide range of applications in agricultural, environmental, industrial, and medical fields (Schaechter and Lederberg, 2004; Cinar et al., 2003). Efficient bioreactors or fermenters provides a controllable environment enabling the biological and biochemical requirements for the development of desired product (Singh et al., 2014). These bioreactors are capable of maintaining and monitor the reaction parameters required for the desired biological activity by controlling the heat transfer, mass transfer, fluid velocity, shear stress, temperature, pH, oxygen, carbon dioxide, and nutrient supply, reaction rate, and cell growth (Mustafa et al., 2018).

In recent years, SSF bioreactors were used for the production of many industrially important enzymes such as protease, pectinase, α-amylase, xylanase, CMCase, cellulase, β-glucosidase, pectinase, phytase, endoglucanase, etc. (Tsouko et al., 2017; Pitol et al., 2016; Derakhti et al., 2012; Diaz et al., 2013; Alam et al., 2009; Dhillon et al., 2013; Biz et al., 2016; Saithi and Tongta, 2016; Farinas et al., 2011). Other important applications of these bioreactors include biomass production (e.g., BCAs/ bio-pesticides, bio-fertilizers, food additives, single cell protein, and micro-algae) and metabolite formation (e.g., organic acids, pigments, antibiotics, ethanol, and aromatic compounds)

TABLE 2.1 SSF Bio-Reactors Used for Production of Bio-Control Agents.

Type of bioreactor	Substrate	Spores	Target pathogen/Pest/Disease	References
Tray	Rice	*B. bassiana*	Coleopteran pests	Xie et al. (2013)
Tray (intermitently mixed)	Wheat bran/maize meal	*Clonostachys rosea*	*Botrytis cinerea*	Zhang et al. (2014)
Plastic bag	Rice grains	*M. anisopliae*	Locusts and grasshoppers	Cherry et al. (1999)
Plastic bag	Rice grains	*Lecanicillium lecanii*	—	Ribeiro-Machado et al. (2010)
Tray	Rice grains	*B. bassiana*	—	Ye et al. (2006)
Plastic tray	Rice media	*B. bassiana and M. anisopliae*	—	Alves and Pereira, (1989)
Packed column bioreactor	Rice and sugarcane bagasse	*M. anisopliae*	—	Arzumanov et al. (2005)
Packed bed	Rice straw and wheat bran	*B. bassiana*	Diamondback moth (*Plutella xylostella*)	Kang et al. (2005)
Packed bed	Oats and inert solids (hemp, perlite, and bagasse)	*Coniothyrium minitans*	—	Weber et al. (1999)
Plastic bag	Rice grains	*B. bassiana*	Coffee berry borer *Hypothenemus hampei*	Posada-Flórez (2008)
Autoclavable bag	Maize	*M. anisopliae*	Oil palm rhinoceros beetle, *Oryctes rhinoceros*	Moslim et al. (2005)
Polyethylene bioreactor	Corn cob	*T. asperellum*	—	De la Cruz-Quiroz et al. (2017)
Erlenmeyer flask	Brewer's spent grain	*B. bassiana*	*Galleria mellonella*	Qui et al. (2019)
Zymotis bioreactor	Bagasse + nutrient solution	*T. harzianum*	—	Roussos et al. (1991)
Plastic bag	Peat:vermi-culite:nutrients	*Penicillium frecuentans*	Brown rot of stone fruits	De Cal et al. (2002)
Erlenmeyer flask	Coffee husk	*T. harzianum*	*F. oxysporum, R. solani, P. capsici*	Das and Sabu (2020)
Erlenmeyer flask	Wheat bran	*P. lilacinum*	*M. incognita*	Das et al. (2020)

(Bhattacharyya et al., 2008; Spier et al., 2011). Table 2.1 shows recent examples of SSF bioreactors used for production of BCAs.

In the fermentation process (SSF or SmF), the bioreactor or fermenter provides best-suited conditions for microbial growth for biological or biochemical activity (Krishna, 2005; Tekere, 2019). It is necessary to understand the main aspects to be taken into account for better performance of the bioreactor, so that the bioreactors can be designed efficiently. Important factors that need to be considered in the design and development of SSF bioreactors is that it should be able to remove the excess metabolic heat within the fermented bed itself as the higher temperature in the system can adversely affect microbial growth and development (Webb and Manan, 2017; Hashemi et al., 2011). Besides, a wide range of factors influence the efficiency of bioprocess in bioreactors, including sampling and transfer systems, inoculation methods, equipment sterilization, fermenting medium composition and nutrient parameters (carbon, nitrogen, phosphorous sources, metals, etc.), substrate type, particle size, aeration, gaseous environment, agitation or shaking arrangements, measurement and control of bioprocess parameters (temperature, pH, moisture content), construction materials of fermentation tank/vessel, and the ability to withstand pressure (Pandey et al., 2001; Aidoo et al., 1982; Gowthaman et al., 2001; Gomez et al., 2016). Different types of bioreactors have been developed and used in SSF processes. SSF bioreactors must be able to hold nutritional media or substrates generally of natural materials such as agro-industrial residues and food wastes and be well sealed to avoid contamination (Krishna, 2005; Webb and Manan, 2017). Several studies have been carried out for the production of value-added products under SSF conditions. *In-vitro* SSF process has generally been carried out in conical flasks, beakers, Petri plates, roux bottles, glass jars, and tubes (as column fermenter), while large-scale fermentation has been performed in tray, drum, and deep-trough type fermenters (Pandey et al., 1999). Generally, the excess heat from the fermenters are removed by forced aeration, but for small-scale purposes, simple trays without forced aeration are sometimes used. Unlike SmF, the choice of an appropriate bioreactor system is difficult in SSF, given the heterogeneity of the solid substrate (Pandey, 2003). Despite the heterogeneity of the solid matrix, the critical factors to be considered for the design of a sophisticated and sound bioreactor includes transport of oxygen and metabolic heat, which involves aeration/agitation, moisture content, temperature, and the type of microorganism and solid substrate used (Webb and Manan, 2017).

2.2.1   TRAY FERMENTERS FOR THE PRODUCTION OF BCAs

Tray bioreactors are the simplest of all types of bioreactors used in SSF. The tray-type SSF bioreactor includes a chamber containing a large number of individual trays stacked on top of one another with suitable gap between them. The trays are usually constructed of wood, bamboo, metal, stainless steel, or plastic and the trays are typically open at the top and have perforated bottoms with mesh to hold the solid substrate to facilitate air circulation (Webb and Manan, 2017). SSF is carried out in this static trays without mechanical agitation. Agitation of the bed is not necessary; if done, is very infrequent, and is typically done manually. A tray fermenter would have unmixed substrate bed without forced aeration under static conditions (Krishna, 2005; Spier et al., 2011; Ge et al., 2017). Usually, the perforated trays loaded with thin layers (typically 5–15 cm) of moist solid substrate are placed in a chamber or room with controlled temperature and humidity for optimal growth of micro-organisms (Spier et al., 2011). In fact, thin layers of substrates avoid overheating and helps in maintaining aerobic conditions. Tray-type SSF bioreactors are commonly used for the production of traditional fermented foods such as soy or tamari, koji, natto, miso, and temph (Arora et al., 2018).

Several studies have been carried out using tray bioreactors for the production of BCAs or bio-pestcides. The use of tray bioreactors results in enhanced production of fungal conidia from filamentous fungi *Beauveria bassiana* and *Metarhizium* sp. (Mendez-Gonzalez et al., 2018). Xie et al. (2013) used a tray bioreactor for conidial production of fungal BCA, *Beauveria bassiana* Bb-202. This entomopathogenic strain has the potential to control the coleopteran pests. In their study, they used rice for the mass production of this fungus in tray bioreactor under SSF. The results of their study show that the highest yield of 3.94×10^{12} conidia kg^{-1} rice was obtained as the substrate of 2 cm thickness was cut into many small pieces (6 cm $\times$ 4 cm $\times$ 2 cm). It indicated that smaller pieces of solid substrates would increase the surface area of solid substrate and the conidia yields were significantly enhanced. They concluded that metabolic heat and gas transfer of substrate in SSF was improved by cutting the solid substrate into smaller pieces.

Ye et al. (2006) developed an upright multi-tray conidiation chamber for the production of aerial conidia of *B. bassiana*. This fermentation chamber was equipped with 25 bottom-meshed metal trays parallel to each other, which had a total capacity of > or =50 kg rice with each tray holding ≥2 kg. They used rice grains as substrate for the growth and spore production of

B. bassiana. SSF were carried out in tray fermenter. In repeated trials, an average yield of 2.4 (1.8–2.7) $\times$ 10^{12} conidia kg^{-1} rice was obtained after the seventh day of fermentation in a fully loaded chamber. Therefore, the novel SSF apparatus would have a high potential for bulk production of fungal conidia of *B. bassiana* and their field application.

Alves and Pereira (1989) developed a technique for culturing the entomo-pathogenic fungi *Metarhizium anisopliae* and *Beauveria bassiana* in plastic trays. They inoculated the fungi (Spore suspension) in rice media inside the plastic bags and the mycelia were transferred to plastic trays and incubated at 27°C under aseptic conditions. At the end of fermentation, the average production yield reached 6–11.4% for *M. anisopliae* with 10^{10} conidia/g and 3.0% for *B. bassiana* with 2.0 $\times$ 10^{11} conidia/g.

Coban and Sargin (2019) described an alternative production method to submerged liquid culture for the production of the BCA, *Trichoderma harzianum* micro propagules, by an integrated-tray bioreactor system. They optimized the different process parameters for the production of micro propagules of *T. harzianum* EGE-K38 in modified czapec medium (MCM) containing 8 g/L glucose using one-factor-at-a-time method and box-behnken design (statistical method). The process variables such as incubation temperature, inoculum size, spore concentration, airflow rate, and medium volume were individually studied as one factor at a time. After optimization, the highest number of micro propagules obtained was 5.2 $\pm$ 0.2 $\times$ 10^9 CFU/mL and dry cell weight was 17 $\pm$ 2 g/L. Their study concluded that the fermentation process carried out in the designed-tray bioreactor offers great potential for large-scale production of fungal BCAs for commercial applications.

Some of the earlier studies dealing with the design and development of static or tray bioreactors mainly focus on the evaluation of important process variables affecting the production of fungal biomass or conidia, as well as developing a reproducible bioprocess method susceptible to a scale-up approach for large-scale conidia production.

2.3   DESIGN AND DEVELOPMENT OF AN SSF BIOREACTOR FOR PRODUCTION OF FUNGAL BCAs

A bioreactor system holds great promise for microbial bioprocess technology, which is used for the manufacturing and extracting the desired biomaterial from a microorganism by fermenting a selected microorganism. Tray bioreactors consist of a chamber in which conditioned air (temperature, air flow, and relative humidity) is circulated around the substrate bed of individual trays (Mitchell et al., 2006). Technologies are currently being developed for the on-site production of fungal BCAs to supply fresh materials for immediate applications to insect- or pest-infested crops or soil. Mass production of BCAs is required for effective utilization in field. With advances in modeling and optimization techniques, production using SSF can be made appropriate for rural farmers.

For the on-site production of BCA, a semi-sterile tray-type SSF bioreactor was designed by our research group through appropriation of technology. The designed fermenter was suitable for on-site production with minimum energy requirement. The novel SSF reactor comprises a perforated external chamber having a removable cover that extends from a top portion of the perforated external chamber. It further comprises a non-perforated internal chamber disposed within the perforated external chamber and defines a gap between an inner surface of the perforated external chamber and an outer surface of the non-perforated internal chamber. The fermenter consists of three perforated trays of varying size (4–5 kg holding capacity) mounted on a perforated central shaft to accommodate solid substrate such as agro-processing residues. These trays are arranged hierarchically based on the size within the non-perforated internal chamber. Porous trays and center shaft allows proper aeration through substrate bed and easy recovery of spores. The water current system in the bioreactor allows easy spore wash for collecting viable spore for aerial applications. Schematic representation of the designed-tray fermenter is given in Figure 2.1.

The advantages of the present invention include, but not limited to, having a simpler design and structure of the SSF bioreactor or tray fermenter. Such simple design and structure enables the bioreactor to be easily deployable on-site. Further, the trays are arranged hierarchically based on their size, which helps in enhanced biomass yield. In addition, this SSF bioreactor does not require an external power supply, control devices, agitators, water circulation mechanisms, etc., and as such is less costly to manufacture. The salient features of this bioreactor includes minimum energy requirement, easy recovery of BCAs, and stress-free operation under semi-sterile condition.

2.4 PRODUCTION OF BCA, *T. HARZINAUM* CH1, IN SSF BIOREACTOR

BCAs are ecofriendly sources for sustainable agriculture. They have attracted great attention during the past decade as a promising alternative to chemical pesticides in integrated pest management. BCAs can control different kinds of pests, although each separate active ingredient is relatively specific for its target pest (Das and Sabu, 2020). The use of biological fungicides like *T. viride* and *T. harzianum,* and biological insecticides like *Beauveria bassiana, Metarhizium anisopliae,* and *Purpureocillium lilacinum* in crop field will reduce the risks of fungal diseases and insect attack (Das and Sabu, 2020; Das et al., 2020). Among the various fungal BCAs, *Trichoderma* spp. are widely exploited fungi and are among the most commonly used microbial fungicides. *Trichoderma* spp. plays a major role as a BCA due to its ability to improve crop fields through multiple roles, such as bio-fungicide and growth-promoting agent (Kumar et al., 2014). *T. harzianum* is mainly used in the production of bio-control biomass because of its ability to utilize a variety of carbon sources, such as rice bran, sugarcane bagasse, chickpea husk, wheat straw, rice straw (Gangadharan and Jeyarajan, 1990), corn bran, wheat bran (Cavalcante et al., 2008), cassava by-product (Jhon and Jeeva, 2014), spent malt (Gopalakrishnan et al., 2003), and coffee husk (Das and Sabu, 2020).

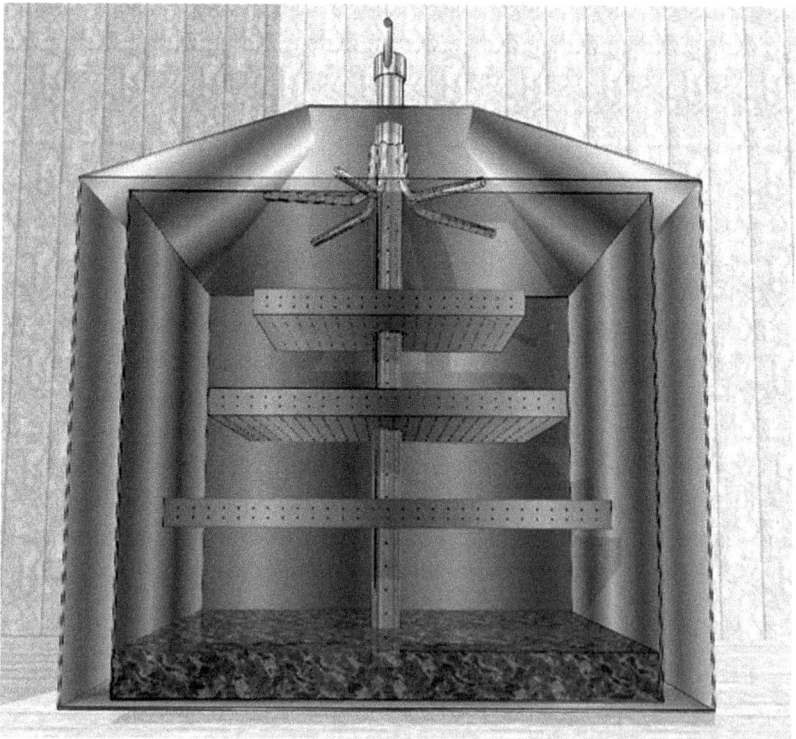

FIGURE 2.1 Diagrammatic representation of the solid-state bioreactor developed for on-site production of BCAs.

In our study, an antagonistic fungi isolated from the forest soil of Palakkad, Kerala, India has been identified as *Trichoderma harzianum* CH1 (GenBank Accession Number: KX756617.1) (Fig. 2.2). The strain is highly effective for the management of various soil-borne phytopathogens mainly pathogens of black pepper and ginger crops, such as *Fusarium oxysporum*, *Rhizoctonia solani*, and *Phytophthora capsici* (Das et al., 2019; Das and Sabu, 2020). The biomass production of *T. harzianum* CH1 was carried out in a newly designed tray fermenter using coffee husk moistened with mineral salt media as substrate by keeping optimized process parameters achieved under flask culture (moisture content 70% (w/v) and external carbon source (jaggery-2 % (w/v), pH-4). Coffee husk, which is a potential agro-waste, is abundantly available in the farming sector and can be recycled and utilized for *Trichoderma* multiplication. Incubation was carried out for 12 days at room temperature and the biomass production was estimated continuously (by n-Acetyl-D-glucosamine assay) in each day (Ramachandran et al., 2005). The optimum time of incubation for maximum biomass production (172.32 mg/gdfs) in the tray fermenter was found to be 7 days (Fig. 2.3). After SSF, 7–8 kg of fermented biomass was obtained from the three trays, which is suitable for field-level application. Fungal spores were separated and extracted in to a suitable medium for aerial applications.

FIGURE 2.2 Newly isolated *Trichoderma harzianum* CH1.

FIGURE 2.3   Mass production of *T. harzianum* CH1 in coffee husk using the newly designed tray fermenter.

This study helps to support the farmers to improve crop production by distributing viable, pre-assessed, and characterized novel strains. Installation and operation of the newly designed customized bioreactor system in their farms helps in on-site production of high-quality BCAs in an economically feasible and sustainable manner using locally available agro-processing residues.

2.5 SUSTAINABLE LIVELIHOOD DEVELOPMENT THROUGH ON-SITE PRODUCTION OF BCAs

Agriculture and its allied sectors is unquestionably the principal source of livelihood in India. Over 70% of the Indian rural households are still dependent on agriculture for their livelihood, out of which 82% of farmers are small and marginal (Food and Agriculture Organization). The unbridled infestation of pests and diseases made the farming a losing game. The hazardous chemicals used for the infestation of pest and diseases have not only miserably failed but also added human illness far and wide due to the high exposure to agro-chemicals. The widespread use of agro-chemicals led to the significant reduction of below-ground microbial population, which maintained the ecological equilibrium and vitality of the soil. The absolute failure of chemical measures to contain the agrarian crisis led to the adoption of organic farming in a big way. But the availability of quality bio-inputs posed a major cause of concern.

For the empowerment of underprivileged farmers, rural transformation and employment are considered important. To empower the farmers, the newly developed economically feasible commercial tray fermenter was successfully utilized through mass multiplication of BCAs. SSF of BCAs was carried out in the designed fermenter with suitable agro-industrial residues, which are locally available. Bioprocess optimization upon SSF in tray fermenter enhanced the production of BCAs contributing to the profit margin in production. Direct application of tray fermenter in field for on-site production may help farmers in economical production of BCAs and improve crop production as well. Awareness and training programmes were also conducted for effective operation of the bioreactor and production and field application of BCAs. The training of the farmers in BCA production and field-level application helped the farmers in changing their farming activities to a sustainable way providing multiple economic and environmental benefits. Ecological benefits such as reducing carbon emissions from agricultural soils, less-polluted soils, air, and water in agricultural fields have been added to the benefits of using BCAs. Instead of depending on external supply of BCAs with compromised quality, the farmers could produce the high-quality, viable BCAs in their farm itself. The strategy helps them to minimize the expenses for procuring BCAs, which, in turn, improved their livelihood.

2.6 CONCLUSIONS

Growing public concerns on overuse of pesticides in agriculture and their effects on the environment has led to researches on use of safe and ecofriendly technology to control the pests and pathogens. The arbitrary use of chemical pesticides led to the significant reduction of below-ground microbial population, which maintained the ecological equilibrium, and vitality of the soil. The absolute failure of chemical measures to contain the agrarian crisis led to the adoption of organic farming in a big way. BCAs are better than hazardous chemical pesticides and it reduces or eliminates the risk associated with the use of toxic pesticides in the agriculture. But the availability of quality bio-inputs (BCAs/bio-pesticides) posed a major cause of concern. On-site production and enhanced utilization of bio inputs will definitely become a suitable substitute for the agrochemicals, which is economically feasible also. Therefore, on-site production of BCAs and their field-level application are required for an eco-friendly agricultural practice. Several studies have been reported on the use of cheap agro-residues as raw materials for the production of BCAs, mainly fungal BCAs using SSF technology. In recent times, SSF bioreactors have undergone rapid modifications and advancement in order to maximize the productivity and increase the commercial use of SSF processes. In our study, a solid-state bioreactor or tray fermenter was designed and developed for onsite production of BCAs. Mass production of *T. harzianum* was successfully achieved in the designed fermenter. The fermented substrate with viable fungal biomass was used directly for field application. The novel bioreactor is a promising apparatus for on-site biomass production of BCAs under SSF where quality and viability of the product are guaranteed. Since the fermenter is operated by the trained farmers, it improves their livelihood by reduction in spending for BCAs from elsewhere.

KEYWORDS

- **bioreactor**
- **solid-state fermentation**
- **bio-control agent**
- **agro-processing residue**
- **microbial biocontrol**

REFERENCES

Aidoo, K. E.; Hendry, R.; Wood, J. B. Solid Substrate Fermentations. *Adv. Applied Microbiol.* **1982**, *28*, 201–237.

Alam, M. Z.; Mamun, A. A.; Qudsieh, I. Y.; Muyibi, S. A.; Salleh, H. M.; Omar, N. M. Solid-State Bioconversion of Oil Palm Empty Fruit Bunches for Cellulase Enzyme Production Using a Rotary Drum Bioreactor. *Biochem. Eng. J.* **2009**, *46*, 61–64.

Alves, S. B.; Pereira, R. M. Production of *Metarhizium anisopliae* (Metsch.) Sorok. and *Beauveria bassiana* (Bals.) Vuill. in Plastic Trays. *Ecossistema* **1989**, *14*,188–192.

Arora, S.; Rani, R.; Ghosh, S. Bioreactors in Solid-State Fermentation Technology: Design, Applications and Engineering Aspects. *J. Biotechnol.* **2018**, *269*, 16–34.

Arzumanov, T.; Jenkins, N.; Roussos, S. Effect of Aeration and Substrate Moisture Content on Sporulation of *Metarhizium anisopliae* Var. *acridum*. *Process Biochem.* **2005**, *40*, 1037–1042.

Bhattacharyya, B. C.; Banerjee, S.; Ghosh, T. K. Bioreactors: Functions in Fermentation Processes. In *Advances in Fermentation Technology*; Pandey, A., Larroche, C., Soccol, C. R., Dussap, C. G. (Eds.); Asiatech Publishers, Inc.: New Delhi, 2008; pp 172–201.

Biz, A.; Finkler, A. T. J.; Pitol, L. O.; Medina, B. S.; Krieger, N.; Mitchell, D. A. Production of Pectinases by Solid-State Fermentation of a Mixture of Citrus Waste and Sugarcane Bagasse in a Pilot-Scale Packed-Bed Bioreactor. *Biochem. Eng. J.* **2016**, *111*, 54–62.

Cavalcante, R. S.; Lima, H. L. S.; Pinto, G. A. S.; Gava, C. A. T.; Rodrigues, S. Effect of Moisture on *Trichoderma conidia* Production on Corn and Wheat Bran by Solid-State Fermentation. *Food Bioprocess Technol.* **2008**, *1*, 100–104.

Cherry, A. J.; Jenkins, N. E.; Heviefo, G.; Bateman, R.; Lomer, C. J. Operational and Economic Analysis of West African Pilot Scale Production Plant for Aerial Conidia of *Metarhizium* Spp. for use Mycoinsecticide against Locusts and Grasshoppers. *Biocntrl. Sci. Technol.* **1999**, *9*, 35–51.

Cinar, A.; Birol, G.; Parulekar, S. J.; Undey, C. *Batch Fermentation: Modeling Monitoring and Control*; CRC Press: Boca Raton, FL, 2003; 648 p.

Coban, I.; Sargin, S. Production of Trichoderma Micropropagules as a Biocontrol Agent in Static Liquid Culture Conditions by Using an Integrated Bioreactor System. *Biocntrl. Sci. Technol.* **2019**, *29*, 1197–1214.

Couto, S. R.; Sanromán, M. A. Application of Solid-State Fermentation to Food Industry: A Review. *J. Food Eng.* **2006**, *76*, 291–302.

Das, M. M.; Haridas, M.; Sabu, A. Biological Control of Black Pepper and Ginger Pathogens, *Fusarium oxysporum*, *Rhizoctonia solani* and *Phytophthora capsici*, using *Trichoderma* Spp. *Biocatal. Agric. Biotechnol.* **2019**, *17*, 177–183.

Das, M.; Sabu, A. Agro-Processing Residues for the Production of Fungal BCAs. In *Valorisation of Agro-Industrial Residues—Volume II: Non-Biological Approaches*; Springer: Cham, 2020; pp 107–126.

Das, M. M.; Haridas, M.; Sabu, A. Process Development for the Enhanced Production of Bio-Nematicide *Purpureocillium lilacinum* KU8 Under Solid-State Fermentation. *Bioresour. Technol.* **2020**, *308*, 123328.

De Cal, A.; Larena, I.; Guijarro, B.; Melgarejo, P. Mass Production of Conidia of *Penicillium frecuentans*, a Biocontrol Agent Against Brown Rot of Stone Fruits. *Biocontrol Sci. Technol.* **2002**, *12*, 715–725.

De la Cruz-Quiroz, R.; Roussos, S.; Hernandez-Castillo, D.; Rodríguez-Herrera, R.; López-López, L.; Castillo, F.; Aguilar, C. N. Solid-State Fermentation in a Bag Bioreactor: Effect of Corn Cob Mixed with Phytopathogen Biomass on Spore and Cellulase Production by *Trichoderma asperellum. Intech: Ferment. Process.* **2017**, *3*, 43–56.

Derakhti, S.; Shojaosadati, S. A.; Hashemi, M.; Khajeh, K. Process Parameters Study of α-Amylase Production in a Packed-Bed Bioreactor Under Solid-State Fermentation with Possibility of Temperature Monitoring. *Prep. Biochem. Biotechnol.* **2012**, *42*, 203–216.

Dhillon, G. S.; Brar, S. K.; Kaur, S.; Verma, M. Bioproduction and Extraction Optimization of Citric Acid from *Aspergillus niger* by Rotating Drum Type Solid-State Bioreactor. *Ind. Crops Prod.* **2013**, *41*, 78–84.

Díaz, A. B.; De Ory, I.; Caro, I.; Blandino, A. Solid-State Fermentation in a Rotating Drum Bioreactor for the Production of Hydrolytic Enzymes. In *ICHEAP-9: 9th International Conference on Chemical and Process Engineering, Pts.* 2009; pp 1–3.

Farinas, C. S.; Vitcosque, G. L.; Fonseca, R. F.; Neto, V. B.; Couri, S. Modeling the Effects of Solid-State Fermentation Operating Conditions on Endoglucanase Production Using an Instrumented Bioreactor. *Ind. Crops Prod.* **2011**, *34*, 1186–1192.

Gaikwad, V.; Panghal, A.; Jadhav, S.; Sharma, P.; Bagal, A.; Jadhav, A.; Chhikara, N. In *Designing of Fermenter and Its Utilization in Food Industries*, 2018.

Gangadharan, K.; Jeyarajan, R. Mass Multiplication of *Trichoderma* Spp. *J. Biol. Cntrl.* **1990**, *4*, 70–71.

Ge, X.; Vasco-Correa, J.; Li, Y. Solid-State Fermentation Bioreactors and Fundamentals. In *Current Developments in Biotechnology and Bioengineering*; Elsevier: Amsterdam, 2017; pp 381–402.

Gomez, S.; Fernandez, F. J.; Vega, M. C. Heterologous Expression of Proteins in Aspergillus. In *New and Future Developments in Microbial Biotechnology and Bioengineering*; Elsevier: Amsterdam, 2016; pp 55–68.

Gopalakrishnan, C.; Ramanujam, B.; Prasad, R. D.; Rao, N. S.; Rabindra, R. J. Use of Brewery Waste Amended Spent Malt as Substrate for Mass Production of *Trichoderma. J. Biol. Control* **2003**, *17*, 167–170.

Gowthaman, M. K.; Krishna, C.; Moo-Young, M. Fungal Solid-State Fermentation—An Overview. In *Applied Mycology and Biotechnology, Vol.1. Agriculture and Food Productions*; Khachatourians, G. G., Arora, D. K. (Eds.); Elsevier Science: The Netherlands, 2001; pp 305–352.

Gupta, S.; Dikshit, A. K. Biopesticides: An Ecofriendly Approach for Pest Control. *J. Biopest.* **2010**, *3*, 186–188.

Hashemi, M.; Mousavi, S. M.; Razavi, S. H.; Shojaosadati, S. A. Mathematical Modeling of Biomass and α-Amylase Production Kinetics by *Bacillus* Sp. in Solid-State Fermentation Based on Solid Dry Weight Variation. *Biochem. Eng. J.* **2011**, *53*, 159–164.

John, N. S.; Jeeva, M. L. Efficacy of Cassava By-Products as Carrier Materials of *Trichoderma harzianum*, a Bio-Control Agent against *Sclerotium rolfsii* Causing Collar Rot in Elephant Foot Yam. *J. Root Crop* **2014**, *40*, 1–6.

Jyoti, S.; Singh, D. *Fungi as Biocontrol Agents in Sustainable Agriculture. Microbes and Environmental Management*; Studium Press: India, 2016; pp 172–194.

Kang, S. W.; Lee, S. H.; Yoon, C. S.; Kim, S. W. Conidia Production by *Beauveria bassiana* (for the Biocontrol of a Diamondback Moth) During Solid-State Fermentation in a Packed-Bed Bioreactor. *Biotechnol. Lett.* **2005**, *27*, 135.

Krishna, C. Solid-State Fermentation Systems—An Overview. *Crit. Rev. Biotechnol.* **2005**, *25*, 1–30.

Kumar, S.; Manibhushan, T.; Archana, R. Trichooderma: Mass Production, Formulation, Quality Control, Delivery and Its Scope in Commercialization in India for the Management of Plant Diseases. *Afr. J. Agric. Res.* **2014,** *9,* 3838–3852.

Li, Y.; Ruiyan, S.; Yu, J.; Kandasamy, S.; Chen, J. Antagonistic and Biocontrol Potential of *Trichoderma asperellum* ZJSX5003 against the Maize Stalk Rot Pathogen *Fusarium gramineurum. Indian J. Microbiol.* **2016,** *56,* 318–327.

Lizardi-Jiménez, M. A.; Hernández-Martínez, R. Solid-State Fermentation (SSF): Diversity of Applications to Valorize Waste and Biomass. *3 Biotech.* **2017,** *7,* 44.

Méndez-González, F.; Loera-Corral, O.; Saucedo-Castañeda, G.; Favela-Torres, E. Bioreactors for the Production of Biological Control Agents Produced by Solid-State Fermentation. In *Current Developments in Biotechnology and Bioengineering*; Elsevier: Amsterdam, 2018; pp 109–121.

Mitchell, D. A.; Berovič, M.; Krieger, N. Introduction to Solid-State Fermentation Bioreactors. In *Solid-State Fermentation Bioreactors*; Springer: Berlin, Heidelberg, 2006; pp 33–44.

Moo Young, M.; Moreira, A.; Tegerdy, R. Principles of Solid-Substrate Fermentation. In *The Filamentous Fungi*; Smith, J., Berry, D., Kriistiansen, B. (Eds.); Edward Arnold: London, 1983; pp 144–177.

Moslim, R.; Hamid, N. H.; Wahid, M. B.; Kamarudin, N.; Ali, S. R. A. In *Mass Production of Metarhizium Anisopliae using Solid-State Fermentation and Wet Harvesting Methods, Proceedings of the PIPOC 2005 International Palm Oil Congress (Agriculture, Biotechnology and Sustainability)*, Kuala, Lumpur, Malaysia: Malaysian Palm Oil Board, 2005; pp 928–943.

Mustafa, M. G.; Khan, M. G. M.; Nguyen, D.; Iqbal, S. Techniques in Biotechnology: Essential for Industry. In *Omics Technologies and Bio-Engineering*; Academic Press: USA, 2018; pp 233–249.

Ooijkaas, L. P. *Fungal Biopesticide Production by Solid-State Fermentation. Growth and Sporulation of Coniothyrium ministrans*, 2000.

Pandey, A. Effect of Particle Size of Substrate of Enzyme Production in Solid-State Fermentation. *Bioresour. Technol.* **1991,** *37,* 169–172.

Pandey, A. Solid-State Fermentation. *Biochem. Eng. J.* **2003,** *13,* 81–84.

Pandey, A.; Selvakumar, P.; Soccol, C. R.; Nigam, P. Solid-State Fermentation for the Production of Industrial Enzymes. *Curr. Sci.* **1999,** *77,* 149–162.

Pandey, A.; Soccol, C. R.; Rodriguez-Leon, J. A.; Nigam, P. *Solid-State Fermentation in Biotechnology-Fundamentals and Applications*; Asiatech Publishers: New Delhi, 2001; pp.100–221.

Pandey, A.; Larroche, C.; Soccol, C. R.; Dussap, C. G. (Eds.). Bioreactors: Functions in Fermentation Processes. *Advances in Fermentation Technology*; Asiatech Publishers, Inc.: New Delhi, Chapter 7, 2008; pp 172–201.

Petrot, I.; Alabouvette, C.; Estefania, H. E.; Franca, S. The Use of Microbial Biocontrol Agents against Soil-Borne Diseases. *Agric. Innov.* **2015,** *37,* 103–133.

Pitol, L. O.; Biz, A.; Mallmann, E.; Krieger, N.; Mitchell, D. A. Production of Pectinases by Solid-State Fermentation in a Pilot-Scale Packed-Bed Bioreactor. *Chem. Eng. J.* **2016,** *283,* 1009–1018.

Posada-Flórez, F. J. Production of *Beauveria bassiana* Fungal Spores on Rice to Control the Coffee Berry Borer, *Hypothenemus hampei*, in Colombia. *J. Insect Sci.* **2008,** *8,* 41.

Prakash, M. G.; Gopal, K. V.; Anandraj, M.; Sharma, Y. R. Evaluation of Substrate for Mass Multiplication of Fungal Biocontrol Agents *Trichoderma harzianum* and *Trichoderma virens. J. Spices Aromat. Crops* **1999,** *8,* 207–210.

Qiu, L.; Li, J.; Li, Z.; Wang, J. Production and Characterization of Biocontrol Fertilizer from Brewer's Spent Grain via Solid-State Fermentation. *Sci. Rep.* **2019,** *9*, 480.

Ramachandran, S.; Roopesh, K.; Nampoothiri, K. M.; Szakacs, G.; Pandey, A. Mixed Substrate Fermentation for the Production of Phytase by *Rhizopus* Spp. Using Oil Cakes as Substrates. *Process Biochem.* **2005,** *40*, 1749–1754.

Ribeiro-Machado, A. C.; Monteiro, A. C.; Belasco, Geraldo-Martins, M. I. E. Production Technology for Entomopathogenic Fungus Using a Biphasic Culture System. *Pesq. Agropec. Bras.* **2010,** *45*, 1157–1163.

Roussos, S.; Olmos, A.; Raimbault, M.; Saucedo-Castañeda, G.; Lonsane, B. K. Strategies for Large Scale Inoculum Development for Solid-State Fermentation System: Conidiospores of *Trichoderma harzianum. Biotechnol. Technol.* **1991,** *5*, 415–420.

Saithi, S.; Tongta, A. Phytase Production of *Aspergillus niger* on Soybean Meal by Solid-State Fermentation Using a Rotating Drum Bioreactor. *Agric. Agric. Sci. Proc.* **2016,** *11*, 25–30.

Sala, A.; Barrena, R.; Artola, A.; Sánchez, A. Current Developments in the Production of Fungal Biological Control Agents by Solid-State Fermentation Using Organic Solid Waste. *Crit. Rev. Environ. Sci. Technol.* **2019,** *49*, 1–40.

Sandhu, S. S.; Sharma, A. K.; Beniwal, V.; Goel, G.; Batra, P.; Kumar, A.; Jaglan, S.; Sharma, A. K.; Malhotra, S. Myco-Biocontrol of Insect Pests: Factors Involved. Mechanism and Regulation. *J. Pathog.* **2012,** *2012*. https://doi.org/10.1155/2012/126819.

Schaechter, M.; Lederberg, J. *The Desk Encyclopedia of Microbiology*; Academic Press, Elsevier: New York, 2004; p 1149.

Singh, J.; Kaushik, N.; Biswas, S. Bioreactors—Technology & Design Analysis. *SciTech J.* **2014,** *1*, 28–36.

Singhania, R. R.; Patel, A. K.; Soccol, C. R.; Pandey, A. Recent Advances in Solid-State Fermentation. *Biochem. Eng. J.* **2009,** *44*, 13–18.

Soccol, C. R.; Vandenberghe, L. P. Overview of Applied Solid-State Fermentation in Brazil. *Biochem. Eng. J.* **2003,** *13*, 205–218.

Spier, M. R.; Vandenberghe, L. P. S.; Medeiros, A. B. P.; Soccol, C. R. Application of Different Types of Bioreactors in Bioprocesses. *Bioreactors: Design, Properties and Applications*; Nova Science Publishers Inc.: New York, 2011; pp 55–90.

Tekere, M. Microbial Bioremediation and Different Bioreactors Designs Applied. *Biotechnology and Bioengineering*; Intech Open, 2019.

Tranier, M. S.; Pognant-Gros, J.; Quiroz, R. D. L. C.; González, C. N. A.; Mateille, T.; Roussos, S. Commercial Biological Control Agents Targeted Against Plant-Parasitic Root-Knot Nematodes. *Braz. Arch. Biol. Technol.* **2014,** *57*, 831–841.

Tsouko, E.; Kachrimanidou, V.; Dos Santos, A. F.; Lima, M. E. D. N. V.; Papanikolaou, S.; de Castro, A. M.; Freire, D. M. G.; Koutinas, A. A. Valorization of By-Products from Palm Oil Mills for the Production of Generic Fermentation Media for Microbial Oil Synthesis. *Appl. Biochem. Biotechnol.* **2017,** *181*, 1241–1256.

Usta, C. Microorganisms in Biological Pest Control—A Review (Bacterial Toxin Application and Effect of Environmental Factors). *Current Progress in Biological Research*; IntechOpen, 2013; pp 287–317.

Webb, C.; Manan, M. A. Design Aspects of Solid-State Fermentation as Applied to Microbial Bioprocessing. *J. Appl. Biotechnol. Bioeng.* **2017,** *4* (1), 511–532.

Weber, F. J.; Tramper, J.; Rinzema, A. A Simplified Material and Energy Balance Approach for Process Development and Scale-Up of *Coniothyrium minitans* Conidia Production by Solid-State Cultivation in a Packed-Bed Reactor. *Biotechnol. Bioeng.* **1999,** *65*, 447–458.

Weiland, P. Principles of Solid-State Fermentation. *Treatment of Lignocellulosic with White Rot Fungi*; Zadrazil, F., Reiniger, P. (Eds.); Elsevier: London, 1988; pp 64–76.

Xie, L.; Chen, H. M.; Yang, J. B. Conidia Production by *Beauveria bassiana* on Rice in Solid-State Fermentation Using Tray Bioreactor. *Advanced Materials Research*; Trans Tech Publications Ltd., 2013; Vol. 610, pp 3478–3482.

Ye, S. D.; Ying, S. H.; Chen, C.; Feng, M. G. New Solid-State Fermentation Chamber for Bulk Production of Aerial Conidia of Fungal Biocontrol Agents on Rice. *Biotechnol. Lett.* **2006,** *28*, 799–804.

Zhang, Y.; Liu, J.; Zhou, Y.; Ge, Y. Spore Production of *Clonostachys rosea* in a New Solid-State Fermentation Reactor. *Appl. Biochem. Biotechnol.* **2014,** *174*, 2951–2959.

CHAPTER 3

Fungal Biomass Supported on Polyurethane Shavings for Removing the Dye Direct Black 38

JULIA MARIANA MÁRQUEZ-REYES[1*],
JUAN PABLO HERNÁNDEZ-RODRÍGUEZ[2],
ORQUÍDEA PÉREZ-GONZÁLEZ[2], CELESTINO GARCÍA-GÓMEZ[3], and
HUMBERTO RODRÍGUEZ-FUENTES[4]

[1]UANL, Universidad Autónoma de Nuevo León, Facultad de Agronomía, Laboratorio de Remediación Ambiental y Análisis de Planta, Suelo, Av. Francisco Villa S/N, Colonia Ex Hacienda El Canadá, General Mariano Escobedo, N. L., México

[2]Facultad de Ciencias Biológicas, Av. Pedro de Alba S/N, Ciudad Universitaria, San Nicolás de los Garza, N.L. México

[3]Universidad Autónoma de Nuevo León, Facultad de Agronomía, Laboratorio de Remediación Ambiental y Análisis de Planta, Suelo. Av. Francisco Villa S/N, Colonia Ex Hacienda El Canadá, General Mariano Escobedo, N. L., México

[4]Universidad Autónoma de Nuevo León, Facultad de Agronomía, Av. Francisco Villa S/N, Colonia Ex Hacienda El Canadá, General Mariano Escobedo, N. L., México

Corresponding author. E-mail: julia.marquezrys@uanl.edu.mx

ABSTRACT

Water pollution by azo dyes has increased due to the lack of treatment systems and poor disposal of contaminated water, causing health problems. Currently, there are biological technologies that use fungi for dye discoloration because they are environmentally friendly, have low operating costs, and generate less waste. The goal of this study was to isolate fungal strains with the ability to discolor the dye direct black 38 (ND38) and evaluate the discoloration percentage in a batch reactor and a continuous reactor packed with polyurethane, both with the presence of *Aspergillus* sp. During this experiment, *Syncephalastrum* sp., *Aspergillus* sp., *Monilia* sp., *Cunninghamella* sp., *Isaria* sp., and *Mycothypha* sp. were isolated. In particular, *Cunninghamella* sp., *Mycotypha* sp., *Aspergillus* sp., and *Syncephalastrum* sp. showed the highest discoloration rates (92.13%, 89.56%, 91.44%, and 91.48%, respectively). The batch reactor with *Aspergillus* sp. discolored the ND38 up to 95%. The continuous reactor packed with polyurethane shavings properly immobilized *Aspergillus* sp. and decolorized 80% of the ND38 in the first 12 h. Additionally, it maintained an average discoloration percentage of 40% during its operating time. *Aspergillus* sp. has a high biotechnological potential for the removal of azo dyes.

3.1   INTRODUCTION

In the last decades, due to the increase of new productive activities, thousands of dyes and pigments have been synthesized and released into the environment through wastewater tributaries without any control (Pandey et al., 2016; Tripathi et al., 2019), polluting water bodies and causing a major problem worldwide (Medhi and Thakur, 2018). In industrial effluents, it has been estimated that up to 20% of the dye used is discarded (Pandey et al., 2018). Of all the dyes produced, azo dyes are generated in large quantities, between 60% and 70% (Bruschweiler and Merlot, 2017); therefore, they comprise half of the pollutants discharged into the effluents (Zhao and Hardin, 2007). Some dyes are stable in extreme environmental conditions and due to their recalcitrant, carcinogenic, and mutagenic nature, generate different adverse effects on the environment, affecting the life of the species that inhabit it (Saeed et al., 2016; Pandey et al., 2018).

At present, different physicochemical methods for dye removal have been developed. These include adsorption, electrocoagulation, precipitation, flocculation, foam flotation, ion exchange, advanced oxidation, membrane filtration, ozonation, and reverse osmosis (Callegari et al., 2017); however, all these processes have a high cost, produce large amounts of hazardous waste, and generate high energy costs (Warade et al., 2016). Therefore, it is necessary to find less-expensive alternatives that have the same or even greater effectiveness than these treatments. Within this area, biological treatments that use microorganisms, such as bacteria, algae, and filamentous fungi, represent an efficient alternative for the removal of this type of contaminants. Biological methods are environmentally friendly since they minimize the use of chemicals and by-products. In addition, they reduce energy costs, making them more profitable (Przyśtas et al., 2018; Sharma and Kaur, 2018).

Fungi have been used to remove, adsorb, and absorb heavy metals and dyes because of their multiple physiological mechanisms (Sen et al., 2016; Bhatia et al., 2017); as well as for their ability to cover large areas, rapid growth, easy handling, and low-separation costs at the end of the dye-removal process (Mishra and Malik, 2014). *Aspergillus* sp. is one of the most-used genera, since it has a high capacity for dye degradation. This capacity has been evaluated in the presence of Basic Blue 9 (Fu and Viraraghavan, 2002), Reactive Black (Jin et al., 2007), Procion Red (Corso and Maganha-de-Almeida, 2009), and Remazol Blue (Tastan et al., 2010).

Dye toxicity is one of the main limitations of biological discoloration, since this decreases the biotransformation performance and generates a slow reaction rate. To counteract these limitations, it is necessary to properly select the potential microorganisms that will carry out the discoloration process (Khandare and Govindwar, 2015). The immobilization of microbial cells is a factor that favors the removal of contaminants (Talha et al., 2018), due to its easy use, greater operational stability of the microorganisms, its greater tolerance to high concentrations of contaminants, and the prolongation of operating time (Kureel et al., 2016). Some of the materials evaluated for cell immobilization are coconut shell, charcoal, sugarcane bagasse, peat, compost, polyurethane foam, alginate, etc. (Patil et al., 2006; Singh et al., 2010a, 2010b). Polyurethane is an inert material that is used for cell production since it favors homogeneous growth, facilitates the formation of biofilms, provides greater surface area, shows high mechanical resistance, is resistant to organic solvents, has low compaction, is easy to handle, and allows the quantification of biomass by direct methods (Silva et al., 2013; Quintana-Quirino et al., 2019).

Batch and continuous reactors are being implemented to remove azo dyes through microbial activity supported on organic and inorganic materials. Continuous bed reactors packed with immobilized cells in different substrates are more efficient and easy-to-use than the rest of the reactor designs (Bharti et al., 2017; Giri et al., 2017).

Different studies have evaluated the removal of azo dyes in the presence of fungal activity using batch-type cultures (Fu and Viraraghavan, 2002; Corso and Maganha-de-Almeida, 2008; Tastan et al., 2010; Rani et al., 2014) and reactors packed with polyurethane foam, which have produced good results for the removal of contaminants (Yang et al., 2010); however, there is little published information on the behavior of continuous reactors packed with fungal activity for the removal of azo dyes. Therefore, the objective of this study was to isolate fungal strains with the ability to discolor the ND38 dye and to evaluate the discoloration percentage in a batch reactor and a continuous reactor packed with polyurethane both with *Aspergillus* sp. activity.

3.2   MATERIALS AND METHODS

3.2.1   SAMPLING AND ISOLATION OF FUNGAL STRAINS

To obtain the biological material, samples from the San Nicolás de los Garza area, Nuevo León, Mexico, and a commercial fertilizer were used. The soil was obtained from a site impacted by mining activity, located in San Nicolás de los Garza; whereas, the fertilizer (peat) was acquired in a specialized establishment. During sample pretreatment, large organic and inorganic residues (e.g., plastics, branches, leaves, etc.) were removed. The samples were dried at room temperature and sieved to a particle size of 10 mm. For the inoculums, 2 g of each solid sample were taken and added to a sterile saline solution in order to make serial dilutions (1×10^{-2}). From the last dilution, inoculums were taken and seeded in potato dextrose agar (PDA) plates and then, they were incubated at 30°C for 5 days. Each fungal colony present in the PDA plates was reseeded until obtaining pure cultures. Subsequently, monosporic cultures were established; these were used to stain and obtain morphological information of each fungus, as well as to carry out the taxonomic identification through the use of a specialized manua (Barnett and Hunter, 1998).

3.2.2   INOCULUM PREPARATION

Sowing was performed in Petri dishes with PDA. Cultures were incubated at 30°C (Thermo Scientific, FT Heratherm, Massachusetts, USA) under dark conditions, until a 15-day mycelial growth was achieved for all isolated strains.

3.2.3   SELECTION OF EFFICIENT FUNGAL STRAINS FOR DYE REMOVAL

Purified isolates were selected according to the highest percentage of discoloration shown. A piece of agar of 1 cm in diameter was inoculated in flasks containing 100 mL of potato dextrose broth (CPD), previously sterilized at 121°C and 103 kPa; then 20 mg/L of ND38 were added. The samples were kept in incubation (Thermo Scientific) at 30°C for 5 days under static conditions. After this period of time, the mycelium was separated from the solution by filtration (Whatman No. 1) and the solution was analyzed by UV-Vis spectrophotometry (Thermo Scientific) at 520 nm.

The discoloration percentage was calculated using the equation proposed by Chen et al. (2018):

$$Discoloration \% = \frac{A_o - A_1}{A_0} * 100 \tag{3.1}$$

where A_0 is the initial absorbance of the control and A_1 is the average discoloration absorbance.

3.2.4 BATCH REACTOR WITH ASPERGILLUS SP.

A total of 4×10^7 spores/mL of *Aspergillus* sp. and different concentrations of ND38 (5 mg/L, 10 mg/L, 20 mg/L, 30 mg/L) were inoculated in flasks containing 100 mL of previously sterilized CPD, maintaining the same operating conditions already described. Every 24 h, 1 mL of solution was collected; this solution was centrifuged (Thermo Scientific) at 5000 rpm for 10 min to separate the mycelium.

3.2.5 CONTINUOUS REACTOR

A borosilicate reactor of 60 cm long, 5 cm of internal diameter, and a total volume of 260 mL was used. The polyurethane was obtained from a solid-waste deposit. This material was washed with running water and cut into 7 mm fragments to produce shavings that were used as a support for fungal development. All the polyurethane used in these experiments was disinfected with ultraviolet radiation inside a laminar flow hood (Labconco, S3612504) for 15 min. Polyurethane shavings were inoculated with *Aspergillus* sp. spores (4×10^7 spores/mL) at 30°C in CPD to establish a mycelial growth of 8 days. Subsequently, 3.8 g were placed inside the reactor. The final operating volume of the reactor was 150 mL and with the help of a peristaltic pump, a flow of 0.3 mL/min of the feed solution (consisting of CPD and 10 mg/L of ND38 dye) was maintained. The hydraulic retention time was 8.3 h. Samples of the effluent were collected and filtered at different times in order to quantify the discoloration percentage at 520 nm during the 172-hour operation period of the reactor. The excess of water was removed from the polyurethane and then it was weighed; subsequently, it was dried for 48 hours at 60°C in order to obtain the dry weight. Then, by weight difference, the amount of biomass developed on the polyurethane shavings during the operation time of the reactor was calculated using the following equation:

$$\frac{Biomass\,mg}{polyurethane\,g} = \frac{\left(B_{final} - P\right) - \left(B_{initial} - P\right)}{P} \tag{3.2}$$

where, B_{final} is the dry biomass quantified at the end of the experiment (mg), $B_{initial}$ is the dry biomass quantified at the beginning of the experiment (mg), and P is the amount of polyurethane used in the reactor.

3.3   RESULTS AND DISCUSSION

3.3.1   *ISOLATION AND MORPHOLOGICAL IDENTIFICATION*

The morphology observed in the mycelial development of each isolated strain coincides with what has been previously reported in the literature (Barnett and Hunter, 1998). The strain 1 isolated from the peat showed straight, branched conidiophores, with enlarged tips and with a rod-shaped sporangiole head, where each sporangiole had a row of almost spherical conidia; these characteristics coincided with the morphology described for *Syncephalastrum* sp. Several strains were isolated from the soil with mining activity. The first strain from this soil had biseriate conidial heads in compact columns, smooth wall-hyaline stipes, vesicles of variable shape (e.g., spherical), metulae occupying half or two-thirds of the vesicle, and globose conidia, characteristics that match the genus *Aspergillus* sp. (Fig. 3.1). The second strain showed rolled hyaline conidia corresponding to the genus *Monilia* sp. The third strain had a non-septate white mycelium, branched conidiophores, and hyaline globose conidia, which are characteristic of *Cunninghamella* sp. The fourth strain showed hyaline, ovoid, and dry conidia, typical of the genus *Isaria* sp. Finally, two isolates of *Mycothypha* sp. were obtained, which presented unbranched sporangiophors and typically elongated vesicles coated with sporangioli.

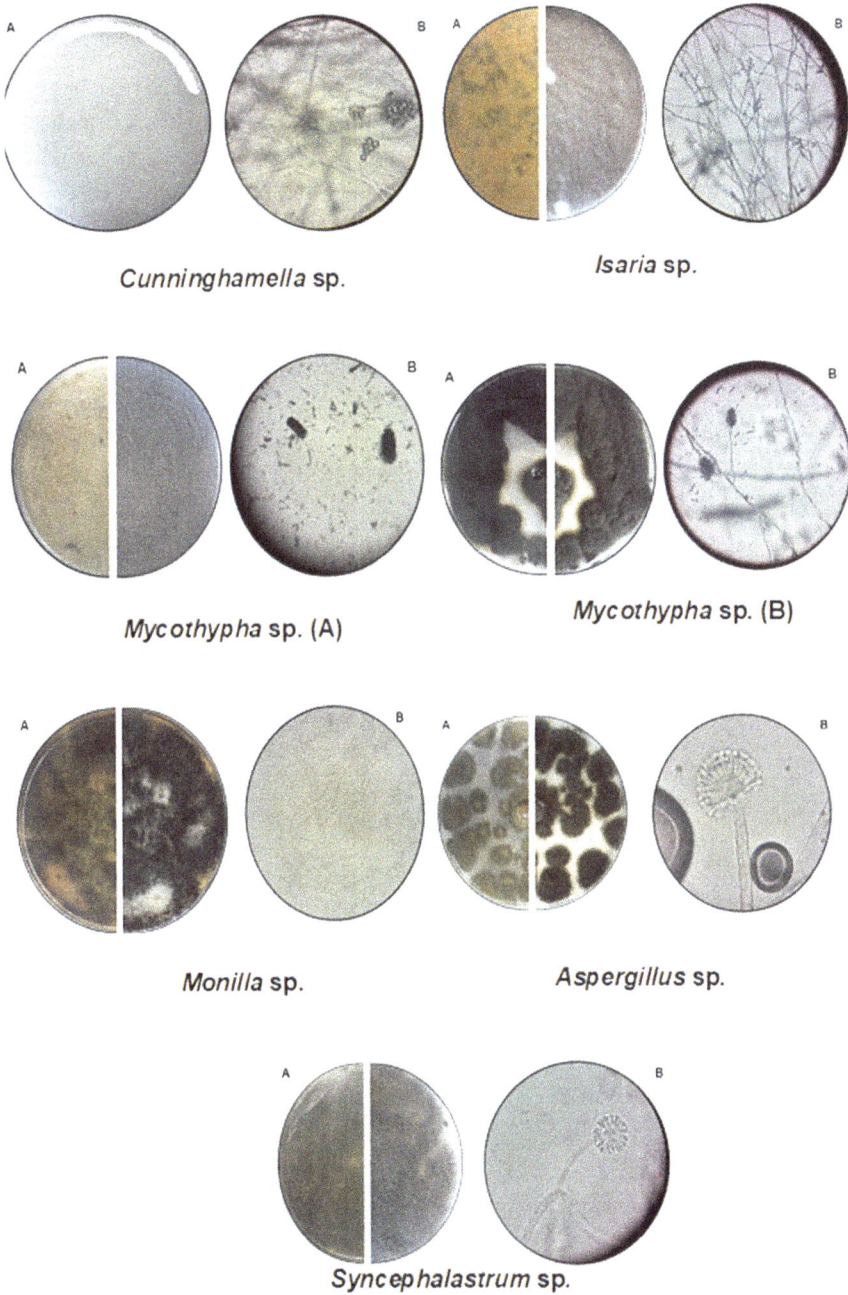

Cunninghamella sp.

Isaria sp.

Mycothypha sp. (A)

Mycothypha sp. (B)

Monilla sp.

Aspergillus sp.

Syncephalastrum sp.

FIGURE 3.1 Images taken of the macroscopic (a) and microscopic (b) part for the characterization of each of the isolated fungal samples.

3.3.2 DISCOLORATION DUE TO FUNGAL ACTIVITY

Figure 3.2 shows the discoloration percentages obtained from each genus of fungi. *Isaria* sp., *Monilia* sp., and *Mycotypha* sp. (A) showed a discoloration percentage below 72%; these genera have not been reported for discoloration processes. Particularly, *Isaria* sp. has been used for biological control (Zimmermann, 2008); while *Monilia* sp. has only been reported in the literature as a cause of plant diseases (Yin et al., 2017). Finally, *Mycotypha* sp. could be related to some type of pathogenicity in humans (Lacroix et al., 2007). On the other hand, *Cunninghamella* sp., *Mycotypha* sp., *Aspergillus* sp., and *Syncephalastrum* sp. showed similar discoloration rates (92.13%, 89.56%, 91.44%, and 91.48%, respectively). *Mycotypha* sp., *Cunninghamella* sp., and *Syncephalastrum* sp. have been used for bioethanol production, herbicide removal, and chitosan production for dye removal (Ambrósio et al., 2012; Jayachandra et al., 2012; Takaki, 2012; Batista et al., 2013). *Aspergillus* sp. has been frequently reported for the discoloration of different types of dyes. This genus has been used with good results in the discoloration of 90% of reactive black 5 (50 μM), 98% of Congo red (150 mg / L), and 94% of true blue (0.02% w/v) (Copete-Pertuz et al., 2019; Khelifi et al., 2009; Ponraj et al., 2011).

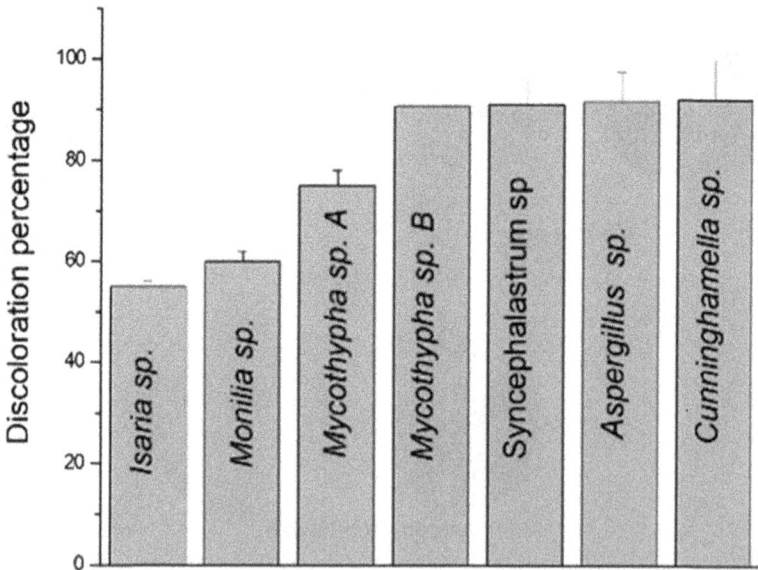

FIGURE 3.2 Percent discoloration of direct black dye 38 by isolated fungi. Averages ± standard deviation are shown.

3.3.3 DISCOLORATION OF ND38 WITH ASPERGILLUS SP.

The discoloration percentage during the first 12 h of exposure was 65%. At the end of the experiment (13 days), it was possible to discolor ND38 from 85% (5 mg/L) to 95% (20 mg/L) (Fig. 3.4).

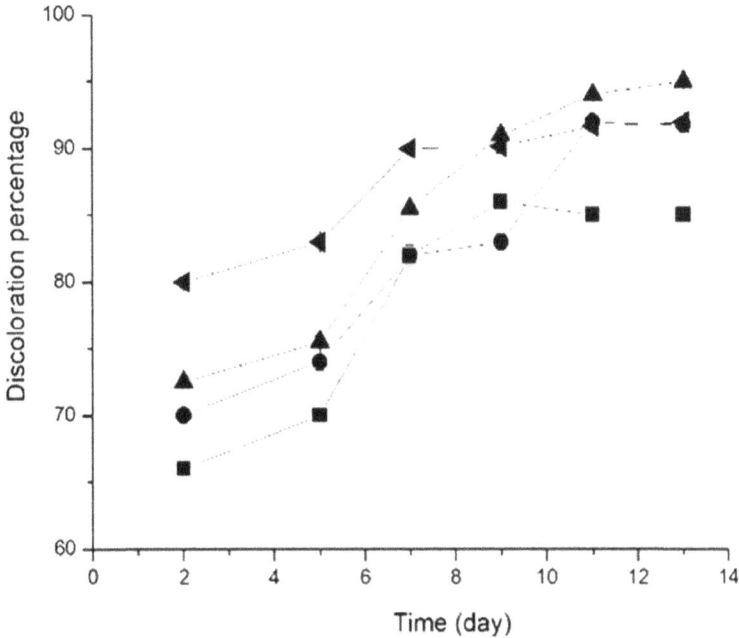

FIGURE 3.3 Percentage of discoloration of direct black dye 38 in a static system with *Aspergillus* sp. activity supported on polyurethane pieces at different concentrations 5 mg/L (■), 10 mg/L (●), 20 mg/L (▲), and 30 mg/L (◄). Here are shown averages of the percentages ± standard deviation.

The discoloration percentages obtained in this work are acceptable compared to those obtained in other studies with azo dyes. For example, the bright red-dye reagent K-2BP (33.9 mg/L) was discolored 77.9% in the presence of *Aspergillus fumigatus* (Wang et al., 2008); while orange G (25 mg/L) was decolorized 78% in the presence of *Aspergilllus niger* (Sivasamy and Sundarabal, 2011). Finally, *Aspergillus niger* combined with *Aspergillus terreus* discolored 100% of MX-5B red (200 µg/mL) (Almedia and Corso, 2014). The results show a greater discoloration during the first experimental hours; for this reason, sorption and degradation phenomena of the dye by biomass are not ruled out. In addition, there are reports that have evaluated the biosorption and biotransformation capacity of azo dyes in *Aspergillus ochraceus* (Kadam et al., 2011), *Aspergillus fumigatus* (Whang and Bao-e, 2007), *Aspergillus niger* (Fu and Viraraghavan, 2002; Bankole et al., 2019), and *Bacillus pseudomycoides* (Naveen et al., 2019).

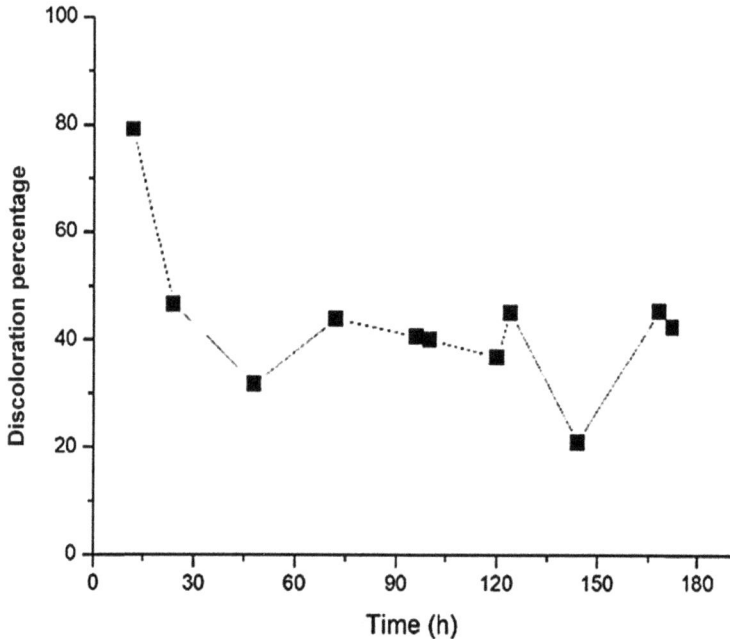

FIGURE 3.4 Percentage of discoloration of direct black dye 38 in the reactor packed with pieces of polyurethane and *Aspergillus* sp. activity. Here are shown averages of the percentages ± standard deviation.

The obtained results seem to indicate that at a higher concentration of the dye in the medium, the percentage of discoloration is increased; however, similar studies have indicated otherwise. For example, *Lysinibacillus* sp. can discolor up to 100% of methanyl yellow at 200 ppm, but this was only able to decolorize 62% at 1000 ppm (Anjaneya et al., 2011); while *Sphingomonas paucimobilis* completely discolored methyl red at 750 ppm, but only 38% at a concentration of 1000 ppm (Ayed et al., 2011).

Most azo dyes have sulfonate substituent groups and a high molecular weight, in addition to being unlikely to move through cell membranes. Therefore, discoloration will depend on the intracellular absorption of the dye and the enzymatic activity (e.g., azoreductases, laccase, peroxidase, tyrosinase, etc.), activated by the genes depending on the microorganism, as well as on the culture conditions (Ramalho et al., 2002; Solís et al., 2012). Additionally, the activation of one or more enzymes can be induced (Solís et al., 2012), which could generate an increase in the discoloration generated by oxidative stress caused by dye concentration.

3.3.4 CONTINUOUS BIOREACTOR

Different types of supports that are used for the immobilization of fungal activity in dye removal have been evaluated in batch and continuous reactors (Espinosa-Ortíz et al., 2016; Sen et al., 2016; Przýstas et al., 2018); however, there are no studies in the literature of a system similar to the one used in this study.

During the first 12 h of operation, 80% of the ND38 was discolored, which coinciding with the results obtained in the batch experiments. Wang and Young-You (2007) reported 89.1% and 93.5% removal of reactive bright red and reactive bright blue (2 g/L), when immobilizing *Aspergillus fumigatus* in chitosan during the first 12 h. Przýstas et al. (2018), reported bright-green (0.1 g/L) discolorations of 82% and 19%, using *Pleurotus ostreatus*, *Gleophyllum odoratum*, and *Polyporus picipes*, immobilized in propylene and sawdust. The obtained results indicate that the discoloration percentages are higher during the first contact hours, due to the biosorption of the dye on the surface of the fungal biomass. Subsequently, the discoloration decreases up to 40% after 30 hours of operation, indicating that the binding sites of the fungal biomass become saturated as well as the electrostatic attraction sites. Sen et al. (2016) indicate that carboxyl, amino, and phosphate groups are responsible for the main dye-binding sites using *Aspergillus niger*; however, the percentage of average discoloration (42%) remained due to fungal activity. The given data suggest that *Aspergillus* sp. was biotransforming the dye through enzymatic activity and extracellular compounds, which interacted with the dye favoring the decolorization process. Copete-Pertuz et al. (2019) have reported that *Aspergillus terreus* and *Trichoderma viride* show laccase and peroxidase activity in the discoloration of reactive black 5. Finally, Arunprasath et al. (2019) concluded that the presence of laccase in the degradation process of malachite green causes the conversion of carboxylic acids into alkanes and alkenes, which can be a part of the fungal-secreted lipase enzyme.

The design system used in this study showed acceptable characteristics in its operation mode. The most-notable feature corresponded to mycelial development, quantifying 400 mg of biomass per one gram of polyurethane.

In the literature, there are only reports of 60 mg biomass/g of polyurethane (Marin-Cervantes et al., 2008). Biomass immobilization improved the discoloration because the surface area increased and the limitations for mass transfer decreased. One disadvantage was the transfer of oxygen in the lower part of the reactor, since it limited the discoloration. This limits the production of reducers that participate in the discoloration process, generates less mycelial development, increases the temperature, and can cause the death of the fungus (Cortés-Sánchez et al., 2014; Pandey et al., 2018; Ferrer-Romero et al., 2019; Menezesa et al., 2019). However, for future experiments, pH conditions, redox potential, temperature, and oxygen concentration should be considered to favor mass transfer and dissipation of metabolic heat generated in the reactor.

3.4　CONCLUSIONS

Aspergillus sp. has the ability to remove the ND38 dye. Polyurethane foam favored the growth and immobilization of the fungus. In batch-type reactors, the discoloration percentages were higher compared to those of the continuous reactor with *Aspergillus* sp. activity and supported on polyurethane shavings. To increase the discoloration efficiency of the continuous system, it is convenient to maintain the rheological characteristics of the fluid that will favor the optimal growth of the fungus.

KEYWORDS

- ***Aspergillus* sp.**
- **azo dye**
- **bioreactor**
- **removal**

REFERENCES

Almedia, E. J. R.; Corso, C. R. Comparative Study of Toxicity of Azo Dye Procion Red MX-5B Following Biosorption and Biodegradation Treatments with the Fungi *Aspergillus niger* and *Aspergillus terreus*. *Chemosphere* **2014,** *112*, 317–322.

Anjaneya, O.; Yogesh Souche, S.; Santoshkumar, M.; Karegoudar, T. B. Decolorization of Sulfonated Azo Dye Metanil Yellow by Newly Isolated Bacterial Strains: *Bacillus* sp. Strain AK1 and *Lysinibacillus* Sp. Strain AK2. *J. Hazard. Mater.* **2011**, *190*, 351–358.

Arunprasath, T.; Sudalai, S.; Meenatchi, R.; Jeyavishnu, K.; Arumugam, A. Biodegradation of Triphenylmethane Dye Malachite Green by a Newly Isolated Fungus Strain. *Biocatal. Agric. Biotechnol.* **2019**, *17*, 672–679.

Ayed, L.; Mahdhi, A.; Cheref, A.; Bakhrouf, A. Decolorization and Degradation of Azo Dye Methyl Red by an Isolated *Sphingomonas paucimobilis*: Biotoxicity and Metabolites Characterization. *Desalination* **2011**, *274*, 272–277.

Ambrósio, S. T.; Vilar Júnior, J. C.; Alves-da-Silva, C. A.; Okada, K.; Nascimento, A. E.; Longo, R. L.; Campos-Takaki, G. M. Biosorption Isotherm Model for the Removal of Reactive Azo Dyes by Inactivated Mycelia of *Cunninghamella elegans* UCP542. *Molecules* **2012**, *17*, 452–462.

Barnett, H. L.; Hunter, B. B. *Illustrated Genera of Imperfect Fungi*, 4th ed.; APS Press: St. Paul, MN, 1998.

Bankole, P. O.; Adekunle, A. A.; Govindwar, S. P. Demethylation and Desulfonation of Textile Industry Dye, Thiazole Yellow G by *Aspergillus niger* LAG. *Biotechnol. Rep.* **2019**, *23*, e00327.

Batista, A. C. L.; Freitas, M. C.; Batista, J. B.; Nascimento, A. E.; Campos-Takaki, G. M. Eco-friendly Chitosan Production by *Syncephalastrum racemosum* and Application to the Removal of Acid Orange 7 (AO7) from Wastewaters. *Molecules* **2013**, *18*, 7646–7660.

Bhatia, D.; Sharma, N. R.; Singh, J.; Kanwar, R. S. Biological Methods for Textile Dye Removal from Wastewater: A Review. *Crit. Rev. Environ. Sci. Technol.* **2017**, *47*, 1–41.

Chen, Y.; Fenga, L.; Lia, H.; Wanga, Y.; Chena, G.; Zhanga, Q. Biodegradation and Detoxification of Direct Black G Textile Dye by a Newly Isolated Thermophilic Microflora. *Bioresour. Technol.* **2018**, *250*, 650–657.

Copete-Pertuza, L. S.; Alandete-Novoa, F.; Plácido, J.; Guillermo, A.; Correa-Londoño, G. A.; Mora-Martínez, A. L. Enhancement of Ligninolytic Enzymes Production and Decolourising Activity in *Leptosphaerulina* Sp. by Co-cultivation with *Trichoderma viride* and *Aspergillus terreus*. *Sci. Total Environ.* **2019**, *646*, 1536–1545.

Corso, C. R.; Maganha-de-Almeida, A. C. Bioremedation of Dyes in Textile Effluents by *Aspergillus oryzae*. *Microb. Ecol.* **2009**, *57*, 384–390.

Cortés-Sánchez, A. J.; Guadarrama, L. M.; Díaz-Ramírez, M. Producción de Biomasa a Partir de *Aspergillus oryzae* en Cultivo Sumergido. *Rev. Cien. Biol. Salud* **2014**, *16*, 11–16.

Espinosa-Ortíz, E. J.; Rene, E. R.; Pakshirajan, K.; Van-Hullebusch, E. D.; Lens, P. N. L. Fungal Pelleted Reactors in Wastewater Treatment: Applications and Perspectives. *Chem. Eng. J.* **2016**, *283*, 553–571.

Fu, Y.; Viraraghavan, T. Removal of Congo Red from an Aqueous Solution by Fungus *Aspergillus niger*. *Adv. Environ. Res.* **2002**, *7*, 239–247.

Ferrer-Romero, J. C.; Mas-Diego, S. M.; Beltrán-Delgado, Y.; Morris-Quevedo, J. H.; Díaz-Fernández, U. Kinetic Study of the Production of Biomass and Phenolic Compounds for *Pleurotus ostreatus* in Submerged Phase. *Rev. Cubana Quím.* **2019**, *31* (1), 15–20.

Giri, B. S.; Goswami, M.; Singh, R. S. Review on Application of Agro-waste Biomass Biochar for Adsorption and Bioremediation Dye. *Med. Biol. Technol. J.* **2017**, *1* (7), 1–3.

Jayachandra, T.; Venugopal, C.; Anu-Appaiah, K. A. Utilization of Phytotoxic Agro-waste Coffee Cherry Husk through Pretreatment by the Ascomycetes Fungi *Mycotypha* for Biomethanation. *Energy Sustain. Dev.* **2012**, *15* (1), 104–108.

Kadam, A. A.; Telke, A. A.; Jagtap, S. S.; Govindwar, S. P. Decolorization of Adsorbed Textile Dyes by Developed Consortium of *Pseudomonas* sp. SUK1 and *Aspergillus ochraceus* NCIM-1146 under Solid State Fermentation. *J. Hazard. Mater.* **2011**, *189*, 486–494.

Khandare, R. V.; Govindwar, S. P. Phytoremediation of Textile Dyes and Effluents: Current Scenario and Future Prospects. *Biotechnol. Adv.* **2015**, *33* (8), 697–1714.

Khelifi, E.; Ayed, L.; Bouallagui, H.; Touhami, Y.; Hamdi, M. Effect of Nitrogen and Carbon Sources on Indigo and Congo Red Decolourization by *Aspergillus alliaceus* Strain 121C. *J. Hazard. Mater.* **2009**, *163*, 1056–1062.

Kureel, M. K.; Geed, S. R.; Giri, B. S.; Shukla, A. K.; Rai, B. N.; Singh, R. S. Removal of Aqueous Benzene in the Immobilized Batch and Continuous Packed Bed Bioreactor by Isolated *Bacillus* Sp. M1. *Res. Effic. Technol.* **2016**, *2*, S87–S95.

Lacroix, C.; Leblanc, T.; Feuilhade-de-Chauvin, M. Isolation of *Mycotypha microspora* from Stool Samples of a Leukemic Child. *J. Mycol. Med.* **2007**, *17* (3), 188–190.

Marin-Cervantes, M. C.; Matsumoto, Y.; Ramírez-Coutiño, L.; Rocha-Pino, Z.; Viniegra, G.; Shirai, K. Effect of Moisture Content in Polyurethane Foams as Support for Solid-Substrate Fermentation of *Lecanicillium lecanii* on the Production Profiles of Chitinases. *Proces. Biochem.* **2008**, *43* (1), 42–32.

Menezesa, O.; Brito, A.; Hallwass, F.; Forencio, L.; Kato, M. T.; Gaazza, S. Coupling Intermittent Micro-aeration to Anaerobic Digestion Improves Tetra-azo Dye Direct Black 22 Treatment in Sequencing Batch Reactors. *Chem. Eng. Res. Des.* **2019**, *146*, 369–378.

Mishra, A.; Malik, A. Novel Fungal Consortium for Bioremediation of Metals and Dyes from Mixed Waste Stream. *Bioresour. Technol.* **2014**, *171*, 217–226.

Naveen, K.; Surbhi, S.; Tithi, M.; Rachana, S.; Simran, T.; Indu, S. T. Biodecolorization of Azo Dye Acid Black 24 by *Bacillus pseudomycoides*: Process Optimization Using Box–Behnken Design Model and Toxicity Assessment. *Bioresour. Technol. Rep.* **2019**, *8*, 100311.

Pandey, R. K.; Rana, B.; Tewari, S.; Sarkar, A.; Dubey, A.; Chandra, D.; Tewari, L. Exploration of Plant-Biomass Degrading Fungi for In Vitro Mycoremediation of Toxic Synthetic Dyes. *Int. J. Curr. Microbiol. Appl. Sci.* **2016**, *5*, 581–592.

Pandey, R. K.; Tewarib, S.; Tewari, L. Lignolytic Mushroom *Lenzites elegans* WDP2: Laccase Production, Characterization, and Bioremediation of Synthetic Dyes. *Ecotoxicol. Environ. Saf.* **2018**, *158*, 50–58.

Patil, N. K.; Veeranagouda, Y.; Vijaykumar, M. H.; Anand-Nayak, S.; Karegoudar, T. B. Enhanced and Potential Degradation of *o*-Phthalate by *Bacillus* Sp. Immobilized Cells in Alginate and Polyurethane. *Int. Biodeter. Biodegr.* **2006**, *57*, 82–87.

Ponraj, M.; Jamunarani, P.; Zambare, V. Isolation and Optimization of Culture Conditions for Decolorization of True Blue Using Dye Decolorizing Fungi. *Asian J. Exp. Biol. Sci.* **2011**, *2* (2), 270–277.

Przýstas, W.; Zabłocka-Godlewska, E.; Grabińska-Sota, E. Efficiency of Decolorization of Different Dyes Using Fungal Biomass Immobilized on Different Solid Supports. *Braz. J. Microbiol.* **2018**, *49* (2), 285–295.

Quintana-Quirinoa, M.; Morales-Osorio, C.; Vigueras-Ramírez, G.; Vázquez-Torres, H.; Shirai, K. Bacterial Cellulose Grows with a Honeycomb Geometry in a Solid-State Culture of *Gluconacetobacter xylinus* Using Polyurethane Foam Support. *Process Biochem.* **2019**, *82*, 1–9.

Ramalho, P. A.; Scholze, H.; Cardoso, M. H.; Ramalho, M. T.; Oliveira-Campos, A. M. Improved Conditions for the Aerobic Reductive Decolourisation of Azo Dyes by *Candida zeylanoides*. *Enzyme Microb. Tech.* **2002**, *31*, 848–854.

Saeed, M.; Haq, A.; Muneer, M.; Adeel, S.; Hamayun, M.; Ismail, M.; Younas, M.; Siddique, M. Degradation of Direct Black 38 Dye Catalyzed by Lab Prepared Nickel Hydroxide in Aqueous Medium. *Glob. NEST J.* **2016**, *18* (2), 309–320.

Sen, S. K.; Raut, S.; Bandyopadhyay, P.; Raut, S. Fungal Decolouration and Degradation of Azo Dyes: A Review. *Fungal Boil. Rev.* **2016**, *30*, 112–133.

Sharma, S.; Kaur, A. Various Methods for Removal of Dyes from Industrial Effluents: A Review. *Indian J. Sci. Technol.* **2018**, *11* (12), 1–21.

Silva, M. F.; Rigo, D.; Mossi, V.; Dallago, R. M.; Henrick, P.; De-Oliveira, G.; Treichel, H. Evaluation of Enzymatic Activity of Commercial Inulinase from *Aspergillus niger* Immobilized in Polyurethane Foam. *Food Bioprod. Process.* **2013**, *91*, 54–59.

Singh, K.; Singh, R. S.; Rai, B. N.; Upadhyay, S. N. Biofiltration of Toluene Using Wood Charcoal as the Biofilter Media. *Bioresour. Technol.* **2010a**, *101*, 3947–3951.

Singh, R. S.; Rai, B. N.; Upadhyay, S. N. Removal of Toluene Vapor from Air Stream Using a Biofilter Packed with Polyurethane Foam. *Process Saf. Environ.* **2010b**, *88*, 366–371.

Sivasamy, A.; Sundarabi, N. Biosorption of an Azo Dye by *Aspergilllus niger* and *Trichoderma* Sp. Fungal Biomass. *Curr. Microbiol.* **2011**, *62*, 351–357.

Solís, M.; Solis, A.; Pérez, H. I.; Manjarrez, N.; Flores, M. Microbial Decolouration of Azo Dyes: A Review. *Process Biochem.* **2012**, *12*, 1723–1748.

Takaki, G. M. A Biosorption Isotherm Model for the Removal of Reactive Azo Dyes by Inactivated Mycelia of *Cunninghamella elegans* UCP542. *Molecules* **2012**, *17*, 452–462.

Talha, M. A.; Goswamib, M.; Giric, B. S.; Sharmad, A.; Raic, B. N.; Singh, R. S. Bioremediation of Congo Red Dye in Immobilized Batch and Continuous Packed Bed Bioreactor by *Brevibacillus parabrevis* Using Coconut Shell Biochar. *Bioresour. Technol.* **2018**, *252*, 37–43.

Tastan, B. E.; Ertugrul, S.; Dönmez, G. Effective Bioremoval of Reactive Dye and Heavy Metals by *Aspergillus versicolor*. *Bioresour. Technol.* **2010**, *101*, 870–876.

Wang, B.; Yong-You, H. U. Comparison of Four Supports for Adsorption of Reactive Dyes by Immobilized *Aspergillus fumigatus* Beads. *Int. J. Environ. Sci.* **2007**, *19* (4), 451–457.

Wang, B.; Hu, Y.; Xie, L.; Peng, K. Biosorption Behavior of Azo Dye by Inactive MC Immobilized *Aspergillus fumigatus* Beads. *Bioresour. Technol.* **2008**, *99*, 794–800.

Warade, A. R.; Gaikwad, R. W.; Sapkal, R. S.; Sapkal, V. S. Study of Removal Techniques for Dyes by Adsorption: A Review. *Int. J. Adv. Resear. Innovat. Ideas Educ.* **2016**, *2*, 3851–3869.

Yang, C.; Suidan, M. T.; Zhu, X.; Kim, B. J. Comparison of Single-layer and Multi-layer Rotating Drum Biofilters for VOC Removal. *Environ. Prog.* **2010**, *22* (2), 87–94.

Yang, C.; Linlin, F.; Hanguang, L.; Yuanxiu, W.; Guotao, C.; Qinghua, Z. Biodegradation and Detoxification of Direct Black G Textile Dye by a Newly Isolated Thermophilic Microflora. *Bioresour. Technol.* **2018**, *250*, 650–657.

Yin, L.; Cai, M.; Diu, S.; Luo, C. Identification of Two *Monilia* Species from Apricot in China. *J. Integr. Agric.* **2017**, *16* (11), 2496–2503.

Zhao, X.; Hardin, I. R. HPLC and Spectrophotometric Analysis of Biodegradation of Azo Dyes by *Pleurotus ostreatus*. *Dyes Pigm.* **2007**, *73* (3), 322–325.

Zimmermann, G. The Entomopathogenic Fungi *Isaria farinosa* (Formerly *Paecilomyces farinosus*) and the *Isaria fumosorosea* Species Complex (Formerly *Paecilomyces fumosoroseus*): Biology, Ecology and Use in Biological Control. *Biocntrl. Sci. Technol.* **2008**, *18*, 865–901.

CHAPTER 4

Purpureocillium lilacinum: A Promising Bionematicide for Sustainable Agriculture

M. MOUSUMI DAS[1], RAUL RODRIGUEZ HERRERA[2], M. HARIDAS[3], and A. SABU[1*]

[1]*Department of Biotechnology and Microbiology, Centre for Bio-innovation and Product Development, Kannur University, Dr. Janaki Ammal Campus, Thalassery, Kannur 670661, Kerala, India*

[2]*Facultad de Ciencias Químicas, Universidad Autónoma de Coahuila, Blvd. V. Carranza y Jose, Cardenas Valdez s/n, Col. Republica Ote., 25280 Saltillo Coahuila, Mexico*

[3]*Department of Biotechnology and Microbiology, Inter University Centre for Bioscience, Kannur University, Dr. Janaki Ammal Campus, Thalassery, Kannur 670661, Kerala, India*

*Corresponding author. E-mail: drsabu@gmail.com

ABSTRACT

Applications of nematophagous fungi in crop fields for the management of plant pests and parasitic nematodes are cost-effective, increase the yield of agricultural products, minimize the usage of agrochemicals, and prevent the environment from the pesticide pollution. In the present day perspective with the increasing cost of synthetic agrochemicals along with increasing incidences of pesticide toxicity, the application of microbial biocontrol agents (BCAs) holds great promise for the crop protection/crop production. Basically, biological control is an eco-friendly pest management strategy that is not harmful to the health of humans and animals. *Purpureocillium lilacinum* is one of the most promising and practicable bionematicide recommended for plant protection against plant diseases in organic agriculture. In recent decades, many BCAs have been used in the crop protection. However, *P. lilacinum* has been recognized for a long time as standard commercial bionematicide for the control of plant-parasitic nematodes infecting economically important crops. This chapter summarizes the importance of *P. lilacinum* as natural antagonists of various nematode parasites and the known mechanisms of action responsible for the nematicidal activity. Mass production of *P. lilacinum* using agro-processing residues under solid-state fermentation is also discussed.

4.1　INTRODUCTION

Indiscriminate and extensive use of synthetic chemicals in agriculture caused incredible harm to the environment and human health (Jeyaratnam, 1990; Pimentel, 2005). There is now massive evidence that many of these agrochemicals pose adverse health effects to humans and other life forms. Improper usage of these chemicals can lead to secondary pest outbreaks, destruction of biodiversity, and poses significant risks to non-target species and the environment (Gross and Rosenheim, 2011; Andrea et al., 2000; Arias-Estévez et al., 2008). Thus, the current scenario demands an environmentally sound, biologically based pest management strategy for controlling the pest, as chemical pesticides are not suitable for the cultivation of crop. Biological control of soil-borne phytopathogens and plant-parasitic nematodes is an ecologically safe, agronomically durable, consistent and economically feasible alternative to hazardous agrochemicals (Gupta and Dikshit, 2010; Das et al., 2019). The emergence of fungal antagonists as an alternative to agrochemicals has made it a promising biocontrol strategy to control the plant diseases and offer great potential for field application without substantially harming the environment (Das and Sabu, 2020). Application of bio-inputs in sufficient quantities only will help to regain the lost vitality of the soil. Such kind of a move would reduce the occurrence of crop diseases and consequently revitalization and rejuvenation of the plantation crops. Biocontrol agents (BCAs) also help to maintain and establish a favorable balance between the plant species along with their natural enemies. Compared with bacterial BCAs, fungal antagonists have made great progress, mainly due to their high reproductive rate, host specificity, adaptability to different environmental conditions, persistence, dispersal efficiency, and of their easy maintenance and production characteristics, which can ensure their survival longer without a host (Sandhu et al., 2012; Das and Sabu, 2020).

To date, numerous antagonistic organisms have been reported to reduce the root-knot nematode (RKN) populations in soil and roots (Stirling, 2011). *Purpureocillium lilacinum* (previously called *Paecilomyces lilacinus* [Thom] Samson) is a promising BCA because of its ability to capture and infect sedentary plant-parasitic nematode, insect, and acarian species by producing a vast array of inhibitory secondary metabolites (Lopez-Lima et al., 2014; Atkins et al., 2005; Park et al., 2004; Kepenekci et al., 2018). Also, it has been investigated as an alternative to the commonly used commercial nematicides (Dube and Smart, 1987; Mendoza et al., 2007). This group of fungi is an important egg and female parasite of sedentary endoparasitic nematode species (Dube and Smart, 1987; Atkins et al., 2005; Khan et al., 2006a). Previous studies reported that *P. lilacinum* adapts well in varied climatic conditions and is much effective in controlling plant-parasitic species of *Meloidogyne* (Grace et al., 2019), *Pratylenchus* (Kepenekci et al., 2018), *Rotylenchulus* (Walters and Barker, 1994), *Globodera* (Sreenivasan, 2017; Davide and Zorilla, 1983), *Nacobbus* (Jatala, 1986), *Heterodera* (Anastasiadis et al., 2008), *Tylenchulus* (Jatala, 1986), and *Radopholus* (Anastasiadis et al., 2008). Besides its nematicidal activity, *P. lilacinum* causes harmful effects on insect herbivores such as, *Ceratitis capitata*, *Setora nitens*, *A. gossypii*, and *Triatoma infestans* (Imoulan, 2011; Wakil et al., 2012; Rao et al., 2012; Marti et al., 2006; Fiedler and Sosnowska, 2007). There are many isolates of *P. lilacinus* and many commercial formulations in particular *P. lilacinus* strain 251 have been adopted recently in agriculture as a BCA against nematodes worldwide (Atkins et al., 2005; Kiewnick and Sikora, 2006).

The most important and fascinating feature of *P. lilacinum* is the study of mechanisms varying for the management of nematode diseases in which RKNs and plant pathogens are getting antagonized by antagonist resulting from different types of interaction between organisms. It includes direct parasitism of the egg stage, juveniles, and females (Kiewnick and Sikora, 2006; Holland et al., 1999). Mycelial penetration into the egg-shell was initiated by appressorium formation and subsequent eggs were completely colonized, which resulted in the condensation of egg contents (Swarnakumari and Kalaiarasan, 2017). The production of hydrolytic enzymes (proteases and chitinases) and toxic metabolites (leucinotoxins and acetic acid) by *P. lilacinum* is also linked with the nematode infection process (Khan et al., 2004; Yang et al., 2015; Kiewnick and Sikora, 2006). Among them, proteases and chitinases dissolve the outer vitelline layer of nematode eggshell, facilitating the penetration of fungal hyphae into the eggs and destroying the early stages of embryonic development (Khan et al., 2006a; Mukhtar et al., 2013).

From the current situation with the increasing cost of agrochemicals and the increasing incidences of pesticide toxicity, the application of microbial pesticides is an ecologically sound strategy for sustaining soil health and protecting plants from various diseases. The ability of *P. lilacinum* to antagonize, parasitize, and kill nematodes and insects, confirms it as an effective biocontrol agent. This chapter discusses the importance of *P. lilacinum* as bionematicide, its biocontrol activity and its mass production and application in plant disease management programs.

4.2   NEED OF BIOCONTROL AGENTS

One of the major problems in agricultural crops including vegetables are plant-parasitic nematodes, phytopathogens, and insects. Among them, plant-parasitic nematodes constitute one of the major limiting factors for cultivation of crops. Nematodes attack the plant roots and weaken their ability to absorb water and nutrients which results in a low yield of organic produce. Also, the nematode-infested plant becomes more susceptible to other stress factors such as heat, water deficit, mineral deficiencies, and microbial pathogens. Therefore, the use of chemical nematicides for the management of nematode pests remains the foremost and effective control measure but with some serious constraints. Commercial chemical nematicides are extremely toxic to the mammals and non-target organisms prevailing in the same niche, contaminate groundwater resources and have residual effects on farm produce (Grace et al., 2019; Udo et al., 2014). In the era of the promotion of biosafety and quality agricultural products, the farmers are resorting to eco-friendly crop protection methods like BCA.

Biocontrol agents are recommended worldwide and there is renewed interest in the use of fungal antagonists for the control of plant diseases. Microbial biocontrol appears to be an environmentally safer and ecologically feasible option for the control of root-knot nematode, *Meloidogyne* sp. with great potential for promoting sustainability in agricultural systems (Rao et al., 2015; Khan et al., 2019). The potentiality of *P. lilacinum* as a biocontrol agent has been found equivalent to commonly used commercial agricultural nematicides. *P. lilacinum* is a proven efficient mycobiocontrol agent in its field application in controlling various genera of nematodes, such as *Meloidogyne* sp., *Globodera* sp., *Pratylenchus* sp., *Rotylenchulus* sp., *Heterodera* sp., *Radopholus* sp., etc. (Khan et al., 2019; Kepenekci et al., 2018; Sreenivasan, 2017; Walters and Barker, 1994; Anastasiadis et al., 2008). Infestations of soil with *P. lilacinus* in green house and field conditions have been reported to reduce the root-knot nematode disease incidence and improve crop yields (Brand et al., 2004).

4.3 *PURPUREOCILLIUM LILACINUM*—AN OVERVIEW

Purpureocillium lilacinum (Thom) Luangsaard, Hywel-Jones, Houbraken and Samson (formerly known as *Paecilomyces lilacinus*) (Sordariomycetes: Hypocreales) is a soil inhabiting fungus, that has been reported as a potential biocontrol agent of RKNs and other nematode pests (Luangsa-Ard et al., 2011; Udo et al., 2014). It has been widely distributed in warm regions of the world. The fungus was isolated firstly from insects in tropical regions (Domsch et al., 1980). *Purpureocillium* species have been reported as endophytic fungi while generally found in all types of soils such as forest soil, grass land, desert soil, cultivated and noncultivated soil, estuarine, sewage sludge, decaying vegetation, and insects. The fungal species are fast growing under the optimum temperature range between 26 and 30°C. *Purpureocillium* displays robust growth over a wide pH range, which makes them to act in a wide variety of agricultural soils, and allows it to grow on a variety of substrates. The fungus used a variety of compounds such as carbon and nitrogen sources for microbial growth and biomass production (Das et al., 2020; Kumar et al., 2017).

Purpureocillium lilacinum, forms a dense, septate, and branched hyaline mycelium which gives rise to conidiophores. Conidiophores are arising mainly from submerged hyphae, bear loose whorls of branches and phialides with occasional formation of synnemata. Conidiophore color morphology varies from species to species but typically yellow to purple (Das et al., 2020). Conidia were formed in divergent chains at the end of phialides, which sometimes became slightly roughened to ellipsoidal fusiform, smooth-walled to slightly roughened (Kepenekci et al., 2015).

4.3.1 *PURPUREOCILLIUM LILACINUM AS BIOCONTROL AGENT*

P. lilacinum is a successful antagonist having biocontrol ability against economically important plant-parasitic nematodes, soil-borne pathogens, and insects. Also, the nematophagous ovicidal fungi have the capacity to infect, colonize, and digest living stages of their nematode hosts (eggs, cysts, and nematode females) in the soil (Huang et al., 2004). This fungal group has played a crucial role in the biological control of RKNs, lesion nematodes, cyst nematodes, burrowing nematodes, etc. Table 4.1 summarizes the biocontrol activity of *P. lilacinum* against nematode pests and insects.

TABLE 4.1   *Purpureocillium lilacinum* for the Management of Various Nematode Pests and Insects.

Antagonist	Plant pathogens/pests/ nematodes	Host	Method of application	References
P. lilacinus	*M. incognita*	Cardamom	Soil application	Eapen and Venugopal (1995)
P. lilacinus	*M. incognita*	Okra	Soil application	Thakur and Devi (2007)
P. lilacinus	*Radopholus similis*	Banana	Soil application	Mendoza et al. (2004)
P. lilacinus	*Tylenchulus. semipenetrans*	*Citrus jambhiri*	Soil application (*P. lilacinus* cultured on rice bran media)	Deka et al. (2002)
P. lilacinus	*Heterodera cajani*	Pigeon pea	Soil application	Siddiqui and Mahmood (1995)
P. lilacinus	*R. similis*	Betel vine	Soil application	Sosamma et al. (1994)
P. lilacinum	*M. incognita*	Mung bean (*Vigna radiata* L.)	Soil application	Khan et al. (2019)
P. lilacinus	*M. javanica*	Tobacco	Soil application (*P. lilacinus* infested wheat seeds)	Hewlett et al. (1988)
P. lilacinum	*M. javanica*	Pineapple	Soil application	Kiriga et al. (2018)
P. lilacinum	Leaf cutter ants (*Acromyrmex lundii*)	Crops	Conidial suspension	Goffre and Folgarait (2015)
P. lilacinum	*Pratylenchus thornei*	Wheat (*Tritiicum* spp.)	Soil application	Kepenekci et al. (2018)
P. lilacinum	*Meloidogyne enterolobii*	Tomato and banana	Spore suspension	Silva et al. (2017)
P. lilacinus strain 251	*Meloidogyne hapla*	Tomato	Soil application	Kiewnick and Sikora (2006)
P. lilacinus	*M. javanica*	Broad bean, Okra	Soil drenching with culture filtrate of *P. lilacinum*	Zareen et al. (1999)
P. lilacinus	*Rotylenchulus reniformis*	Tomato	Soil application	Walters and Barker (1994)

TABLE 4.1 *(Continued)*

Antagonist	Plant pathogens/pests/ nematodes	Host	Method of application	References
P. lilacinus + organic materials	*M. incognita*	Chilli	Soil application	Ahmad and Khan (2004)
P. lilacinus + Monacrosporium lysipagum	*M. javanica, Heterodera avenae, R. similis*	Tomato, cereals, barely, banana	Soil application	Anastasiadis et al. (2008)
P. lilacinus + Beauveria bassiana	Cotton aphid (*Aphis gossypii*)	Cotton	Seed treatment	Lopez et al. (2014)
P. lilacinus	*M. incognita*	Potato	Soil application	Jatala et al. (1980)
P. lilacinus integrated with aldicarb, Datura stramonium, Tagetes minuta, Ricinus communis, chicken manure	*M. javanica*	Tomato	Soil application	Oduor-Owino (2003)
P. lilacinus integrated with neem cake, groundnut cake, linseed cake, castor cake, mahua cake	*M. javanica*	Brinjal	Soil application	Ashraf and Khan (2010)
P. lilacinus + neem cake	*M. incognita*	Tomato	Soil treatment	Parvatha et al. (1997)
P. lilacinum	*M. incognita*	Tomato	Soil treatment	Singh et al. (2013)
P. lilacinus	*M. javanica*	Tomato	Soil treatment	Esfahani and Pour (2006)
P. lilacinus	*R. reniformis*	Chickpea	Soil treatment	Anver and Alam (1999)
P. lilacinus + fruit waste	*R. reniformis*	Chick pea	Soil treatment	Ashraf and Khan (2008)
P. lilacinus and P. chlamydosporia	*M. incognita*	Brinjal	Soil treatment	Cannayane and Rajendran (2001)
P. lilacinum + Pseudomonas fluorescens	*Meloidogyne hapla*	Carrot	Soil drenching and seed treatment	Sreenivasan (2018)

TABLE 4.1   *(Continued)*

Antagonist	Plant pathogens/pests/ nematodes	Host	Method of application	References
P. lilacinum + velum	*Meloidogyne incognita*	Tomato	Soil application	Dahlin et al. (2019)
P. lilacinum + vermicompost	*Meloidogyne incognita*	Tuberose (*Polianthes tuberosa* L.)	Soil application	Grace et al. (2019)
P. lilacinum + Pseudomonas fluorescens	*Globodera rostochiensis* and *Globodera pallida*	Potato	Soil drenching and seed treatment	Sreenivasan (2017)
P. lilacinus + leaf extract of Lantana camara	*Meloidogyne incognita*	Tomato	Soil application	Udo et al. (2014)
P. lilacinus	*M. incognita*	Okra	Soil application	Saikia and Roy (1994)

4.3.2 PURPUREOCILLIUM LILACINUM: CURRENT STATUS

The fungus *P. lilacinum* is formerly known as *Paecilomyces lilacinus*, having undergone a recent taxonomic revision (Luangsa-Ard et al., 2011). There is a resurgent interest in nematophagous fungi because they are considered to offer an ecologically safer and inexpensive alternative to hazardous chemical nematicides. An increasing trend has been observed over the past three decades in various aspects of bionematicide formulation and its application in crop fields. Biopesticides represented approximately 3.5% (valued at 1.6 billion USD in 2009) of the total pesticide market (Lehr, 2010). Among these, a major share of the market is contributed by bioinsecticides, but there are also microbial fungicides, bionematicides and bioherbicides (Wilson and Jackson, 2013). Early research has speculated that *P. lilacinum* was applied to crop field using various organic materials such as oil cakes, wheat bran, leaf residues, and gram seeds as carrier (Cannayane and Sivakumar, 2001; Siddiqui and Mahmood, 1996). Nowadays, the fungus *P. lilacinum* is marketed for the management of plant pathogenic nematodes in several countries.

The nematopagous fungus *P. lilacinum* was initially marketed as an agriculture bionematicide by the Australian Technology Innovation Corporation, but the commercial success has been driven by PROPHYTA GmbH, Malchow, Germany (now named Bayer CropScience Biologics GmbH), which obtained the rights and permissions to develop and market the effective bionematicide (BioAct WG) in 2001. The company has also developed a commercially successful wettable powder formulation (BioAct WP). The commercial *P. lilacinum* strain 251 (PL251) has been registered with the US EPA (United States Environmental Protection Agency) as an agricultural bionematicide and which is the commercially formulated product to control plant-parasitic nematodes of various crops. These strains are currently sold under the trade names "BioAct" by Intrachem Bio Italia and MeloCon® by Certis USA (Wilson and Jackson, 2013). Also, it has been commercialized by Asiatic Technologies Incorporation, Philippines and sold in the form of wettable powder in the name of BIOCON. There are a number of commercial *Purpureocillium*-based bionematicides are available in the market (Table 4.2).

4.3.3 MECHANISM OF ACTION

The merit of potential fungal BCAs with different modes of action against plant-parasitic nematodes should be continuously studied for more details about their molecular mechanisms of toxicological processes. *P. lilacinum*,

TABLE 4.2 *P. lilacinum*-Based Commercial Products against Plant-Parasitic Nematodes.

Active ingredient	Commercial product/ Company	Target nematodes	Crops	References
P. lilacinum	BioAct® WP Bayer Crop Science, USA	*Meloidogyne* spp.	Vegetables	Yadav (2017), Abd-Elgawad and Askary (2018), Moosavi and Zare (2012), Wilson and Jackson (2013)
P. lilacinum	BioAct® WG Bayer Crop Science, USA	*Meloidogyne* spp.	Tobacco, pineapples, tree nuts, peaches, citrus, strawberries, ornamentals, grape vines, bananas, turf grass	Yadav (2017), Abd-Elgawad and Askary (2018), Moosavi and Zare (2012), Wilson and Jackson (2013)
P. lilacinum	MeloCon®WG Prophyta GmbH, Germany Certis, USA	*R. similis, Rotylenchulus reniformis, Heterodera* spp. *and Globodera* spp., *Nacobbus* spp., *Belonolaimus* spp., *Helicotylenchus* spp., and *Pratylenchus* spp.	Fruit, vegetables, vine, tuber, and ornamental crops	Abd-Elgawad and Askary (2018), Wilson and Jackson (2013)
P. lilacinum	NemOut	Reniform, root-knot and lance types of nematodes	Cotton and pea nuts	Wilson and Jackson (2013)
P. lilacinum	PL Gold BASF Worldwide, Becker Underwood, South Africa	*Meloidogyne* spp.	Banana, tomato	Yadav (2017)
P. lilacinum	BIOCON Asiatic Technologies Inc. Manila, Philippines	Root-knot and cyst nematodes	Unspecified	Yadav (2017), Davide (1990)
P. lilacinum	Shakti Paecil, Shakti Biotech, India	Root-knot nematodes, cyst nematodes, lesion nematodes	Unspecified	Abd-Elgawad and Askary (2018)

TABLE 4.2 *(Continued)*

Active ingredient	Commercial product/ Company	Target nematodes	Crops	References
P. lilacinum	PAECILO® Agri Life, India	Root-knot nematode, cyst nematodes, reniform nematodes, burrowing nematodes, citrus and lesion nematodes	Cereals, pulses, millets, oil seeds, fiber crops, vegetables, sugar crops, etc.	http://www.agrilife.in/biopesti_microrigin_pacel.htm Abd-Elgawad and Askary (2018)
P. lilacinum	Paecilon Enpro Bio Sciences Private Limited, India	Soil nematodes	Pomegranate	Abd-Elgawad and Askary (2018)
P. lilacinum	Nematofree, International Panaacea Limited, India	*Meloidogyne* spp. *Rotylenchulus* spp. Citrus nematodes *Radopholus* spp.	Onion, tomato, chilli, capsicum, brinjal, okra, papaya, and tuberose Capsicum and Papaya Acid lime Banana	Abd-Elgawad and Askary (2018)
P. lilacinum	Gmax bioguard Greenmax Agro Tech, India	*Meloidogyne* spp.. *Radopholus similis* reniform nematodes, *Rotylenchulus reniformis, Tylenchulus semipenetrans*	Banana, citrus, horticultural crops	Abd-Elgawad and Askary (2018)
P. lilacinum	Yorker Agriland Biotech, India	Root-knot nematodes, cyst nematodes, reniform nematodes, stunt nematodes and black lesion nematodes	Vegetables, trees, fruits, sugar cane, oil seeds, tobacco, tea, cotton, coffee, etc.	Yadav (2017), Abd-Elgawad and Askary (2018)
P. lilacinum	PI Plus® (*P. lilacinus* strain 251), wettable powder, Biological Control Products, South Africa	Cyst and root-knot nematodes	Vegetables, citrus, banana	Yadav (2017), Abd-Elgawad and Askary (2018)

TABLE 4.2 *(Continued)*

Active ingredient	Commercial product/ Company	Target nematodes	Crops	References
P. lilacinum	BIOSTAT[r] LAM International	*Radopholus* sp., *Meloidogyne* sp., *Tylenchus* sp., *Pratylenchus* sp., *Ditylenchus* sp., *Rotylenchulus* sp., and *Helicotylenchus* sp.	Vegetables	Tranier et al. (2014)
Arthrobotrys sp., *Dactyllela* spp., *Paecilomyces* sp., *Glomus* sp., *Bacillus* sp., and *Pseudomonas* sp.	REM G[r] Green Solutions, Italy	Soil nematodes	Tomato	Tranier et al. (2014)

parasitizes and subsequently kills various plant-parasitic nematodes. The ovicidal activity of nematophagous fungus *P. lilacinus* on nematode eggs occurs by mechanical and/or enzymatic action. The nematode egg infection begins with the attachment of hyphal filaments on the egg surface and the formation of an appressorium. From the appressorium, the fungus adheres to the nematode eggshell and develops an infection peg to penetrate the eggshell. This results in the colonization of internal contents and disintegration of the egg cell wall. Consequently, embryonic development and egg hatching were completely inhibited (Swarnakumari and Kalaiarasan, 2017). The nematophagous fungus uses its conidia to infect root-knot nematode such as *Meloidogyne* spp. that germinate rapidly on the cuticle and then penetrate the body of the nematode through appressoria. After mycelial penetration, the nutrients are taken up by the fungi and reproduce massively by invading the host cell until its death (Ortiz Paz et al., 2015).

Extracellular hydrolytic enzymes, such as proteases and chitinases from fungi are involved in the infection mechanism of nematode eggs. Serine protease might play a pivotal role in the penetration of the fungus through cuticle or egg-shell of nematodes and causes degradation of eggshell components, and inhibits hatching (Bonants et al., 1995). Fungal chitinases play a major role in the antagonism to nematode eggs as it breaks down the nematode eggshell to assist *eggshell penetration* by the fungus. The release of ammonia during chitin decomposition is reported to be toxic to the second-stage juveniles of RKNs. The fungal-infected eggs swell and buckle. As mycelial penetration continues, the vitelline layer of the egg breaks and loses its integrity (Khan et al., 2004; Abd-Elgawad and Askary, 2018). Also, the filamentous fungus has been reported to produce peptidyl antibiotics such as leucinostatin, lilacin, and a mycotoxin paecilotoxin (Abd-Elgawad and Askary, 2018). The culture filtrate of *P. lilacinus* contains nematicidal metabolite acetic acid, which affects the movement of nematode (Djiam et al., 1991).

4.3.4 ANTAGONISTIC AND BIOCONTROL POTENTIAL OF PURPUREOCILLIUM LILACINUM

4.3.4.1 EFFECT OF P. LILACINUM ON ROOT-KNOT NEMATODES (RKNS)

Root-knot nematodes (RKNs, *Meloidogyne* spp.) are considered as one of the most destructive pests of vegetable crops (Sasser, 1980). The pest causes about 5% average yield losses in the world and could be more in tropic and sub-tropic agriculture (Taylor and Sasser, 1978). Many nematode species have been shown to predispose vegetable crops to infection by soil-borne bacterial or fungal pathogens or to transmit virus diseases, which contribute to more indirect crop yield losses. Among plant-parasitic nematodes, the RKN is the most damaging one which poses a great threat to agriculture. The most economically devastating species of RKNs are *Meloidogyne incognita*, *M. javanica*, *M. hapla*, and *M. arenaria*. These predominant species of RKNs are extremely polyphagous and multiply on thousands of cultivated and wild plants. Current management of RKNs includes nematode-resistant varieties, crop rotation, and soil cultivation, multiple cropping systems, organic soil amendments, and BCAs. The use of BCAs appears as an ecologically safer RKN management option with great potential for promoting sustainable agriculture. The progress made in the control of RKNs with *P. lilacinus* in in-vitro and in-vivo conditions has been quite impressive and several studies have been carried out using *P. lilacinus* and *P. lilacinum* for biological control of RKNs.

Jatala (1985) used this fungus in controlling *M. incognita* on potatoes. It was observed that the nematopagous fungus greatly reduced the tuber and root-galling caused by RKNs, which result in greater plant vigor. Dahlin et al. (2019) conducted greenhouse experiments to study the combined effects of the chemical nematicide fluopyram and fungal antagonist (*P. lilacinum* strain 251) on *M. incognita* infested tomatoes. This study revealed that the combination of chemical pesticide (Velum) to downregulate the nematode population followed by the application of biological nematicide is more successful to control the *M. incognita* population compared to each treatment alone and enhanced the tomato yields.

Ahmad and Khan (2004) evaluated the integration of organic additives with the fungal antagonist, *P. lilacinus* for the management of root-knot nematode, *M. incognita* infecting chilli (*Capsicum annuum* var. Pusa jwala). They found that integration of biocontrol fungus *P. lilacinus* with *Calotropis procera* leaves significantly reduced the multiplication of *M. incognita* population infecting chilli by parasitizing females, egg masses, and eggs. Also, the gall formation on the roots was greatly reduced, resulting in improved plant growth and yield.

Grace et al. (2019) tested the efficacy of liquid formulation of *P. lilacinum* against *Meloidogyne incognita* infecting tuberose (*Polianthes tuberosa* L.). Greenhouse and field studies on biocontrol effects of *P. lilacinum* on *M. incognita* showed a significant inhibition of nematode reproduction and root galling, resulted in increased growth and yield of tuberose. In greenhouse conditions, corms treated along with soil application of 2 t/ha of vermicom-post enriched with *P. lilacinum* resulted in a maximum increase in the plant growth and yield parameters like root length, spike length, spike weight, root weight, bulb weight, and reduced nematode gall index.

Khan et al. (2019) evaluated the nematicidal efficacy of biocontrol agent *P. lilacinum* against *M. incognita* on mung bean (*Vigna radiata* L.). Their study found that application of *P. lilacinum* alone in a clay pot filled with sterilized soil (1 kg) mixed with farmyard manure resulted in a significant enhancement in growth and physiological parameters of mung bean cv. "PDM-139" and reduced the nematode eggs and egg masses production. Sequential and concomitant inoculation of *P. lilacinum* with *M. incognita* also improved plant growth parameters of mung bean.

Sikora and Kiewnick (2006) investigated the potential of nematophagous fungus *P. lilacinus* strain 251 to reduce *Meloidogyne hapla* Chitwood infesta-tion on tomato at unfavorable temperature range (25–33°C) in the glasshouse conditions. The study revealed that tomato plants repeatedly treated with parasitic fungus showed significantly higher fruit yield (23–102%). The egg pathogenic fungus PL251 has already reported high biocontrol efficacy toward *M. incognita* on tomato plants in growth chamber and glasshouse experiments (Kiewnick and Sikora, 2004).

Sarven et al. (2019) reported that the high dose of 1×10^6 CFU/g soil of *P. lilacinum* PLSAU-1 is necessary for the successful management of *Meloidogyne* infection in brinjal roots. Hewlett et al. (1988) described a study in which a high proportion of *M. javanica* galls colonized by *P. lilacinus* in Hairy vetch (*Vicia villosa* Roth). Singh et al. (2013) demonstrated the field effectiveness of *P. lilacinum* (HYBDPL-04 strain) when applied as a soil treatment against *M. incognita* eggs and juveniles' population in tomato plants. As a result, the isolates of *P. lilacinum* in their study were found to improve plant health, biomass as well as tomato yield, and reduce nematode multiplication. Glasshouse experiments were conducted to evaluate the individual and combined effects of *Lantana camara* aqueous leaf extract and a bio formulated *P. lilacinus* against *M. incognita* race I on tomato. The results of their study showed that double application of bionematicide (during transplantation and 2 weeks after transplantation) in combination with *L. camara* leaf extract (0.80 g ml^{-1}) reduces galls and number of egg masses per root system and significantly improved growth and accumulation of dry matter (Udo et al., 2014). Brand et al. (2004) demonstrated that *P. lilacinus* spores produced under solid-state fermentation (SSF) using agro-residues have potential nematicidal activity against the *M. incognita*. Pandey and

Trivedi (1992) applied *P. lilacinus* to *Capsicum annuum* seedlings infected with root-knot nematode, *M. incognita* in a pot experiment. The results of their study revealed that there was a significant reduction in number of galls, hatching of eggs per egg mass, and final nematode population in the soil.

Hazarika et al. (1998) showed that field level application of *P. lilacinus* mass cultured in mustard oil cake gave better control against *M. incognita* in betel vine (*Piper bettle*).

Perveen and Ghaffaar (1998) investigated that fungal biocontrol agent *P. lilacinus* completely controlled the root-knot nematode *M. javanica* on tomato root in soil containing an inoculum of 2000 eggs/250 g soil with no inoculum of *Fusarium oxysporium*.

Khanna (2000) demonstrated that application of bioagent *P. lilacinus* in *M. incogniita* infested fields resulted in the enhancement of plant growth parameters in eggplants. However, in the sustainable farming system, the application of *P. lilacinus* (3.6 g/600 g soil) was found to be the optimum for controlling severity of root galls and egg mass production, and improving plant growth.

Kadam and Khan (2015) demonstrated that seed treatment with biofor-mulations (*P. lilacinus*, *Pochonia chlamydosporia*, and *Pseudomonas fluo-rescens* [10 g/kg seed]) proved as the most economical for the management of root-knot nematode infecting okra in West Bengal, India. However, the application of *P. lilacinus* (10 g/kg seed) and Farm Yard Manure (FYM) at 10 t/ha gave the highest fruit yield.

4.3.4.2   EFFECT OF P. LILACINUM ON ROOT-LESION NEMATODES (RLNS)

Root lesion nematodes (RLN; *Pratylenchus species*), are major pests with broad host ranges (Nicol, 2002). RLNs poses the most severe threat to quantitative and qualitative production of several economically important crops such as, rice, maize, wheat, cotton, coffee, tea, sugar cane, potato, banana, pine apple, vegetables, tropical fruits, and ornamentals. There are 97 validly described RLNs (Handoo et al., 2008), two of which are known to be especially damaging: *Pratylenchus thornei* and *P. neglectus* (Smiley, 2010). *P. thornei* is known to be the most destructive pest of wheat, soybean, chickpea, sunflower, and opium (Dasgupta et al., 2010). These nematodes have the ability to penetrate and infect roots, rhizomes, pods, and tubers. Several chemical nematicides have been developed to combat RLN infec-tions. Unfortunately, the chemical nematicides used to control root-lesion nematode, *Pratylenchus* species have been considered less effective, toxic to the beneficial soil microflora, and a potential threat to the environment (Thomason, 1987). As a common soil inhabitant, *P. lilacinum* has been reported to inhibit RLN, *Pratylenchus thornei*.

Kepenekci et al. (2018) used three inoculum levels of *P. lilacinum* ((10^6, 10^7, and 10^8 conidia cultures ml^{-1}) for the control of *P. thornei* on wheat (*Triticum* spp.) under glass house conditions. Among the various treatments, applications with the higher dose of biocontrol agent (10^8 conidia of *P. lilacinum* ml^{-1}) showed maximum improvement in dry and fresh weight of shoots and reduced the adverse effect of *P. thornei* population.

4.3.4.3 EFFECT OF PURPUREOCILLIUM LILACINUM ON CYST NEMATODES

Cyst nematodes, (*Heterodera* and *Globodera* spp.) are of worldwide concern as a parasite of several economically important crops (Khan, 2015). In contrast to RKNs, the mode of action of cyst nematode is similar to that of RKNs, where second stage juveniles (J2) infect the host and develop to adult stages within host tissue. Cyst nematodes penetrate roots with their stylet and induce the formation of a multinucleate feeding site in the infected plant roots called syncytium. The formation of multinucleate syncytium involves dissolution of cell walls and protoplast fusion of numerous adjacent cells (Khan, 2015). Some nematode species like *Globodera rostochiensis* and *G. pallida* in potato, *Heterodera glycines* in soyabean, *Heterodera avenae*, and *H. filipjevi* in cereals, and *Heterodera cajani* in pigeon pea are serious problems.

Previous studies reported that enzymatic and exopathic toxic compounds of *P. lilacinus* inhibited *Globodera pallida* egg hatching under laboratory and greenhouse conditions. The biocontrol effects of egg parasitic fungus, *P. lilacinus* on *G. pallida* indicated up to 30% without direct hyphal penetration while the egg hatching was stimulated in the premature nematode eggs containing second-stage juveniles (Jatala et al., 1985).

Khan et al. (2006b) reported that *P. lilacinus* in combination with *Monacrosporium lysipagum* reduced 65% of *Heterodera avenae* cysts on barley in controlled pot experiments. Jacobs and Crump (2003) investigated the combined effect of *P. lilacinus* with fungicides and chemical nematicides as part of an integrated pest management (IPM) program to manage the potato cyst nematodes. Treatment with *P. lilacinus* formulated with alginate pellets greatly reduced the disease symptoms and multiplication of potato cyst nematodes in soil. Gul (1991) evaluated the biocontrol potential of *P. lilacinus* against golden nematode, *Globodera rostochiensis* in pot experiments using two application methods: the direct pouring of fungal-infected wheat grains to the soil and coating potato tubers with fungus. Tuber yield and plant height were significantly highest with tuber coating while fresh shoot weight enhanced with dose. Sreenivasan (2017) also reported that combined application of *Pseudomonas fluorescens* and *P. lilacinum* in potato fields reduced the *Globodera* population effectively.

4.3.4.4 EFFECT OF PURPUREOCILLIUM LILACINUM ON BURROWING NEMATODES

The burrowing nematode, *Radopholus similis* is a quarantine nematode pest worldwide and is capable of causing diseases on many vegetables, fruits, spices, and plantation crops (Khan, 2015). *R. similis* damages agricultural crops of economic importance include banana, citrus, betel vine, coconut, areca nut, ginger, black pepper, etc. This migratory endoparasite enters the plant roots and feeds host tissues and cause decay. Due to its intracellular migration, the parasitic nematode destroys the cell cytoplasm, disintegrate the nucleus and ruptures the cell wall, resulting in the formation of cavities or burrows inside the root tissue. All developmental stages of the nematode are able to exploit all the plant parts but most exclusively they infect roots. Females lay eggs in plant root tissues and the newly hatched second stage juveniles starts feeding and develop inside the plant tissue. Therefore, the entire life cycle of this nematode may be completed within the root tissue. *R. similis* completes its life cycle in 20–25 days at temperatures ranging from 24 to 32°C (Bernard et al., 2017). Khan et al. (2006b) reported a study in which *P. lilacinus* and *M. lysipagum* are used to control nematode infection in banana. In their study, they have inoculated *Radopholus similis* populations to 2-month-old banana plantlets. After 2 weeks of nematode inoculation, the soil was treated with 5.6×10^8 *P. lilacinum* and 6×10^5 *M. lysipagum* conidia per plant to control nematode infestation. They found that the combined application of *P. lilacinum* and nematode-trapping fungi significantly reduced *Radopholus similis* populations by 92% in the soil and by 54% in the roots during the first harvest. In the second harvest, there was a 99% reduction of *R. similis* population in the soil and 94% in the roots were observed.

4.3.4.5   BIOCONTROL OF RENIFORM NEMATODES USING P. LILACINUM

Reniform nematode (*Rotylenchulus reniformis*) is one of the most serious pests of transplanted vegetable crops (Khan, 2015). The term reniform refers to the kidney-shaped body of the adult female. *Rotylenchulus reniformis* (Linford and Oliviera) are sedentary semi-endoparasitic species in which the females enter the root cortex, establish a permanent-feeding site within the host root from which it can extract the nutrients required for reproduction. About 10 valid species of *Rotylenchulus* are known and distributed worldwide. Among these nematodes, *Rotylenchulus reniformis* is the most economically important species (Robinson et al., 1997). *Paecilomyces lilacinus* has been effective in controlling *R. reniformis* infecting vegetable crops. Walters and Barker (1994) described a study in which rice-cultured *P. lilacinus* has adverse effects on the development of *R. reniformis* population on tomatoes under the micro plot and greenhouse conditions. Also, Reddy and Khan (1988) reported that *P. lilacinus* significantly reduced the *R. reniformis* infestation in "Pusa Ruby" tomato (*Lycopersicon esculentum*).

4.3.5   GROWTH AND SPORULATION OF P. LILACINUM ON AGRICULTURAL CROPS

The use of fungal BCAs for the management of parasitic nematodes and plant diseases is an eco-friendly strategy to minimize the hazardous effects of synthetic chemicals in humans, animals, and to the environment (Das and Sabu, 2020). Production and enhanced utilization of BCAs will definitely become a suitable substitute for the chemical pesticides and insecticides, which are now facing serious resistance from the farmers and common public. Therefore, the application of BCAs will definitely help the farmers in changing their farming activities to the sustainable way providing multiple economic and environmental benefits. This will gradually pave the way to clean agriculture. Several commercial bio-inputs are already available in the market and others will be available in the near future. Commercial microbial bio-inputs must be inexpensive, yield high concentrations of viable conidia/ spores, and provide consistent field efficacy to control target pests and pathogens. A great alternative to develop low-cost bio-input is the utilization of agro-industrial residues (Robl et al., 2009; Das and Sabu, 2020).

P. lilacinum has been found equivalent to the commercially used chemical nematicides (Kerry, 1990) and also as a biological agent for the control of greenhouse insects and mite pests (Fiedler and Sosnowska, 2007). Diverse raw materials have been used for the enhanced production of nematophagous fungi, such as coffee husk, wheat bran, beer waste, cassava bagasse, sugarcane bagasse, and spent tea waste (Das et al., 2020; Brand et al., 2004). Several studies have been carried out to utilize the agro-industrial residues for the production of value-added products. Over the past decades, many researchers have succeeded in using agricultural residues for the growth and spore production of *P. lilacinum*, mainly using SSF technology. The development of bioprocesses for the production of biopesticides and bionematicides from the agro-industrial residues through SSF has received much attention, since it is able to provide high-quality bioproducts without any toxicity associated to the chemical pesticides (Das and Sabu, 2020).

Some of the earlier studies on fermentation of agro-residues have been focused on the production of bionematicides. Brand et al. (2004) used coffee husk, defatted soybean cake, and cassava bagasse as a substrate for the growth and spore production of *P. lilanus* and sugarcane bagasse was used as inert support. SSF was carried out in 250 ml Erlenmeyer flasks at 28°C for 10 days. After SSF, the fermented products were evaluated for their nematicidal activity in pot experiments containing one seedling of plant *Coleus* inoculated with *M. incognita* race I. Antagonistic fungi significantly reduced the number of nematodes and improved the growth of nematode-infected plants. Among the fermented substrates tested, defatted soybean cake was found most effective with the least number of RKNs in plant; while the reduction of nematodes with coffee husk was 80% and with cassava bagasse was around 60%.

Leena et al. (2003) used several byproducts from sugar industry and agro-industrial wastes for the mass production of entomopathogenic fungi, *Paecilomyces farinosus* (hotmskiold) and *P. lilacinus* (Thom.) Samson. Among the agro-industrial wastes tested, the sugarcane pressmud significantly enhanced the growth and sporulation of both fungal species. Robl et al. (2009) mentioned the utilization of refuse potato for mycelial growth and conidia/spore formation of *P. lilacinus* strain Endo 69. Brand et al. (2010) reported that coffee husk-based biocompost supported the growth of egg parasitic fungi, *P. lilacinus*, which infects the *M. incognita* coffee nematode (Brand et al., 2010).

Walia et al. (1991) conducted a pot experiment to evaluate the biocontrol efficacy of *P. lilacinus* in okra. They proved that the effectiveness of the fungus in reducing root galling was better than chemical nematicide. They applied *P. lilacinus* in soil 10 days after sowing at 1 g/kg soil (5×10^8 spores), or as seed treatment (3×10^7 spores/seed) or in different combinations to sandy soil infected with *M. javanica*. In their study, plant growth was significantly increased after soil treatment with *P. lilacinus* grown on wheat bran. The methods (soil/seed treatment) or time (pre/post sowing) of fungus application were found to be equally effective and found to be better than chemical nematicide carbofuran (1 kg a.i/ha) in nematode gall suppression.

Application of *P. lilacinum* multiplied on rice grains in *M. incognita* infected aubergine (*Solanum melongena* L.) plants reduced the number of root galls, eggs per egg mass, and final population of nematodes in soil (Trivedi, 1990).

Goswami et al. (2006) conducted greenhouse experiments to study the efficacy of two bioagents along with mustard oil cake and furadan against *M. incognita* infecting tomato. Integrated application of fungal BCAs such as *P. lilacinus* and *Trichoderma viride* along with mustard cake and furadan in the management of root-knot nematode, *M. incognita*, resulted in enhanced plant growth and reduced the number of galls per plant.

Integration of oil cakes of neem and mustard with fungal biocontrol agent *P. lilacinus* significantly improved plant growth and nodulation in chickpea. According to the authors, combined treatment of *P. lilacinus* at 10 ml (3.5×10^7 spores/ml) and (1.0 g/kg) neem cake significantly suppressed the multiplication of *M. incognita*, reduced the number of nematode galls, and improved plant growth compared to plants inoculated with nematode alone (Bhat et al., 1998).

In our previous research, we have isolated a novel *P. lilacinum* strain KU8 from forest rhizosphere soil and characterized for its nematicidal activity against *M. incognita* infecting ginger. We also established a bioprocess for enhancing fungal biomass production using agro-processing residues under SSF. SSF is the most suitable fermentation technique for the biomass production of fungal BCAs. Conidiospores produced must be cost-effective, viable, virulent, and show good stability for large-scale applications at a competitive price (Brand et al., 2010). In our work, we investigated the use of different agro-processing residues for cultivating *P. lilacinum* and developed a statistical model for optimizing biomass production. We screened various agro-processing residues (coffee husk, sugarcane bagasse, wheat bran, beer waste, and spent tea waste) for obtaining maximum fungal biomass of bionematicide, *P. lilacinum* strain KU8 under SSF. All the agro-residues supported the growth and sporulation of fungi along with bioconversion of waste materials. Among the selected substrates wheat bran had the highest fungal biomass production (35.63 mg/gdfs) compared to beer waste (34.23 mg/gdfs), sugarcane bagasse (28.23 mg/gdfs), coffee husk (27.69 mg/gdfs), and spent tea waste (29.34 mg/gdfs) under randomly selected conditions (Figs. 4.1 and 4.2).

FIGURE 4.1   Growth of *P. lilacinum* KU8 on wheat bran.

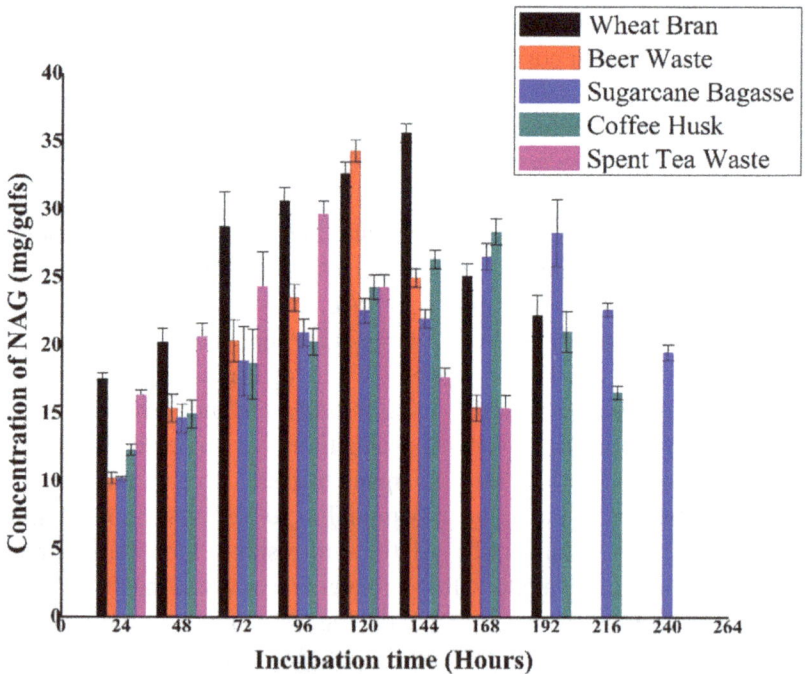

FIGURE 4.2   Production of *P. lilacinum* KU8 biomass at different incubation periods in wheat bran, beer waste, sugarcane bagasse, coffee husk, and spent tea waste. Error bar represents standard deviation (n = 3).

Further optimization of bioprocess parameters by central composite design (CCD) using response surface methodology (RSM) yielded 107.46 mg/gdfs of fungal biomass in wheat bran. Bioprocess parameters such as moisture content, yeast extract, and incubation time were used for optimization studies. A combination of moisture content (67.98% v/w), yeast extract (2.29%), and incubation time (142.2 hours) was found optimum for maximum production. Biomass yield of 107.46 mg/gdfs was obtained under the optimized conditions which was 1.35 fold higher as compared to a classical method and 3.01 fold higher than the unoptimized conditions (Das et al., 2020). From our study, it was clear that adoption of bioconversion of agro-residues for bionematicide production may become eco-enterprising for the rural farming communities for enhancing their livelihood along with improving crop productivity and soil health.

Despite the availability of many commercially available bionematicides, the high cost and low viable biomass content of bionematicides constitute the major constraint to its widespread use in agriculture. Thus, our study suggests that optimization of bioprocess parameters for *P. lilacinum* is very important as it would improve the product quality and reduce the overall manufacturing cost of the bionematicide. Bionematicides not only help to control the plant-parasitic nematodes of economic crops but also facilitate the agro-industrial waste utilization and production of value-added products through bioprocess.

4.4 CONCLUSIONS

Chemical treatment with fumigants or nematicides is frequently used for the management of plant-parasitic nematodes but is generally considered dangerous to humans, animals, and the surrounding environment. The situation triggered intensive research to develop eco-friendly alternatives against nematode pests and various diseases.

Biological control of plant diseases and nematodes is an ecologically safer, sustainable, and cost-effective alternative to hazardous agrochemicals. Several targets specific fungal BCAs are effective against parasitic nematodes and they mitigate the negative environmental effects of chemical inputs. *P. lilacinum* is a proven efficient fungal BCA in its field application for the management of parasitic nematodes such as root-knot, root-lesion, cyst, reniform, and burrowing nematodes. The application of fungal BCAs in nematode management is an eco-friendly biological alternative to chemical nematicides and it does not leave any harmful effect on the environment.

Perhaps the most important criteria in selecting a biocontrol agent for commercial application is the availability of cost-effective production and stabilization technology for the development of an effective bioformulation. The production of fungal biomass by SSF has arisen as a sustainable bioprocess method, which is slowly becoming preferred over conventional submerged fermentation (SmF) considering the advantage of using of agro-industrial wastes as substrates. BCAs produced by SSF are cost-effective and present good stability, better quality, and high-density propagules with higher conidia content than SmF. *P. lilacinum* sporulates well on the solid substrates mainly agro-processing residues such as wheat bran, coffee husk, cassava bagasse, and oil cakes (Brand et al., 2004; Das and Sabu, 2020).

This chapter provided fundamental knowledge on *P. lilacinum*, its biocontrol potential against various nematodes, the known mode of action, and mass production using cheap agro-processing residues. The future of the utilization of bionematicide is bright but depends on the technological advancements and market opportunities. The commercial importance of *P. lilacinum* as bionematicide is dependent on the development of a low-cost, viable product which provides consistent efficacy. Further studies on the development of nematicide compounds from *P. lilacinum* KU8 are progressing, for the preparations of bionematicide formulations to control nematode pests, pathogens, and soil-borne diseases.

KEYWORDS

- *P. lilacinum*
- biocontrol agent
- bionematicide
- *Meloidogyne* sp.
- agro-processing residue
- solid-state fermentation

REFERENCES

Abd-Elgawad, M. M.; Askary, T. H. Fungal and Bacterial Nematicides in Integrated Nematode Management Strategies. *Egypt J. Biol. Pest Cntrl.* **2018,** *28,* 74.

AgriLife Biosolutions for Soil and Crops. http://www.agrilife.in/biopesti_microrigin_pacelo. htm (accessed May 10, 2020).

Ahmad, S. F.; Khan, T. A. Management of Root-Knot Nematode, *Meloidogyne incognita*, by Integration of *Paecilomyces lilacinus* with Organic Materials in Chilli. *Arch. Phytopathol. Pl. Prot.* **2004**, *37*, 35–40.

Anastasiadis, I. A.; Giannakou, I. O.; Prophetou-Athanasiadou, D. A.; Gowen, S. R. The Combined Effect of the Application of a Biocontrol Agent *Paecilomyces lilacinus*, with Various Practices for the Control of Root-Knot Nematodes. *Crop Prot.* **2008**, *27*, 352–361.

Andrea, M. M.; Peres, T. B.; Luchini, L. C.; Pettinelli, J. A. Impact of Long-Term Pesticide Application on Some Soil Biological Parameters. *J. Environ. Sci. Health B* **2000**, *35*, 297–307.

Anver, S.; Alam, M. M. Control of *Meloidogyne incognita* and *Rotylenchulus reniformis* Singly and Concomitantly on Chickpea and Pigeon Pea. *Arch. Phytopathol. Pl. Prot.* **1999**, *32*, 161–172.

Arias-Estévez, M.; López-Periago, E.; Martínez-Carballo, E.; Simal-Gándara, J.; Mejuto, J. C.; García-Río, L. The Mobility and Degradation of Pesticides in Soils and the Pollution of Groundwater Resources. *Agric. Ecosyst. Environ.* **2008**, *123*, 247–260.

Ashraf, M. S.; Khan, T. A. Biomanagement of Reniform Nematode, *Rotylenchulus Reniformis* by Fruit Wastes and *Paecilomyces lilacinus* on Chickpea. *World J. Agric. Sci.* **2008**, *4*, 492–494.

Ashraf, M. S.; Khan, T. A. Integrated Approach for the Management of *Meloidogyne javanica* on Eggplant Using Oil Cakes and Biocontrol Agents. *Arch. Phytopathol. Pl. Prot.* **2010**, *43*, 609–614.

Atkins, S. D.; Clark, I. M.; Pande, S.; Hirsch, P. R.; Kerry, B. K. The use of Real-Time PCR and Species-Specific Primers for the Identification and Monitoring of *Paecilomyces lilacinus*. *FEMS Microbiol. Ecol.* **2005**, *51*, 257–264.

Bernard, G. C.; Egnin, M.; Bonsi, C. The Impact of Plant-Parasitic Nematodes on Agriculture and Methods of Control. *Nematology—Concepts, Diagnosis and Control*; 2017.

Bhat, M. Y.; Hisamuddin, Fazal, M. Combined Application of *Paecilomyces lilacinus* and Oil Cakes for the Protection of Chickpea against *Meloidogyne incognita*. In *Nematological Challenges and Opportunities in 21ˢᵗ Century. Proceedings of the Third International Symposium of Afro-Asian Society of Nematologists (TISAAN), Sugarcane Breeding Institute (ICAR)*, Coimbatore, India, April 16–19, 1998.

Bonants, P. J.; Fitters, P. F.; Thijs, H.; den Belder, E.; Waalwijk, C.; Henfling, J. W. D. A Basic Serine Protease from *Paecilomyces lilacinus* with Biological Activity Against *Meloidogyne hapla* Eggs. *Microbiology* **1995**, *141*, 775–784.

Brand, D.; Roussos, S.; Pandey, A.; Zilioli, P. C.; Pohl, J.; occol, C. R. Development of a Bionematicide with *Paecilomyces lilacinus* to Control *Meloidogyne incognita*. *Appl. Biochem. Biotechol.* **2004**, *118*, 81–88.

Brand, D.; Soccol, C. R.; Sabu, A.; Roussos, S. Production of Fungal Biological Control Agents Through Solid-State Fermentation: A Case Study on *Paecilomyces lilacinus* against Root-Knot Nematodes. *Micol. Apl. Int.* **2010**, *22*, 31–48.

Cannayane, I.; Rajendran, G. Application of Biocontrol Agents and Oil Cakes for the Management of *Meloidogyne Incognita* in Brinjal (*Solanum melongena* L.). *Curr. Nematol.* **2001**, *12*, 51–55.

Cannayane, I.; Sivakumar, C. V. Nematode Egg-Parasitic Fungus 1: *Paecilomyces lilacinus*—A Review. *Agric. Rev.* **2001**, *22*, 79–86.

Dahlin, P.; Eder, R.; Consoli, E.; Krauss, J.; Kiewnick, S. Integrated Control of Meloidogyne Incognita in Tomatoes using Fluopyram and *Purpureocillium lilacinum* Strain 251. *Crop Prot.* **2019,** *124,* 104874.

Das, M.; Sabu, A. Agro-Processing Residues for the Production of Fungal Bio-Control Agents. *In Valorisation of Agro-Industrial Residues—Volume II: Non-Biological Approaches*; Springer: Cham, 2020; pp 107–126.

Das, M. M.; Haridas, M.; Sabu, A. Biological Control of Black Pepper and Ginger Pathogens, *Fusarium oxysporum, Rhizoctonia solani* and *Phytophthora capsici*, Using *Trichoderma* spp. *Biocatal. Agric. Biotechnol.* **2019,** *17,* 177–183.

Das, M. M.; Haridas, M.; Sabu, A. Process Development for the Enhanced Production of Bio-Nematicide *Purpureocillium lilacinum* KU8 under Solid-State Fermentation. *Bioresour. Technol.* **2020,** *308,* 123–328.

Dasgupta, M. K.; Mukheriee, B.; Bhattacharya, C.; Khan, M. R.; Ghosh, S. Nematode Infestation in Tea. In *Nematode Infestation, Part II: Industrial Crops*; Khan, M. R., Jairajpuri, M. S. (Eds.); The National Academy of Sciences: India, 2010; pp. 213–255, 464.

Davide, R. G. Biological Control of Nematodes using *Paecilomyces lilacinus* in the Philippines. Integrated Pest Management for Tropical Root and Tuber Crops. In *Proceedings of the Global Status of and Prospects for Integrated Pest Management of Root and Tuber Crops in the Tropics*, Ibadan, Nigeria, October 25–30, 1987, 1990; pp 156–163.

Davide, R. G.; Zorilla, R. A. Evaluation of a Fungus, *Paecilomyces lilacinus* (Thom.) Samson, for the Biological Control of the Potato Cyst Nematode, *Globodera rostochie* Woll. as Compared with Some Nematicides. *Philipp. Agric.* **1983,** *66,* 397–404.

Deka, R.; Sinha, A. K.; Neog, P. P. Effect of *Paecilomyces lilacinus* and Botanicals against *Tylenchulus semipenetrans* on *Citrus jambhiri*. *Indian J. Nematol.* **2002,** *32,* 183–233.

Djiam, C.; Dijarowski, L.; Ponchet, M.; Arpin, N.; Favrebonin, J. Acetic Acid: A Selective Nematicidal Metabolite from Culture Filtrates of *Paecilomyces lilacinus* (Thom) Samson and *Trichoderma longibrachiatum* Rifai. *Nematologica* **1991,** *37,* 101–112.

Domsch, K. H.; Gams, W.; Anderson, T. H. *Compendium of soil fungi*; Academic Press: London, UK, 1980; p 860.

Dube, B.; Smart, G. Biological Control of *Meloidogyne incognita* by *Paecilomyces lilacinus* and *Pasteuria penetrans*. *J. Nematol.* **1987,** *19,* 222–227.

Eapen, S. J.; Venugopal, M. N. Field Evaluation of *Trichoderma* Spp. and *Paecilomyces lilacinus* for Control of Root Knot Nematodes and Fungal Diseases of Cardamom Nurseries. *Indian J. Nematol.* **1995,** *25,* 15–16.

Esfahani, M. N.; Pour, B. A. The Effects of *Paecilomyces lilacinus* on the Pathogenesis of *Meloidogyne javanica* and Tomato Plant Growth Parameters. *Iran Agric. Res.* **2006,** *24,* 67–75.

Fiedler, Z.; Sosnowska, D. Nematophagous Fungus *Paecilomyces lilacinus* (Thom) Samson Is also a Biological Agent for Control of Greenhouse Insects and Mite Pests. *BioControl* **2007,** *52,* 547–558.

Goffre, D.; Folgarait, P. J. *Purpureocillium lilacinum*, Potential Agent for Biological Control of the Leaf-Cutting Ant, *Acromyrmex lundii*. *J. Invertebr. Pathol.* **2015,** *130,* 107–115.

Goswami, B. K.; Pandey, R. K.; Rathour, K. S.; Bhattacharya, C.; Singh, L. Integrated Application of Some Compatible Biocontrol Agents Along with Mustard Oil, Seed Cake and Furadon on *Meloidogyne incognita* Infecting Tomato Plants. *J. Zhejiang Univ. Sci. B* **2006,** *11,* 873–875.

Grace, G. N.; Shivananda, T. N.; Rao, M. S.; Umamaheswari, R. Management of Nematodes Using Liquid Formulations of *Purpureocillium lilacinum* in Tuberose. *J. Entomol. Zool. Stud.* **2019**, *7*, 720–724.

Gross, K.; Rosenheim, J. A. Quantifying Secondary Pest Outbreaks in Cotton and Their Monetary Cost with Causal Inference Statistics. *Ecol. Appl.* **2011**, *21*, 2770–2780.

Gul, A. Biological Control of Golden Cyst Nematode of Potato with *Paecilomyces lilacinus* (Thom) Samson. *Sarhad J. Agric.* **1991**, *7*, 377–382.

Gupta, S.; Dikshit A. K. Biopesticides: An Ecofriendly Approach for Pest Control. *J. Biopest.* **2010**, *3*, 186–188.

Handoo, Z. A.; Carta, L. K.; Skantar, A. M. Taxonomy, Morphology and Phylogenetics of Coffee-Associated Root-Lesion Nematodes, *Pratylenchus* spp. In *Plant-Parasitic Nematodes of Coffee*; Springer: Dordrecht, 2008; pp 29–50.

Hazarika, K.; Dutta, P. K.; Saikia, M. K. Field Efficacy of *Paecilomyces lilacinus* against Root Knot Nematode in Betel Vine. *J. Phytol. Res.* **1998**, *1*, 171–173.

Hewlett, T. E.; Dickson, D. W.; Mitchell, D. J.; Kannwischer-Mitchell, M. E. Evaluation of *Paecilomyces lilacinus* as a Biocontrol Agent of *Meloidogyne Javanica* on Tobacco. *J. Nematol.* **1988**, *20*, 578–584.

Holland, R. J.; Williams, K. L.; Khan, A. Infection of *Meloidogyne javanica* by *Paecilomyces lilacinus*. *Nematology* **1999**, *1*, 131–139.

Huang, X.; Zhao, N.; Zhang, K. Extracellular Enzymes Serving as Virulence Factors in Nematophagous Fungi Involved in Infection of the Host. *Res. Microbiol.* **2004**, *155*, 811–816.

Imoulan, A. Natural Occurrence of Soil-Borne Entomopathogenic Fungi in the Moroccan Endemic Forest of *Argania spinosa* and Their Pathogenicity to *Ceratitis capitata*. *J. Microbiol. Biotechnol.* **2011**, *27*, 2619–2628.

Jacobs, H.; Crump, D. H. Interactions between Nematophagous Fungi and Consequences for Their Potential as Biological Agents for the Control of Potato Cyst Nematodes. *Mycol. Res.* **2003**, *107*, 47–56.

Jatala, P. Biological Control of Nematodes. In *Advanced Treatise on Meloidogyne, Biology and Control*; Sasser, J. N., Carter, C. C. (Eds.); North Carolina State Graphics: North Carolina, 1985; Vol I, pp 303–308.

Jatala, P. Biological Control of Plant-Parasitic Nematodes. *Annu. Rev. Phytopathol.* **1986**, *24*, 453–489.

Jatala, P.; Kaltenbach, R.; Bocangel, M.; Devaux, A. J.; Campos, R. Field Application of *Paecilomyces lilacinus* for Controlling *Meloidogyne incognita* on Potatoes. *J. Nematol.* **1980**, *12*, 226–227.

Jatala, P.; Franco, J.; Gonzalez, A.; OHara, C. M. Hatching Stimulation and Inhibition of *Globodera pallida* Eggs by the Enzymatic and Exopathic Toxic Compounds of Some Biocontrol Fungi. *J. Nematol.* **1985**, *17* (4), 501–501.

Jeyaratnam, J. Acute Pesticide Poisoning: A Major Global Health Problem. *World Health Stat. Q.* **1990**, *43*, 139–144.

Kadam, V.; Khan, M. R. Biomanagement of Root-Knot Nematode (*Meloidogyne incognita*) Infecting Okra in West Bengal, India. *Indian J. Nematol.* **2015**, *45*, 178–183.

Kepenekci, I.; Atay, T.; Oksal, E.; Saglam, H. D.; Tulek, A.; Evlice, E. Identification of Turkish Isolate of the Entomopathogenic Fungi, *Purpureocillium lilacinum* (Syn: *Paecilomyces lilacinus*) and Its Effect on Potato Pests, *Phthorimaea operculella* (Zeller) (Lepidoptera: Gelechiidae) and *Leptinotarsa decemlineata* (Say) (Coleoptera: Chrysomelidae). *Egypt J. Biol. Pest Cntrl.* **2015**, *25*, 121–127.

Kepenekçi, İ.; Toktay, H.; Oksal, E.; Bozbuğa, R.; İmren, M. Effect of *Purpureocillium lilacinum* on Root Lesion Nematode, *Pratylenchus thornei*. *J. Agric. Sci.* **2018**, *24*, 323–328.

Kerry, B. R. An assessment of Progress toward Microbial Control of Plant-Parasitic Nematode. *J. Nematol.* **1990**, *22*, 621–631.

Khan, A.; Tariq, M.; Asif, M.; Khan, F.; Ansari, T.; Siddiqui, M. A. Integrated Management of *Meloidogyne incognita* Infecting *Vigna radiata* L. Using Biocontrol Agent *Purpureocillium lilacinum*. *Trends Appl. Sci. Res.* **2019**, *14*, 119–124.

Khan, A.; Williams, K. L.; Nevalainen, H. K. Effects of *Paecilomyces lilacinus* Protease and Chitinase on the Eggshell Structures and Hatching of *Meloidogyne javanica* Juveniles. *Biol. Cntrl.* **2004**, *31*, 346–352.

Khan, A.; Williams, K. L.; Nevalainen, H. K. Infection of Plant-Parasitic Nematodes by *Paecilomyces lilacinus* and *Monacrosporium lysipagum*. *BioControl* **2006a**, *51*, 659–678.

Khan, A.; Williams, K. L.; Nevalainen, H. K. M. Control of Plant-Parasitic Nematodes by *Paecilomyces lilacinus* and *Monacrosporium lysipagum* in Pot Trials. *BioControl* **2006b**, *51*, 643–658.

Khan, M. R. Nematode Diseases of Crops in India. In *Recent Advances in the Diagnosis and Management of Plant Diseases*; Springer: New Delhi, 2015; pp 183–224.

Khanna, A. S. Biomanagement of *Meloidogyne incognita* (Kofoid & White) Chitwood by *Paecilomyces lilacinus* (Thom.) Samson on Eggplant. *Pest Manage. Econ. Zool.* **2000**, *8*, 133–136.

Kiewnick, S.; Sikora, R. A. Optimizing the Biological Control of Plant-Parasitic Nematodes with *Paecilomyces lilacinus* Strain 251. *Phytopathology* **2004**, *94*, S51.

Kiriga, A. W.; Haukeland, S.; Kariuki, G. M.; Coyne, D. L.; Beek, N. V. Effect of *Trichoderma* Spp. and *Purpureocillium lilacinum* on *Meloidogyne javanica* in Commercial Pineapple Production in Kenya. *Biol. Cntrl.* **2018**, *119*, 27–32.

Kumar, S. N.; Namasivayam, S. K. R.; Kamil, T. M., Ravi, T. Optimization of Process Parameters on the Viable Inocula of Entomopathogenic Fungi by Response Surface Methodology for the Fungal Consortium Preparation. *J. Biopest.* **2017**, *10*, 77–82.

Leena, M. D.; Easwaramoorthy, S.; Nirmala, R. In Vitro Production of Entomopathogenic Fungi *Paecilomyces farinosus* (Hotmskiold) and *Paecilomyces lilacinus* (Thom.) Samson Using Byproducts of Sugar Industry and other Agro-Industrial Byproducts and Wastes. *Sugar Technol.* **2003**, *5*, 231–236.

Lehr, P. *Biopesticides: The Global Market*, Report Code CHM029B, BCC Research, 2010.

Lopez-Lima, D.; Carrion, G.; Núñez-Sánchez, Á. E. Isolation of Fungi Associated with *Criconemoides* Sp. and Their Potential Use in the Biological Control of Ectoparasitic and Semiendoparasitic Nematodes in Sugar Cane. *Austr. J. Crop Sci.* **2014**, *8*, 389–396.

Luangsa-Ard, J.; Houbraken, J.; Van Doorn, T.; Hong, S. B.; Borman, A. M.; Hywel-Jones, N. L.; Samson, R. A. *Purpureocillium*, a New Genus for the Medically Important. *FEMS Microbiol. Lett.* **2011**, *321*, 141–149.

Marti, G. A.; Lastra, C. C.; Pelizza, S. A.; Garcıa, J. J. Isolation of *Paecilomyces lilacinus* (Thom) Samson (Ascomycota: Hypocreales) from the Chagas Disease Vector, *Triatoma infestans* Klug (Hemiptera: Reduviidae) in an Endemic Area in Argentina. *Mycopathologia* **2006**, *162*, 369–372.

Mendoza, A.; Sikora, R. A.; Kiewnick, S. Efficacy of *Paecilomyces lilacinus* (Strain 251) for the Control of *Radopholus similis* in Banana. *Comm. Agric. Appl. Biol. Sci.* **2004**, *69*, 365–372.

Mendoza, A. R.; Sikora, R. A.; Kiewnick, S. Influence of *Paecilomyces lilacinus* Strain 251 on the Biological Control of the Burrowing Nematode *Radopholus similis* in Banana. *Nematropica* **2007**, *37*, 203–213.

Moosavi, M. R.; Zare, R. Fungi as Biological Control Agents of Plant-Parasitic Nematodes. In *Plant Defence: Biological Control*; Springer: Dordrecht, 2012; pp 67–107.

Mukhtar, T.; Arshad Hussain, M.; Zameer Kayani, M. Biocontrol Potential of *Pasteuria penetrans*, *Pochonia chlamydosporia*, *Paecilomyces lilacinus* and *Trichoderma harzianum* against *Meloidogyne incognita* in Okra. *Phytopathol. Mediterr.* **2013**, *52*, 66–76.

Nicol, J. M. Important Nematode Pests of Cereals. In *Bread Wheat: Improvement and Production*; Curtis, B. C., Rajaram, S., Macpherson, G. (Eds.); FAO Plant Production and Protection Series: Roma, 2002; p 567.

Oduor-Owino, P. Integrated Management of Root-Knot Nematodes Using Agrochemicals, Organic Matter and the Antagonistic Fungus, *Paecilomyces lilacinus* in Natural Field Soil. *Nematol. Medit.* **2003**, *31*, 121–123.

Ortiz Paz, R. A.; Guzmán Piedrahita, Ó. A.; Leguizamón Caycedo, J. *In Vitro* Effect of *Purpureocillium lilacinum* (Thom) Luangsa-Ard et al. and *Pochonia chlamydosporia* var. catenulata (Kamyschko ex Barron & Onions) Zare & Gams on the root-knot nematodes [*P. chlamydosporia* (Kofoid & White) Chitwood and *Meloidogyne mayaguensis* Rammh & Hirschmann]. *Bol. Cien. Centr. Museos. Museo Hist. Nat.* **2015**, *19*, 154–172.

Pandey, R.; Trivedi, P. C. Fungus *Paecilomyces lilacinus* in Controlling Root-Knot Nematode and their Histological Interactions in *Capsicum annuum*. In *Proceedings—National Academy of Sciences India Section B* 1992; pp 275–275.

Park, J. O.; Hargreaves, J. R.; Mc Conville, E. J.; Stirling, G. R.; Ghisalberti, E. L.; Sivasithamparam, K. Production of Leucinostatins and Nematicidal Activity of Australian Isolates of *Paecilomyces lilacinus* (Thom) Samson. *Lett. Appl. Microbiol.* **2004**, *38*, 271–276.

Parvatha, R.; Nagesh, M.; Devappa, V. Effect of Integration of *Pasteuria penetrans*, *Paecilomyces lilacinus* and Neem Cake for the Management of Root-Knot Nematodes Infecting Tomato. *Pest Manage. Hortic. Ecosyst.* **1997**, *3*, 100–104.

Perveen, S.; Ghaffar, A. Use of *Paecilomyces lilacinus* in the Control of *Fusarium oxysporum* Root Rot and *Meloidogyne javanica* Root Knot Infection on Tomato. *Pak. J. Nematol.* **1998**, *16*, 71–76.

Pimentel, D. Environmental and Economic Cost of the Application of Pesticides Primarily in the United States. *Environ. Dev. Sustain.* **2005**, *7*, 229–252.

Rao, N. B. V. C.; Snehalatharani, A.; Emmanuel, N. New Record of *Paecilomyces lilacinus* (Deuteromycota: Hyphomycetes) as an Entomopathogenic Fungi on Slug Caterpillar of Coconut. *Insect Environ.* **2012**, *17*, 151–153.

Rao, M. S.; Umamaheswari. R.; Chakravarty, A. K.; Grace, G. N.; Kamalnath, M.; Prabu, P.; et al. A Frontier Area of Research on Liquid Bio-Pesticides: The Way Forward for Sustainable Agriculture in India. *Curr. Sci.* **2015**, *108*, 1590–1592.

Reddy, P. P.; Khan, R. M. Evaluation of *Paecilomyces lilacinus* for the Biological Control of *Rotylenchulus reniformis* Infecting Tomato, Compared with Carbofuran. *Nematol. Mediterr.* **1988**, *16*, 113–115.

Robinson, A. F.; Inserra, R. N.; Caswell-Chen, E. P.; Vovlas, N.; Troccoli, A. *Rotylenchulus* Species: Identification, Distribution, Host Ranges, and Crop Plant Resistance. *Nematropica* **1997**, *27*, 127–180.

Robl, D.; Sung, L. B.; Novakovich, J. H.; Marangoni, P. R.; Zawadneak, M. A. C.; Dalzoto, P. R.; Gabardo, J.; Pimentel, I. C. Spore Production in *Paecilomyces lilacinus* (Thom.) Samson Strains on Agro-Industrial Residues. *Braz. J. Microbiol.* **2009**, *40*, 296–300.

Saikia, M. K.; Roy, A. K. Efficacy of *Paecilomyces lilacinus* on the Reduction of Attack of *Meloidogyne incognita* on Okra. *Indian J. Nematol.* **1994**, *24*, 163–167.

Sandhu, S. S.; Sharma, A. K.; Beniwal, V.; Goel, G.; Batra, P.; Kumar, A.; Jaglan, S.; Sharma, A. K.; Malhotra, S. Myco-Biocontrol of Insect Pests: Factors Involved. Mechanism and Regulation. *J. Pathog.* **2012**, *2012*, 10 p. https://doi.org/10.1155/2012/126819.

Sarven, M. S.; Aminuzzaman, F. M.; Huq, M. E.. Dose-Response Relations between *Purpureocillium lilacinum* PLSAU-1 and *Meloidogyne incognita* Infecting Brinjal Plant on Plant Growth and Nematode Management: A Greenhouse Study. *Egypt. J. Biol. Pest Cntrl.* **2019**, *29*, 26.

Sasser, J. N. Root Knot Nematodes: A Global Menace to Crop Production. *Plant Dis.* **1980**, *64*, 36–41.

Seenivasan, N. Effect of Concomitant Application of *Pseudomonas fluorescens* and *Purpureocillium lilacinum* in Carrot Fields Infested with *Meloidogyne hapla*. *Arch. Phytopathol. Plant Prot.* **2018**, *51*, 30–40.

Siddiqui, Z. A.; Mahmood, I. Some Observations on the Management of the Wilt Disease Complex of Pigeon Pea by Treatment with a Vesicular Arbuscular Fungus and Biocontrol Agents for Nematodes. *Bioresour. Technol.* **1995**, *54*, 227–230.

Siddiqui, Z. A.; Mahmood, I. Biological Control of Plant-Parasitic Nematodes by Fungi: A Review. *Bioresour. Technol.* **1996**, *58*, 229–239.

Sikora, R.; Kiewnick, S. Evaluation of *Paecilomyces lilacinus* Strain 251 for the Biological Control of the Northern Root-Knot Nematode *Meloidogyne hapla* Chitwood. *Nematology* **2006**, *8*, 69–78.

Silva, S. D.; Carneiro, R. M.; Faria, M.; Souza, D. A.; Monnerat, R. G.; Lopes, R. B. Evaluation of *Pochonia chlamydosporia* and *Purpureocillium lilacinum* for Suppression of *Meloidogyne enterolobii* on Tomato and Banana. *J. Nematol.* **2017**, *49*, 77.

Singh, S.; Pandey, R. K.; Goswami, B. K. Bio-Control Activity of *Purpureocillium lilacinum* Strains in Managing Root-Knot Disease of Tomato Caused by *Meloidogyne incognita*. *Biocontrol Sci Technol.* **2013**, *23*, 1469–1489.

Smiley, R. W. Root-Lesion Nematodes: Biology and Management in Pacific Northwest Wheat Cropping Systems. *PNW Extens. Bull.* **2010**, *617*, 9.

Sosamma, V. N.; Geetha, S. M.; Koshy, P. K. Effect of the Fungus, *Paecilomyces lilacinus* on the Burrowing Nematode *Radopholus similis* Infesting Betel Vine. *Indian J. Nematol.* **1994**, *24*, 50–53.

Sreenivasan, N. Combined Application of *Pseudomonas fluorescens* and *Purpureocillium lilacinum* Liquid Formulations to Manage *Globodera* Spp. on Potato. *J. Crop Prot.* **2017**, *6*, 529–537.

Stirling, G. R. Suppressive Biological Factors Influence Populations of Root Lesion Nematode (*Pratylenchus thornei*) on Wheat in Vertosols from the Northern Grain-Growing Region of Australia. *Austr. Plant. Pathol.* **2011**, *40*, 416–429.

Swarnakumari, N.; Kalaiarasan, P. Mechanism of Nematode Infection by Fungal Antagonists, *Purpureocillium lilacinum* (Thom) Samson and *Pochonia chlamydosporia* (Goddard) Zare & Gams 2001. *Pest Manage. Hortic. Ecosyst.* **2017**, *23*, 165–169.

Taylor, A. L.; Sasser, J. N. *Biology, Identification and Control of Root Knot Nematode (Meloidogyne Spp.)*; Department of Plant Pathology Raleigh, North Carolina, St. University and USAID, North Carolina State Graphics: Raleigh, NC, 1978, 111 p.

Thakur, N. S. A.; Devi, G. Management of *Meloidogyne incognita* Attacking Okra by Nematophagous Fungi, *Arthrobotrys oligospora* and *Paecilomyces lilacinus*. *Agric. Sci. Digest* **2007**, *27*, 50–52.

Thomason, I. J. Challenges Facing Nematology: Environmental Risk with nematicide and the Need for New Approaches. In *Vistas on Nematology, Society of Nematologists*; Veech, J. A., Dickson, D. W., Eds.; Hyattsville: USA, 1987; pp 469–476.

Tranier, M. S.; Pognant-Gros, J.; Quiroz, R. D. L. C.; González, C. N. A.; Mateille, T.; Roussos, S. Commercial Biological Control Agents Targeted Against Plant-Parasitic Root-Knot Nematodes. *Braz. Arch. Biol. Technol.* **2014,** *57,* 831–841.

Trivedi, P. C. Evaluation of Fungus, *P. lilacinus* for the Biological Control of Root-Knot Nematode, *M. incognita* on *Solanum melonqena*. In *Proceedings of the 3rd International Conference on Plant Protection in the Tropics*; Genting Highlands: Malaysia, 1990; Vol 6, pp 29–33.

Udo, I. A.; Osai, E. O.; Ukeh, D. A. Management of Root-Knot Disease on Tomato with Bio-Formulated *Paecilomyces lilacinus* and Leaf Extract of *Lantana camara*. *Braz. Arch. Biol. Technol.* **2014,** *57,* 486–492.

Wakil, W.; Ashfaq, M.; Ghazanfar, M. U.; Kwon, Y. J.; Ullah, E.; Islam, S.; Ali, K. Testing *Paecilomyces lilacinus*, Diatomaceous Earth and *Azadirachta indica* Alone and in Combination against Cotton Aphid (*Aphis gossypii* Glover) (Insecta: Homoptera: Aphididae). *Afr. J. Biotechnol.* **2012,** *11,* 821–828.

Walia, R. K.; Bansal, R. K.; Bhatti, D. S. Effect of *Paecilomyces lilacinus* Application Time and Method in Controlling *M. javanica* on Okra. *Nematol. Medit.* **1991,** *19,* 247–249.

Walters, S. A.; Barker, K. R. Efficacy of *Paecilomyces lilacinus* in Suppressing *Rotylenchulus reniformis* on Tomato. *J. Nematol.* **1994,** *26,* 600.

Wilson, M. J.; Jackson, T. A. Progress in the Commercialization of Bionematicides. *BioControl* **2013,** *58,* 715–722.

Yadav, U. Recent Trends in Nematode Management Practices: The Indian Context. *Int. Res. J. Eng. Technol.* **2017,** *4,* 482–489.

Yang, F.; Abdelnabby, H.; Xiao, Y. The Role of a Phospholipase (PLD) in Virulence of *Purpureocillium lilacinum* (*Paecilomyces lilacinum*). *Microb. Pathol.* **2015,** *85,* 11–20.

Zareen, A.; Zaki, M. J.; Ghaffar, A. Effect of Culture Filtrate of Fungi in the Control of *Meloidogyne javanica* Root-Knot Nematodes on Okra and Broad Bean. *Pak. J. Biol. Sci.* **1999,** *2,* 1441–1444.

CHAPTER 5

Essential Oils: An Overview of Extraction Methods, Applications, and Perspectives

MIREYA VÁZQUEZ-AGUILAR[1], ISRAEL BAUTISTA-HERNÁNDEZ[1], ROMEO ROJAS[1], CECILIA CASTRO-LÓPEZ[2], and GUILLERMO CRISTIAN GUADALUPE MARTÍNEZ-ÁVILA[1*]

[1]*Laboratory of Chemistry and Biochemistry, School of Agronomy, Autonomous University of Nuevo Leon, General Escobedo, Nuevo León, 66050, México*

[2]*Laboratory of Chemistry and Biotechnology of Dairy Products, Research Center in Food & Development A.C. (CIAD, A.C.), Sonora, 83304, México*

*Corresponding author. E-mail: guillermo.martinezavl@uanl.edu.mx

ABSTRACT

Plant materials and by-products are important sources of essential oils. These oils have extreme relevance for agronomic, food, and pharmacological industries due to their great variety of biological activities. Nowadays, several conventional methods and emerging technologies have been investigated for the extraction of different compounds in order to study their yield, physicochemical, and functional properties in different areas. In this sense, this document reports recent information about importance and benefits of essential oils, their functional properties on foodstuffs, and their application in agronomic and cosmetic areas. In addition, more suitable methodologies applied for their extraction and characterization are presented. Finally, an overview of patents required to protect several essential oil–based products was reviewed.

5.1 INTRODUCTION

Essential oils are liquid fractions secreted in different plant tissues. These are composed by secondary metabolites that are produced by plants in order to protect themselves from pests, invasion of other plants, as well as to attract pollinating insects (León et al., 2015). In recent years, it has been a growing interest from the agri-food sector to find the main compounds present in essential to be applied in different industries, for which a large content of oxygenates, volatile and non-volatile residues, and mixtures of hydrocarbons are reported; within these compounds, those that are found in great abundance are derived from terpenes. The combination of these compounds provides different functions, for example, once obtained the essential oil from plants of interest, it is used in the food industry for production of fragrance and flavor (León et al., 2015). Over time, not only aromatic and flavor food functions have been assigned, antibacterial, antifungal, insecticidal, and antiparasitic properties (Smigielski et al., 2014), additives in foods acting as preservatives, preventing the growth of pathogens and damaging the microorganisms of the product have also been found (Torres et al., 2017).

One of the great advantages in the application of essential oils is that they are recognized as safe agents (GRAS) by the Drug Administration and United States Foods (FDA) (Smigielski et al., 2014; Torres et al., 2017), which demonstrate that they are substances in which toxicological studies have been carried out and have proven that they do not generate any risk to human health, in comparison with many chemical additives used in the food industry that cause long-term health damage, so these oils would be an alternative to the use of chemical additives. They are also safe alternatives against the environmental pollution, safety, and public health problems since it has been found that exist an inadequate use of agrochemicals that damage the environment and the health of the human being (Smigielski et al., 2014; Vaillant et al., 2015). Based on these premises, the aim of this work is to report recent information about importance and benefits of essential oils, their functional properties on foodstuffs and their application in agronomic and cosmetic areas. In addition, more suitable extraction methods applied for their extraction and characterization are presented. Finally, an overview of patents required to protect several essential oil–based products was reviewed.

5.2 CHEMICAL COMPOSITION OF ESSENTIAL OILS

To be possible to apply a new product or substance that will be used for human consumption, it is always important to carry out detailed and validated protocols to obtain reliable results and not cause any future damage to human health. In addition to guaranteeing health, it is important to carry out the chemical characterization in order to know the structure and properties of the compounds present in the material, in order to apply pharmaceutical, biotechnological, or nutritional use in the future (Abbasipour et al., 2017).

Chemical characterization by mass spectrometry is the most used and helps in the identification of compounds from plant material (Ramesh, 2017) among them phenolic compounds in essential oil, have been shown to have an antimicrobial and antioxidant activity, so they have been used in food industry (Mark et al., 2019). However, it is important to recognize that the characterization of compounds in plant materials is challenged due to the chemical diversity they possess and the need for standards for most chemical characterization methods in addition to the detection of high molecular weight compounds (Wen et al., 2018). To overcome these challenges, other methods have been implemented to improve the characterization techniques, know the compounds present in some substance, and analyze the toxic residues. Several technologies and combinations of them have been developed, such as liquid chromatography from which HPLC and UPLC derive that are used for separate and quantify compounds that were dissolved in a solution and that can also be combined with gas-chromatography–mass spectrometry (Kiani et al., 2018).

Other techniques such as Isotope Ratio Mass Spectrometry and Near Infrared Spectroscopy, which applied to several essential oils and become cross-validation model, in which analog essential oil compounds have been detected, such as the cinnamon (*Cinnamomum zeylanicum*) and clove oil (*Syzygium aromaticum*) (Bun et al., 2016). The most used method is the gas chromatography coupled to mass spectrometry (GC-MS), which analyzes volatile and semi-volatiles molecules trough that is possible to acquire data and identify components by comparing references that report known spectra (Deda et al., 2019). Through this technology, different compounds of the family of terpenes have been detected in essential oils extracted from different plant species that have diverse applications, most of them as environmental alternatives in the agroindustry, the food industry, cosmetic industry and in medicine that can be analyzed in Table 5.1 where it is possible to observe that all *Citrus* species contain myrcene and D-Limonene, in *Daucus carota L.* and *Citrus sp.* is found a derivative from pinene, the α-pinene.

The differences observed in the functions of the essential oils as well as in the compounds found will depend on several factors. Not all essential oils have the same compounds, there are quantitative and qualitative variations in the content since this will depend on the plant species, according to their genotype; for example, (Da Silva et al., 2017) observed that in *Psidium guajava* there are significant differences in the chemical composition because of the genotypes, based on the chromatographic profile of the essential oil, where in the C10, C13, and SEC genotype, compounds such as trans-caryophyllene were obtained, and in C3 and C6 had alpha-humelene. Also, other variations depend on environmental factors such as soil, the stage of maturity for example it was observed that in *Eucalyptus globulus* oil from Morrocan has a low content of 1,8-cineole in May, than in July, increasing a 19.8% (Tisserand and Young, 2014). Also, it depends on the type of climate, the method of extraction, solvents used, as shown in the Figure 5.1, and other unconventional factors such as fungi where (Fokom et al., 2019) showed that the inoculation of fungus (arbuscular mycorrhizae) in lemongrass (*Cymbopogon citratus*) that occur naturally during plant growth generate a symbiosis and together with the ideal harvest period are key in obtaining quality essential oil having a high proportion of compounds that contribute to antioxidant properties like myrcene, isogeranial, neral, and 6-methyl-5-hepten-2-one. All these effects could also affect the biological activity of the oil.

Plant material

Conventional and classical techniques ← .... → **Innovative techniques**

Extraction methods

- Hydro-distillation
- Steam distillation
- Organic solvent extraction

- Supercritical fluid extraction (SCFE)
- Ultrasound-assisted extraction (UAE)
- Microwave-assisted extraction (MAE)
- Enzyme-assisted extraction

Essential oil (EO)

Applications

Agronomic   **Medicine**   **Cosmetic**   **Food**

FIGURE 5.1 Production flow diagram of essential oils.

It has been observed that the compounds found in essential oils have related chemical structures and many of them contain an aroma that characterizes them, for example limonene, linalool and pinenes, which

TABLE 5.1 Applications of the Essential Oils in Different Plant Species.

Plant	Tissue source	Main components of essential oil	Functional activity/application	Target organism/Food product	References
Daucus carota L	Seeds	Carotol, α-pinene, and geranyl acetate	Antimicrobial activity	*Bacillus subtilis* and *Candida* sp	Smigielski et al. (2014)
Citrus sp.	Commercial oil	D-Limonene, Myrcene, α-Pinene, and Linalool	Antimicrobial and antioxidant activity	*Listeria monocytogenes*	Torres et al. (2017)
Citrus sinensis	Peel	D-Limonene, Myrcene, α-Pinene, and Terpinen-4-ol	Activity against weed germination and growth	*Helianthus annuus, Portulaca oleracea, Lupinus albus* and *Malva parviflora*	El Sawi et al. (in press)
Citrus aurantium		D-limonene, Myrcene, Lynayl acetate, and α-Pinene terpinene.	Natural pesticide		
Citrus reticulata		D-limonene, γ-terpinene, α-terpineol, α-phellandrene, and β-pinene	Repellent activity		
Cannabis sativa L	Residue of industrial hemp	--------	Antifungal activity	*Musca domestica* and *Myzus pesicae*	Pavelac et al. (2018)
Lantana camara	Leaves	α-Humelene, cis-caryophyllene, germacrene-D, bicyclogermacrene, aromadendrene, and β-curcumine	Treatment against parasitic disease	*Callosobruchus maculatus*	Zandi et al. (2012)
Allium sativum	Commercial oil	Allicin	Increase in the lipidic profile in meat	*Candida*	Mendoza et al. (2017)
Mentha piperita		--------	Antioxidant activity	*Anisakis*	Romero et al. (2014)
Lippia origanoides		--------	Antimicrobial and antifungal activity	Chicken meat	Madrid et al. (2018)

TABLE 5.1 *(Continued)*

Plant	Tissue source	Main components of essential oil	Functional activity/application	Target organism/Food product	References
Prunus armeniaca	Seeds	Oleic acid, linoleic acid, 1-methyl-2-pyrrolidone, and palmitic acid	Increase humidity resistance and mechanic properties of films	Bread slices *B. subtilis, E. Coli,* and *Rhizopus stolonifera* Chitosan films	Priyadarshi et al. (2018)
Lippia origanoides	Commercial oil	--------	Antioxidant activity Provide adequate texture (pasty) in meat products	Meat Children with dysphagia and cerebral palsy	Barbosa et al. (2017)
Deep blue and Copaiba oil	—	Methyl salicylate, menthol, and β-caryophyllene	Anti-inflammatory and analgesic	People with hand arthritis	Bahr et al. (2018)
Curcuma aureginosa	Rhizome	Germacrone	Inhibitor of axillary hair growth	Women with androgenic axillary hair-growth	Srivilai et al. (2016)
Cedrus deodara	—	—	Insecticide	Adult mealworms (*Tenebrio molitor*)	Buneri et al. (2017)
Teucrium polium L.	Aerial parts	Lycopersene, dodecane, and 1,5-dimethyldecahydronaphthalene	Insecticide	Red flour beetle (*Tribolium castaneum Herbst*)	Ebadollahi et al. (2019)
C. acaulis D. ambrosi-oides Mentha longifolia	Roots Flowering aerial parts Aerial parts	Polyacetylene carlina oxide, ascaridole, and piperitenoneoxide	Insecticide	Horn (*Prostephanus truncates*) khapra (*Trogoderma granarium*)	Kavallieratos et al. (2020)
Ocimum gratissimum	Leaves mixed with young branches	Thymol	Insecticide	Cotton leafworm or Egyptian (*Spodoptera litoralis*) Housefly (*Musca domestica*)	Benellia et al. (2019)
Spheranthus amaranthroids	Leaves	D-Carvone	Insecticide	Tobacco cutworm (*Spodoptera litura*)	Murfadunnisa et al. (2019)

TABLE 5.1 *(Continued)*

Plant	Tissue source	Main components of essential oil	Functional activity/ application	Target organism/Food product	References
Vitex negundo	Leaves	β-Caryophyllene	Herbicide	*Avena fatua* L. *Echinochloa crus-galli* L.	Issaa et al. (2020)
E. citriodora O. basilicum M. arvensis	Leaves	Citronellal, Methyl chavicol/ linalool, Menthol/menthone	Herbicide	*Angallis arvensis Cyperus rotundus Cynodon dactylon*	Khare et al. (2019)
Tagetes erecta L., *Asteraceae*	Leaves	a-Terpinolene, dihydrotagetone	Cytotoxicity	Murine melanoma Human colon carcinoma cell lines	Oliveiraa et al. (2015)
Myrica rubra	Leaves	B-Caryophyllene, a-Humulene	Cytotoxicity	Ileocecal adenocarcinoma cell lines	Langhasovaa et al. (2014)

are found in large quantities (Tisserand and Young, 2014). Most of the essential oils used to produce aroma or flavor are those that come from citrus, the oxygenated fraction is what gives them the intense flavor and consists mainly of alcohols, aldehydes, ketones and esters, and other volatile compounds. Also, there are non-volatile chemical compounds such as hydrocarbons, sterols, fatty acids, waxes, carotenoids, coumarins, psoralens, and flavonoids (Torres et al., 2017).

Essential oils also have other compounds called terpenes like the monoterpenes which explains the inhibition of germination in "weeds" since they have phytotoxic properties that cause anatomical and physiological damages in plants because the mechanism of these compounds causes the accumulation of lipid globules in the cytoplasm which reduces organelles as mitochondria. Terpenoids have also been found which influence the defense mechanisms of plants, so they represent a potential use in agriculture (El Sawi et al., in press).

Other compounds have been detected such as carvacrol, eugenol and thymol, which are important for inhibit the growth of microorganisms by breaking cell membranes (Bhavaniramya et al., accepted manuscript), and D-Limonene, mainly found in orange essential oil, which is one of the components that contributes to the antioxidant activity of this oil, along with other monoterpenes (Torres et al., 2017; El Sawi et al., in press).

This is how knowing the chemical compounds that a product contains its importance to be able to find applications that meet the needs of the human being, either to isolate the compounds, use the plant material to replace for other synthetic compounds or to avoid consuming because of the damage that could cause in the health.

5.3   POTENTIAL APPLICATIONS OF ESSENTIAL OILS

Scientific research has always been of interest to industries since it has persistently sought to meet the different demands of the market; quantity, quality, low cost, and in recent years, be friendly to the environment. Also, the ingredients of natural origin have been wisely implemented in the industry, since they do not imply complications in the process of substances of synthetic origin, likewise the label "Natural origin" provides a prestige degree to the product highlighting the industry's commitment to preservation of the environment. That is why it is of great importance to know the various applications of essential oils in four of the industries that are fundamental for world trade, the agricultural industry, the pharmaceutical industry (medicine), food industry, and the cosmetic industry.

5.3.1 AGRICULTURE FIELD

In agriculture, there have always been problems related to pests at production and storage steps that continuously cause negative impacts (mainly economic) for farmers (United Nations). The traditional solutions introduce the application of conventional pesticides (produced synthetically) that for the latest decades helps to the stability and increment of production. But also, its application has a negative impact as negative interaction with no-target organisms, poor waste managements, appearance of new resistant pest, and risks to health. That is why the search and application of new alternatives to chemical pesticides has intensified in the last years (Rodríguez et al., 2014) and one of these alternatives is the use of products based in natural extracts.

The use of essential oils has potential as alternative to control of pest based on benefits for environmental and human health. Some benefits of interest are the highly volatile capacity, low toxicity for mammals, different sites of action and less side effects on non-target organisms (Santos et al., 2019; Souza et al. 2019). In addition, several studies evaluated the efficiency of its application against important pests; for example, (Zandi et al. 2012) shows that the compounds found in *Lantana camara* essential oil have a repellent activity against *Callosobruchus maculatus*, a pest with negative effects on the economy of grain legumes. The study explained a possible mechanism of action by their lipophilic nature, and this mechanism of action can be explained by their lipophilic nature, as this condition causes them to be absorbed by the insect's cuticle and interfere with their metabolic and biochemical functions until they are killed (Tandon and Mittal, 2018).

Also, the EO application against coleoptera pest is an important topic to grain industry as a result of the severe impact on quality and quantities in products. Several species like mealworm beetle (*Tenebrio molitor*), horn (*Prostephanus truncates*), khapra (*Trogoderma granarium*), and confused flour beetle (*Tribolium confusum*) have been reported as severe and common problem in storage. Studies carried out by Buneri et al. (2017) compared traditional application insecticides (carbosulfan and imidacloprid) and the *Cedrus deodara* oil against mealworm beetle (*Tenebrio molitor*); the results showed a higher LC50 value of 3.41% (EO) and lower values for traditional methods (LC50: 0.086–0.0231%, respectively); the information not only supports the efficiency of traditional application but also provides us a perspective of insecticide activity in EO.

Other studies focused only in the insecticide activity and the optimal doses for increment the mortality ratios; for example, the application of *Teucrium polium L.* essential oil showed a mortality of 97.97% on red flour beetle (*Tribolium castaneum Herbst*) to concentration of 20 μL/L and 72 min of action (Ebadollahi and Taghinezhad, 2019); Kavallieratos et al. (2020) evaluated eight different EO (*Dysphania ambrosioides, Carlina acaulis, Trachyspermum ammi*, and *Mentha longifolia*) against horn (*Prostephanus truncates*) and khapra (*Trogoderma granarium*); the study established *Trogoderma granarium* mortality of *C. acaulis* essential oil in 97% to 500 mg/L after 3 days of application, also the *D. ambrosioides* and *Mentha longifolia* essential oil eliminated all *T. granarium* adults to 500 mg/L after 4–2 days, respectively. The results demonstrated high insecticide activity by natural components against common pests with a problematic impact in the economy.

One aspect to evaluated in EO application is the interaction between non-target organism, with the aim to elucidate the possible real impact in the environment. A non-target organism adopted as a model for the diagnosis of soil ecosystem health and for evaluating the potential environmental impact is (Bourdineaud et al., 2019). Studies carry out on earthworms analyzed the application of traditional products and natural extracts focused in the non-target organism impact; for example, *Ocimum gratissimum* and *Spheranthus amaranthroids* essential oil were evaluated against several pest as tobacco cutworm (*Spodoptera litura),* cotton leafworm or Egyptian (*Spodoptera littoralis)* and the housefly (*Musca domestica*), respectively. The results obtained showed a lower impact or non-impact in the application of *Spheranthus amaranthroids* EO (1000–1500 mg/L) on earthworm; otherwise, the application of monocrotophos (pesticide) has a morality rate around 80% at 20 μg after 24 h. The Ocimum gratissimum essential oil has a non-mortality impact on *E. fetida* adults, but a-cypermethrin (pesticide) demonstrated mortality ratios of 100 ± 0.0 - 85 ± 3.8% on the 10th day (20.0–10.0 mg/kg) (Benellia et al., 2019; Murfadunnisa et al., 2019).

Another important aspect of the implementation of EO is its antifungal and antimicrobial potential; the EO activity has been found in *Melaleuca quinquenervia L.*, a plant native to Australia with effects on phytopathogenic fungi as *Rhizoctonia solani Kühn, Phytophthora nicotianae Breda Haan, Fusarium Solani Mart*, and others (Vaillant et al., 2015; Leyva et al., 2016). The activity against phytopathogenic organisms lies in compounds with negative interaction on cell machinery, such as trans-cinnamaldehyde, which has been observed as a growth controller in *E. coli* and *S. tifimirium*, by decreasing the level of ATP inside the cell, also the Carvone has a roll in the mechanism by breaking down the outer membrane of the cell but without affecting the level of ATP (Bhavaniramya et al., accepted manuscript).

The insects, fungi, and bacteria are not the only threats on a crop production and storage; there are unwanted plants that grow along with others and take advantage of their nutrients (weeds). It has also been applied essential oils of some species that inhibit the germination and growth of these weeds. Under weeds perspective studies focused in phytotoxicity, cytotoxic and genotoxic evaluation to determinate viability and the efficiency against weeds. For example, Rani et al. (2019) implemented *Vitex negundo essential* oil against two agricultural weeds *Avena fatua L.* and *Echinochloa crusgalli (L.),* obtained 65% and 76% reduction in germination at 1 mg/mL and non-germination at 2.5 mg/mL EO; the repressing effect could be the result of interaction between β-caryophyllene (present in EO) and other major compound in EO. In addition, they reported DNA damage in *Allium cepa L.* (model) and a mitotic index lower when increase de EO concentration (cells being unable to enter the prophase). Also, khare et al. (2019) reported a reduction biomass ratio in weeds (*Angallis arvensis, Cyperus rotundus,* and *Cynodon dactylon*) between 22–69%, 5–70%, and 51–78%, respectively, and a shoot length reduction between 22–69%, 7–92% and 63–86%, on the same order. All the reduction effect was a complement for visible injuries like chlorosis and necrosis. Both studies support the argue that EO could provide herbicidal protection to interest organism but also open the opportunity for new researches with the aim to understand in a better way the mechanism and interactions of EO and environmental aspects.

Another new point of interest it is the application of residues that provide extra value to an agro-industrial waste; a research carried out by El Sawi et al. (2019) shows the use of residues from the citrus juice industry as the peel of *C. sinensis, C. aurantium,* and *C. reticulata* like primary material for EO obtention and consequentially an application against weeds, the data reported best inhibition result in concentration of 3%(v/v) of *C. aurantium* with a 100% of inhibition on seed germination, also the root length was affected by the EO; the application reduce the length and some cases reduce until 60% of root length. At the same time there are plants that are not fully exploited, such as the inflorescences of *Cannabis sativa L.,* a plant used to obtain industrial hemp, studies have shown that essential oil from its inflorescences can be used as pesticides against phytophagous insects such as aphids (*Myzus persicae*), and houseflies (*Musca domestica L*), since these inflorescences secrete secondary metabolites such as cannabinoids and volatile terpenes that damage them (Pavelac et al., 2018).

5.3.2   MEDICINE FIELD

Another important aspect about essential oil is its possible application in medicine. The emergence of microbial strains resistant to conventional antibiotics has sparked an interest in new effective and safe treatment methods. An alternative of interest is the application of essential oil on treatments, as a result of its antimicrobial activity. Studies carried out by (Lin and Changzhu, 2019) evaluate the ability of oregano essential oil as an antimicrobial agent against methicillin-resistant *Staphylococcus* aureus (MRSA). In addition, this study establishes the mechanisms of action of essential oil that include (1) damage to the cell membrane, (2) inhibition of the respiratory chain, (3) the capacity of one of the major components (Carvacrol) to form a chimera with DNA, and (4) inhibition of the expression of the pathogenicity factor pvl.

In addition, others studies analyzed the EO potential as antiseptic and antiparasitic, for example, Mendoza et al. (in press) used *Allium sativum* oil to control *Candida* species in dental prostheses, which was result in more effective than fluconazole, a medication used to counteract this fungus, also (Romero et al., 2014) used the essential oil of *Mentha piperita* for treatment against *anisakisasis*, a condition of parasitic nature that occurs through the intake of fish containing the larva of Anisakis, which produces gastrointestinal symptoms and allergic reactions difficult to cure.

One important aspect is that in recent years there is an increasing resistance of mammalian tumor cells to chemotherapy and treatment present barriers, such a severe side effects, multi-drug resistance, ineffective responses, and expensive services. As a response, the search of new medication has potentiated the most interesting subjects in the field of natural products and essential oil could be an auxiliary in some treatments.

Several studies evaluated the application of essential oil against tumor cells and showed a promising result; for example, Oliveira et al. (2015) evaluated several essential oil cytotoxicity against cancer cell lines (murine melanoma, human colon carcinoma, human breast adenocarcinoma, and human cervical adenocarcinoma) with concentration ranging from 3.12 to 400 µg/mL for 24 h. The best result corresponds to *Tagetes erecta L., Asteraceae* essential oil, the *Tagetes erecta L., Asteraceae* EO showed a IC_{50} value significantly lower (7.47 ± 1.08 and 6.93 ± 0.77 µg/mL) for murine melanoma and human colon carcinoma cell lines than normal cell line IC_{50} value (19.50 ± 5.96 µg/ml). Also, the essential oil showed a selectivity index of 2.61 a 2.81, the values means that the compound is more than twice more cytotoxic to the tumor cell line as compared with the normal cell line. Other study applied *Myrica rubra* essential oil against human colon and ileocecal adenocarcinoma cell lines and found a major sensibility in Caco2, HCT8, and HT29 cells (IC_{50} values 30–49 µg/mL) than SW480 and SW620 cells (IC_{50} values 116 and 132 µg/mL, respectively). In addition, Richardson (2019) conducted a study in which the cytotoxicity and genotoxicity of the major compounds of essential oils was evaluated; their study found a potential application of thymol and geraniol as a genoprotector of human colon cells.

5.3.3 FOOD INDUSTRY

Although essential oils are already used in the food industry, studies have recently been conducted to find other food applications and not only to provide flavor and aroma, for example, the application of the antifungal capacity of these in food which has been studied by Pinho (2018), who evaluated the potential of nano-emulsions of encapsulated oregano essential oil as an antifungal agent in Minas Padrao cheese, the study showed the potential against the growth of *Cladosporium sp, Fusarium sp.* and *Penicillium sp.*

In turn, the application of oil to improve the quality of a food product has been evaluated (Madrid et al., 2018), added oregano essential oil (*Lippia origanoides*) to the broiler feed, which resulted as a growth promoter improving the fatty acid composition in chicken meat, thus being an alternative in the replacement of antibiotics used as growth promoters; besides that oregano oil showed an improvement in taste, secretion of digestive enzymes, improvement in gastric motility, anti-inflammatory, antioxidant, and antimicrobial activity.

It is important to know that in the food industry synthetic antioxidants are used, such as butylated hydroxyanisole (BHA) and butylated hydroxytoluene (BHT), which have had negative side effects, so it has been interesting to find natural antioxidants, in where (Torres et al., 2017; Barbosa et al., 2017) analyzes that the orange essential oil is one of the plant species that could be an alternative in the application in food as additives with antioxidant effect.

In addition to using compounds that have antioxidant activity in food, another strategy in the food industry is to add them to the packaging. (Priyadarshi et al., 2018) demonstrates that it is possible to use apricot kernel (*Prunus armeniaca*) essential oil in combination with chitosan films, which results in a greater antioxidant potential and an efficient inhibition of bacteria such as *B. subtilis* and *E. coli* that it could be due to one of the components found in this oil, N-methyl-2-pyrrolidone because it has been shown to interact with the bacterial membranes and dissolves their lipids, which results in their decomposition, leaving out all intracellular fluids that will cause the death of the bacteria.

Therefore, it is possible to say that essential oils have a great potential in the commercial application as a food packaging material or as an ingredient for food preservation, being thus a natural and environmentally friendly way which it is of great importance for the food industry these essential oils as an alternative, since they reduce the use of synthetic compounds in food that cause future damage to human health.

5.3.4   COSMETIC INDUSTRY

The ingredients of natural origin are implemented wisely in the cosmetic industry, since they do not imply complications in the manufacturing process such as substances of synthetic origin. In addition, the label "Natural origin" provides a degree of prestige to the product that highlights the industry's commitment to the preservation of the environment. In the cosmetic industry, plants oils are generally implemented in cosmetic products as functional ingredients for their antioxidant, anticancer, or photoprotective activities (Sarkic and Stappen, 2018).

Also, its properties have been implemented to preserve products as a result of the application of inhibitory substances, due to the fact that under the appropriate physicochemical conditions the process of oxidation and microbial growth would have the capacity to compromise the quality of the product (Noureddine Halla et al., 2018). The application of essential oil in the cosmetic industry falls not only in the conservative properties it offers to the product but also in the properties offered to the consumer, which include anti-inflammatories activity, moisturizing capacity, and skin protection (Hallaa et al., 2018). An example, Maurya et al. (2018) evaluated the anti-inflammatory capacity of *Citrus limetta Risso*-essential oil in in vitro and in vivo models, obtaining favorable results in the decrease of parameters of interest (edema, ear weight, ear thickness, and oxidative stress level) resulting from the inhibition of the constituents in the production of proinflammatory cytokines, demonstrating the potential for future formulations.

Likewise, the study and development of knowledge about the various properties of essential oils promotes the protection of intellectual property with the registration of patents that involve since preparation methods to products with the action of an essential oil. For example, the method of preparation of radix aucklandiae essential oil (CN107412021) that involves as advantages the high content of essential oil with a stable performance with a potential application in inflammation and acne removal. On the other hand, cosmetic products such as a liniment with wormwood essential oil (CN109718136) or a day cream with *Rose Damascena* essential oil (CN109431904) take advantage of properties like permeability for the benefit of skin protection performance.

The release of new cosmetic products with a higher content of components from natural sources promotes the development of future lines of research with the objective of characterizing and assessing the potential of extracts and essential oils.

5.4 EXTRACTION METHODS APPLIED FOR ESSENTIAL OIL OBTENTION

The increase popularity about the application of personal care products with natural components has had a positive impact on the demand of aromatherapy industry products, including the essential oils market. Also, the essential oil market shows enormous growth and predictions reach US $ 3226.2 million (Ren, 2019). As a result, the development of new methodologies focuses on offering higher extraction speeds, lower operating temperatures, and the application of solvents with a lower environmental impact, while also solving the high energy consumption that has negative impacts on the environment (emissions of CO_2) and production costs (Zhang and Zhao, 2014).

5.4.1 CONVENTIONAL METHODS

5.4.1.1 HYDRODISTILLATION (HD)

In general, traditional extraction methods are simpler and more economic that innovate systems, one of the favorite methods as a result of practicality is hydrodistillation that can be utilized by big or small industries (Chouhana et al., 2019). Also, variants come from the same principle differing in the region assigned for the plant material in the extraction complex (steam-hydrodistillation and steam distillation). Hydrodistillation method consists on introducing the plant material into water, subsequently boiling, to finally condense and decant the essential oil (El asbahani et al., 2018). Alternatively, steam-hydrodistillation involves separated the plant material and boiling water on the same compartment; otherwise, on steam distillation direct steam is injected into the plant material.

Some vantages are the lower water miscibility with terpenic molecules, non-toxic residues on products and the lowers prices of primary material; otherwise, the process involves high amounts of energy, long operational times according to plant material (1–24 h) and the system energy could add unwanted alteration in the terpene molecules (oxidation products) by high temperatures and polar molecules losses (Chouhana et al., 2019; El asbahani et al., 2015; Yousefi et al., 2019).

The hydrodistillation process generate large volumes of residues. As an alternative, several studies evaluated the bioactive potential of HD waste to give an extra value to the process. Santana-Méridas et al. (2014) obtained promising results of *Rosmarinus officinalis L.* solid residues extracts as an antifeedant agent against *L. decemlineata, S. littoralis,* and *M. persicae,* and a high antioxidant activities, comparable to an extract from red grape pomace that was used as a natural antioxidant standard. Most recently, the evaluation of antioxidant activity of EO distillation residues of *Nepeta* species revealed that water extract is a strong radical scavenger (Baranauskienėa et al., 2019.

5.4.1.2   ORGANIC SOLVENT EXTRACTION (SOXHLET)

The organic solvent extraction or Soxhlet is a conventional method for recover organic compounds in plants that is carried out by Soxhlet apparatus invented in 1879 by Franz von Soxhlet. In the extraction process an extracting unit is loaded with plant material inside of perforated bags and by an evaporation–condensation process the samples is repeatedly washed with a selectively process. Some attractive vantages are related with Soxhlet. First, the systems could work on lower temperature under the boiling point of water according to solvent qualities, also, it is a simple and cheap process, and it is not necessary a filtration process at the end in spite of the plant material is in contact to extracting solvent. A negative aspect is the presence of solvent residues on products and the health risk for consumer (food applications) also the high amount of time and energy applied (Hesham et al., 2016; Oreopoulou et al., 2019).

5.4.2   INNOVATIVE METHODS

According to literature, the essential oil extraction is a complex process as a result of the vegetal structure tissue that limits the extraction by water or other extractant agent and consequentially the intracellular metabolites release; also, it is important considerate the negative alterations involved on traditional methods. Under this perspective studies have accelerated the search of new "Innovative" techniques that allow the recovery of essential oils with stable components and in the natural proportion with shorter operating times (Smigielski et al., 2014; El asbahani et al., 2015).

5.4.2.1 SUPERCRITICAL FLUIDS EXTRACTION

Extraction by supercritical fluids has been used for the extraction of flavoring and aromatizing compounds in complex matrix as an alternative to organic solvents. The SCFE is an innovative and environmentally friendly process that uses renewable fluids on (or near) its supercritical conditions (temperature and pressure) as extracting agents (Vichi, 2010).

Under supercritical conditions, the solvent shares properties between liquids and gases that add greater efficiency for the recovery of compounds. The process involves two steps: (1) extraction and (2) separation of solvent. In the first step, a mix of supercritical solvent and a cosolvent (if it is necessary) extract the solute under convection and diffusion principles. Then, a pressure reduction involves a lower dissolvent power on the fluid and the solute separates for solvent (stage 2) (Peter, 2012; Prado et al., 2015). The application of CO_2 as a solvent is the result of chemical properties as its critical pressure (7.38 MPa) and temperature (304.2 K). The application of SFCE has some advantages such as: The solvent can be easily removed due to the adjustment of conditions, the solvent implemented is reusable in the system, its considers a green technology and safe for health, higher yields, and better quality of solutes (Hesham, 2016; Khalil et al., 2017; Chin Chew, 2020).

5.4.2.2 MICROWAVE AND ULTRASOUND ASSISTED EXTRACTION

The coupling of microwave technology (MAE) to conventional methods such as hydrodistillation has been implemented in recent years as an alternative with favorable prospects. Its application considers an electromagnetic field in a time from 300 MHz to 300 GhZ, according to a range of greater application from 0.915 to 2.45 GHz, where the greatest amount of energy is very weak to break a hydrogen bond (0.00001 eV) (Prado et al., 2015). The extraction principle is based on small traces of moisture on plant material, moisture gets heat and starts to evaporate into the cell until it breaks, which helps release the compounds into the medium (Mandal et al., 2015). The process has three sequential stages: (1) Separation of solutes from active sites by increasing temperature and pressure, (2) diffusion of the solvent by the matrix, and (3) release of the solutes from the matrix to the solvent (Oreopoulou et al., 2019). Interest in the methodology has increased the development of new associations with greater benefits to extraction situation. For example, new methodologies are: vacuum microwave hydrodistillation (VMHD) (Oreopoulou et al., 2019) and solvent-free microwave extraction (SFME) (Benmoussa et al., 2016).

The ultrasound assisted extraction (UAE) is one of the most implemented green extraction methods in bioactive compounds research, due to high extractions yields, simplicity, and small operational times. Similarly, UAE has been used as a complement to other techniques such as Soxhlet or hydrodistillation, with the aim to add acoustic energy to the process and improve the extract yields. The impact on plant matrix is the formation of bubbles by cavitation phenomena, and their implosion results in a cutting force that breaks the cell, facilitating the release of compounds in the medium (Mandal et al., 2015).

5.4.3   OTHER ASPECTS IN TRADITIONAL AND INNOVATIVE EXTRACTION METHODS

The stability of the compounds by application of the new technologies is evidenced in the studies carried out by Guerrero et al. (2017) evaluated different *Rosmarinus officinalis* essential oil extraction by CO_2-supercritical extraction, hydro-distillation and steam distillation. The best yields obtained was reported to CO_2-supercritical extraction (2.53%, w/w), also, its antioxidant evaluation by the ABTS radical test (2,2'-azinobis- (3-ethyl-benzothiazolino-6-sulfonic acid)) registered a range between (37.55 ± 0.86–29.67 ± 0.52 mg Trolox/g of essential oil) higher that values obtained for hydro-distillation and steam distillation (2–3 mg Trolox/g of essential oil). Also, the data support the potential application of new green solvent extraction methods as a result of better yield that traditional, and the differences between both extraction methods could be result of chemical modification by extraction conditions as temperature and operational time.

The diversity of the extraction procedures (Fig. 5.1) could be an alternative to improve extraction process, by combinations of methods that allow solving certain critical points such as: lower extractions yields, a better stability in biochemical composition of the extract or improve the extraction of determinate compounds of interest (Saucedo et al., 2018).

Other alternative, it is the application of innovative technologies implemented enzymatic reactions (catalysis), which have the advantage of avoiding unwanted changes result of high temperatures, very extreme pH values, precise specificity, and region selectivity (Saucedo et al., 2018; Nadar et al., 2018). For example, Hosni et al. (Polmanna et al., 2019) implemented a methodology that involves enzymatic reactions (enzyme-assisted extraction/ hydrodistillation) to obtain essential oils from *Laurus nobilis L.*, the result showed better extraction yields of enzyme treatments(cellulase, hemicel-lulose, and xylanase) between 0.54 and 1.24% (w/w) with respect to the control (0.37% w/w), the yield with a lower value corresponds to the three enzymes treatment and it is attributed to interactions between degradation products that form oligomers with negative effect on degradation process. As a result, four points to consider for an efficient application of enzymes in extraction processes are: consider the composition/morphology of the plant material, the catalysis of the enzyme, the optimal enzymatic operational parameters and finally promote a synergism between enzymes (Chouhana et al., 2019; Hesham et al., 2016).

However, as seen in Table 5.2, standard methodologies are still used in the extraction of essential oils, yields in traditional methods have a variant range between 0.1 and 43% with the complication of involving harmful substances that hinder the subsequent use of oil for industry. On the other hand, innovative methods imply yields ranging from 1.41 to 8.63 with a lower amount of energy, free of solvents harmful to the environment and with shorter times. Although traditional methods have been implemented in the industry for more years, the application of new extraction routes is a growing research field with the objective of obtaining higher extraction yields with better product quality.

5.4.4 RECOMMENDATION FOR EXTRACTING AND IMPLEMENTING OILS

The activity of the essential oils will depend a lot on the care of these, since when containing a high amount of terpenic hydrocarbons; it makes them very unstable before the light and the heat, and their solubility in alcohol diminishes, reason why it is recommended to discard these compounds for

TABLE 5.2   Performance and Specifications of Traditional and Innovative Extraction Methods.

Extraction method	Plant material	Sample quantity	Particle size	Extraction time	Temperature	Pressure/flow	Solvent	Yield	Ultrasound or microwave energy	References
Conventional or classical										
Hydrodistillation	*Carex megerianan* kunth	20 g	10 mm	480 min	—	—	Distilled water	0.1%	Not apply	Cui et al. (2018)
	Damask rose	100 g	—	237 min	—	—	Distilled water	0.033%	Not apply	Manouchehria et al. (2018)
Steam distillation	Lavender (*Lavandula* spp.)	30 g	—	113 min	—	—	Distilled water	3.3%	Not apply	Gavahian and Chu (2018)
	Mallee	10 g	—	1–180 min	105°C	1.3 kg/min/m²	Distilled water	—	Not apply	Syamsuddin et al. (2014)
Soxhlet	Eucalyptus	5 g	—	1, 2, and 8 h	78°C	—	Hexane	36.33%	Not apply	Zhao and Zhang (2014)
			—	0.5, 1, 2, and 8 h	90°C	—	Ethanol	7.9%	Not apply	
	Moringa peregrina	10 g	841 μm	11 h	60°C	—	Hexane	43%	Not apply	Mohammadpour et al. (2019)
Super critical fluid extraction	Rosemary	25 g	600 μm	180 min	40–50°C	10.34 and 17.24 MPa/126 mL/min	CO_2	1.41–2.53%	Not apply	Hernández et al. (2017)
	Chrysopogon zizanioides	11 g	450 μm	150 min	40, 50, and 60°C	14, 17, and 20 MPa/1.97×10^{-3} Kg/min	CO_2	1.35–2.23%	Not apply	Santos et al. (2019)
Ultrasound-assisted extraction	White pepper (*Piper nigrum* L.)	10 g	420 μm	7 min	100°C	—	Distilled water	3.4%	50 W (40 KHz)	Wang et al. (2018)

TABLE 5.2 *(Continued)*

Extraction method	Plant material	Sample quantity	Particle size	Extraction time	Temperature	Pressure/flow	Solvent	Yield	Ultrasound or microwave energy	References
Ultrasound-assisted extraction/ Microwave-assisted extraction	White pepper (*Piper nigrum* L.)	10 g	420 μm	7 min	100°C	–	Distilled water	4.1%	50 W (40 KHz)/500 W	Wang et al. (2018)
Microwave-assisted hydrodistillation	*Laurus nobilis* L.	50 g	500 μm	45 min	–	–	Distilled water	1.01%	400 W	Tabana et al. (2018)
	Myristicae arillus	40 g	103 μm	56 min	–	–	Distilled water	8.63%	800 W	Megawati et al. (2019)
Enzyme-assisted extraction/ hydro-distillation	*Laurus nobilis* L.	100 g	500 μm	60 min	40°C	–	Distilled water	1.25%	Not apply	Boulila et al. (2015)

obtaining the final product. There are several technologies to reduce terpene hydrocarbons such as vacuum distillation, solvent extraction, and adsorption chromatography. However, recent studies have found a more innovative method to retain antioxidant activity and protect the active compounds from environmental factors such as light, oxygen and temperature of essential oils, the use of cyclodextrins that are cyclic oligosaccharides, which not only preserve compounds sensitive if not also improve the aqueous solubility of essential oils and thus increase the capacity of the products (Torres et al., 2017). For example, yarrow essential oil (*Achillea millefolium*) has been used as a remedy for digestive problems, amenorrhea, and anti-inflammatory effects, these oils were encapsulated with hydroxypropyl-β-cyclodextrin (HPβCD) by lyophilization, which shows a positive effect since protects active compounds by 27–30% more than the control (Rakmai et al., 2017). Therefore, it is recommended to investigate which compounds might be present in the oil to add or combine other techniques such as encapsulation to retain the compounds of interest.

5.5 PERSPECTIVES OF ESSENTIAL OILS AND INTELLECTUAL PROTECTION

In recent years, consumer interest has shifted more towards herbal products compared to classical or synthetic products due to better accessibility and compatibility with the human body (Ren et al., 2019). As a result of the new trends, multiple studies have been developed to carry out evaluations on essential oils from various plant tissues in order to identify the bioactive compounds present, as well as establish the potential uses or hazards involved.

Essential oils are involved in different industries, which exploit the chemical qualities seeking to add them to their products. The cosmetic industry incorporates essential oils as a result of its biological activity (analgesic, antiseptic, antimicrobial, and diuretic) and mainly because of the pleasant aroma they release (Maurya et al., 2018). In the same way, the pharmaceutical industry implements essential oils for their pharmaceutical potential; studies carried out this year have demonstrated the effectiveness of the essential oil of oregano as an antimicrobial agent against Meticillin-resistant *Staphylococcus aureus* (MRSA) (Lin and Changzhu, 2019). Likewise, the food and agronomic industry seeks to add resistance to products by adding essential oils; recent studies have demonstrated the effectiveness of the essential oil of *O. gratissimum* as an insecticidal agent against *Spodoptera littoralis*, housefly, and *Culex quinquefasciatus* (Benellia et al., 2019).

According to the search in the databases "Google patents" and WIPO-PATENTSCOPE (Accessed September 2, 2019) using the keywords "Essential oil," "food technology," "cosmetic," "medicine" and "agronomic", 12 patents were registered in the period that includes the years 2018–2019 that involve the implementation of essential oil as the main agent in the elaboration of the product Table 5.3. In addition, patents related to essential oils extraction methodologies were also found (CN108949363A; CN105132184B) which describe an increase in the effectiveness of the extraction procedure, a simplicity in development and ease of operation, implementing green-oil free solvent extraction. However, the specific characterization of the oils implemented in the products is not mentioned in the registered patents, likewise in our knowledge there are no reported patents for the isolation of compounds of interest present in the essential oils, which opens a window to the investigation of isolation and purification technologies.

5.6   CONCLUDING REMARKS

Taken together, the information above presented essential oils are a group of bioactive molecules that have proven to be good agents against several microorganisms and as with multipurpose for application in food, agronomic, cosmetic, and pharmaceutical industries. Although numerous studies have been conducted about chemical characterization, there is not enough information available on the effect of processing and storing essential oil–based products to gain a better understanding of the role of these compounds in the product matrix and its quality. Finally, research should also focus on the methods and conditions for extracting these molecules with more suitable single or combined method.

ACKNOWLEDGMENTS

Authors thank to Sectorial fund for research, development and forest technological innovation CONAFOR-CONACYT projects 2018-B-S-65769 and 2018-2-B-S-131466.

TABLE 5.3   Patents Registered in the Last Two Years with Essential Oils as Main Component.

Patent number	Title	Main core	Scope	Publication data	Country
108925626	Nano-essential oil emulsion and application in fresh keeping of fresh cut fruits and vegetables	The invention belongs to the technical field of fruit and vegetable fresh keeping. It consists in a nano-essential oil emulsion with an application in the fresh keeping of fresh cut fruits and vegetables	Food technology	04/12/2018	China
WO/2017/106984	Degradable packaging film for fruits and vegetables	The present invention relates to a degradable film for packaging fruits and vegetables, which comprises a polyolefin-based polymer matrix that incorporates an antimicrobial (biocidal or fungicidal) active ingredient of an essential oil	Food technology	29/06/2019	Chile
109393014	Preservation fresh keeping agent for plant-source *Mangifera indica*	Plant-source essential oil is used as a main component, and cooperates with the vitamin C, the fatty alcohol polyoxymethylene ether and the like, so that the fresh keeping agent has good effect of fresh keeping of the *Mangifera indica*	Food technology	01/03/2019	China
109157435	Bactericidal and bacteriostatic hand lotion containing pepper and Mongolian thyme herb essential oil	The invention discloses a bactericidal and bacteriostatic hand lotion containing pepper and Mongolian thyme herb essential oil	Cosmetic	08/01/2019	China
108309835	Strong antioxidant skincare product containing peony seed antioxidant peptide, and preparation method of strong antioxidant skincare product	The invention belongs to the technical field of cosmetics, and in particular relates to a strong antioxidant skin care product containing peony seed antioxidant peptide, and a preparation method of the strong antioxidant skin care product	Cosmetic	24/07/2018	China

TABLE 5.3 *(Continued)*

Patent number	Title	Main core	Scope	Publication data	Country
WO/2019/162468	Use of an essential oil of *Lemon savory* for improving the biomechanical properties of the skin	The present invention relates to the cosmetic use of at least one essential oil of lemon savory, in particular *Satureja montana* L. ssp. *citriodora*, or a cosmetic composition comprising said essential oil, as an active agent for improving the biomechanical properties of the skin	Cosmetic	29/08/2019	France
20190183955	Composition and method for treating plantar fasciitis in humans	A topical, non-prescription formulation of essential oils combined with skin permeation enhancers for the clinical treatment of plantar fasciitis in humans	Medicine	20/06/2019	United States
WO/2019/118444	Crosslinked particles, composition comprising the crosslinked particles, method for the manufactures thereof, and method of treating an infection	A method of making the crosslinked particles includes emulsifying a mixture of an essential oil and an aqueous solution. The crosslinked particles and compositions including the crosslinked particles can be particularly useful in the treatment of bacterial and fungal infections	Medicine	20/06/2019	United States
109820756	Antipruritic essential oil	The antipruritic essential oil not only has a good effect of beauty and skin care, but also has an antipruritic function, enhances the body immunity by long-term use, accelerates metabolism, clears away heat and removes toxicity, and has high use, safety, and no side effects	Medicine	21/05/2019	China
109820002	Insecticide composition for controlling stored grain pest population	The invention discloses an insecticide composition for controlling the stored grain pest population. The insecticide composition comprises active components of *Illicium verum* essential oil, deltamethrin, and methoprene	Agronomic	31/05/2019	China

TABLE 5.3 *(Continued)*

Patent number	Title	Main core	Scope	Publication data	Country
109380423	Insecticide for cotton in spring and summer and preparation method thereof	The invention discloses an insecticide for cotton in spring and summer	Agronomic	26/02/2019	China
WO/2019/061510	Insecticide specially adapted for urban greening and preparation method therefor	The insecticide is prepared with traditional Chinese medicine and plant extracts and is completely harmless to people. The invention uses lemon essential oil to adjust the smell in order to protect urban street environments and cover up odor	Agronomic	04/04/2019	China

KEYWORDS

- **essential oils**
- **functional applications**
- **extraction methods**
- **intellectual protection**

REFERENCES

Abbasipour, M.; Afshari, M.; Arnold, L.; Bagherzadeh, R.; Bhattarai, N.; Cernik, M.; Chae, T.; Fashandi, H.; Fong, H.; Ge, Y.; Gorji, M.; Jasper, S.; Karki, T.; Khajavi, R.; Ko, F.; Ma, X.; Moucka, F. The Electrospinning Process. In *Electrospun Nanofibers*, 2017; pp 1–8. doi: 10.1016/C2014-0-04496-6

Bahr, T.; Allred, K.; Martinez, D.; Rodriguez, D.; Winterton, P. Effects of a Massage-Like Essential Oil Application Procedure Using Copaiba and Deep Blue Oils in Individuals with Hand Arthritis. *Complement. Therap. Clin. Prac.* **2018,** *33,* 170–176. doi: 10.1016/j. ctcp.2018.10.004

Baranauskienėa, R.; Bendžiuvienėa, V.; Ragažinskienėb, O.; Venskutonisa, P. Essential Oil Composition of Five Nepeta Species Cultivated in Lithuania and Evaluation of Their Bioactivities, Toxicity and Antioxidant Potential of Hydrodistillation Residues. *J. Food Chem. Toxicol.* **2019,** *129,* 269–280. https://doi.org/10.1016/j.fct.2019.04.039

Barbosa, E. M.; Pelaes, A. C.; Oliveira, J.; Alexandre, S.; Nascimento, K.; Scaramal, Madrona.; Gratón, J. M.; Nunes, I. Development and Quality Evaluation of Infant Food with Oregano Essential Oil for Children Diagnosed with Cerebral Palsy. *Food Sci. Techonol.* **2017,** accepted Manuscript. doi: 10.1016/j.lwt.2017.06.016

Benellia, G.; Pavelab, R.; Maggid, F.; Wandjoud, J. G.; Fofiee, N. B.; Koné-Bambae, D.; Sagratinid, G.; Vittorid, S.; Capriolid, G. Insecticidal Activity of the Essential Oil and Polar Extracts from *Ocimum gratissimum* Grown in Ivory Coast: Efficacy on Insect Pests and Vectors and Impact on Non-Target Species. *J. Indust. Crops Prod.* **2019,** *132,* 377–385. https://doi.org/10.1016/j.indcrop.2019.02.047

Benmoussa, H.; Farhat, A.; Romdhane, M.; Bouajila, J. Enhanced Solvent-Free Microwave Extraction of *Foeniculum vulgare* Mill. Essential Oil Seeds Using Double Walled Reactor. *Arab. J. Chem.* **2016.** doi: http://dx.doi.org/10.1016/j.arabjc.2016.02.010

Bhavaniramya, S.; Vishnupriya, S.; Al-Aboody, M. S.; Vijayakumar, R.; Baskaran, D. Role of Essential Oils in Food Safety: Antimicrobial and Antioxidant Applications. *Grain Oil Sci. Technol.* accepted manuscript. doi: 10.1016/j.gaost.2019.03.001

Boulila, A.; Hassen, I.; Haouari, L.; Mejri, F.; Amor, I. B.; Casabianca, H.; Hosni, K. Enzyme-Assisted Extraction of Bioactive Compounds from Bay Leaves (*Laurus nobilis* L.). *Indust. Crops Prod.* **2015,** *74,* 485–493. doi: 10.1016/j.indcrop.2015.05.050

Bourdineaud, J. P.; Štambuk, A.; Šrut, M.; Brkanac, S. R.; Ivanković, D.; Lisjak, D.; Klobučar, R. S.; Dragun, Z.; Bačić, N, Klobučar, G. I. Gold and Silver Nanoparticles Effects to the

Earthworm *Eisenia fetida*—the Importance of Tissue Over Soil Concentrations. *J. Drug Chem. Toxicol.* 2019. https://doi.org/10.1080/01480545.2019.1567757

Bun, T.; Fei, E.; Ahmed, A.; Ho, J. Methods for the Characterization, Authentication, and Adulteration of Essential Oils. In *Essential Oils in Food Preservation, Flavor and Safety*, **2016**; pp 11–17. doi: 10.1016/B978-0-12-416641-7.00002-X

Buneri, I. D.;Yousuf, M.; Attaullah, M.; Afridi, S.; Anjum, S. I.; Rana, H.; Ahmad, N.; Amin, M.;Tahir, M.; Ansari, M. J. A Comparative Toxic Effect of *Cedrus deodara* Oil on Larval Protein Contents and Its Behavioral Effect on Larvae of Mealworm Beetle (*Tenebrio molitor*) (Coleoptera: Tenebrionidae). *J. Saudi J. Biol. Sci.* **2017**. http://dx.doi.org/10.1016/j.sjbs.2017.06.005

Chin Chew, S. Cold-Pressed Rapeseed (*Brassica napus*) Oil: Chemistry and Functionality. *Food Res. Int.* **2020**. doi: https://doi.org/10.1016/j.foodres.2020.108997

Chouhana, K. B. S.; Tandey, R.; Sen, K. K.; Mehtab, R.; Mandala, V. Critical Analysis of Microwave Hydrodiffusion and Gravity as a Green Tool for Extraction of Essential Oils: Time to Replace Traditional Distillation. *Trends Food Sci. Technol.* **2019,** *92*, 12–21. doi: 10.1016/j.tifs.2019.08.006

Cui, H.; Pana, H. W.; Wanga, P. H.; Yanga, X. D.; Zhaia, W. C.; Dong, Y.; Zhoua, H. L. Essential Oils from Carex meyeriana Kunth: Optimization of Hydrodistillation Extraction by Response Surface Methodology and Evaluation of Its Antioxidant and Antimicrobial Activities. *Indust. Crops Prod.* **2018,** *124*, 669–676. doi: 10.1016/j.indcrop.2018.08.041

Da Silva, T.; Da Silva, M.; Menini, L.; Lima, J. R. C.; Alves, L.; Cecon, P. R.; Ferrerira, A. Essential Oil of *Psidium guajava*: Influence of Genotypes and Environment. *Scientia Horticulturae* **2017,** *216*, 38–44. doi: 10.1016/j.scienta.2016.12.026.

Deda, O.; Gika, H.; Raikos, N.; Theodoridis, G. Chapter 4-GC-MS-Based Metabolic Phenotyping. In *The Handbook of Metabolic Phenotyping*, **2019**; pp 137–169.

Ebadollahi, A.; Taghinezhad, E. Modeling and Optimization of the Insecticidal Effects of *Teucrium polium* L. Essential Oil against Red Flour Beetle (*Tribolium castaneum Herbst*) Using Response Surface Methodology. *J. Info. Process. Agric.* **2019**. https://doi.org/10.1016/j.inpa.2019.08.004

El asbahani, A.; Miladi, K.; Badri, W.; Sala, M.; Aït Addi, E. E.; Casabanca, H.; El Mousadik, A.; Hartmann, D.; Jilale, A.; Renaud, F. N. R.; Elaissari, A. Essential Oils: From Extraction to Encapsulation. *Int. J. Pharm.* **2015,** *483*, 220–243. doi: 10.1016/j.ijpharm.2014.12.069

El Sawi, S.; Mohamed, I.; Kowthar, R.; Samia, D. Allelophatic Potential of Essential Oils Isolated from Peels of Three Citrus Species. *Ann. Agric. Sci.* in press. doi: 10.1016/j.aoas.2019.04.003

Fokom, R.; Adamou, S.; Essono, D.; Ngwasiri, D. P.; Eke, P.; Teugwa, C.; Tchoumbougnang, F.; Fekam, B.; Amvam, P. H.; Nwaga, D.; Sharma, A. K. Growth, Essential Oil Content, Chemical Composition and Antioxidant Properties of Lemongrass as Affected by Harvest Period and Arbuscular Mycorrhizal Fungi in Field Conditions. *Indust. Crops Prod.* **2019,** *138*. doi:10.1016/j.indcrop.2019.111477

Gavahian, M.; Chu, Y. H. Ohmic Accelerated Steam Distillation of Essential Oil from Lavender in Comparison with Conventional Steam Distillation. *Innov. Food Sci. Emerg. Technol.* **2018,** *50*, 34–41. doi: 10.1016/j.ifset.2018.10.006

Hallaa, N.; Helenoa, S. A.; Costac, P.; Fernandesa, I. P.; Calhelhaa, R. C.; Boucheritd, K.; Rodrigues, A. E.; Ferreira, I.; Barreiroa, M. F. Chemical Profile and Bioactive Properties of the Essential Oil Isolated from *Ammodaucus leucotrichus* Fruits Growing in Sahara and Its Evaluation as a Cosmeceutical Ingredient. *Indust. Crops Prod.* **2018,** *119*, 249–254. doi:10.1016/j.indcrop.2018.04.043

Hernández, L. A.; Victoria, E. J.; Trejo, A.; Beltran, G. J. CO_2-Supercritical Extraction, Hydrodistillation and Steam Distillation of Essential Oil of Rosemary (*Rosmarinus officinalis*). *J. Food Eng.* **2017,** *200,* 81–86. Doi:10.1016/j.jfoodeng.2016.12.022

Hesham, H. A.; Rassem, A. H.; Nour, R. M. Y. Techniques for Extraction of Essential Oils from Plants: A Review. *Aust. J. Basic & Appl. Sci.* **2016,** *10* (16), 117–127.

Issaa, M.; Chandelb, S.; Singha, H. P.; Batishb, D. R.; Kohlib, R. K.; Yadavd, S. S.; Kumarie, A. Appraisal of Phytotoxic, Cytotoxic and Genotoxic Potential of Essential Oil of a Medicinal Plant *Vitex negundo. J. Indust. Crops Prod.* **2020,** *145.* https://doi.org/10.1016/j. indcrop.2019.112083.

Kavallieratos, N. G.; Boukouvala, M. C.; Ntalli, N.; Skourti, A.; Karagianni, E. S.; Nika, E. P.; Kontodimas, D. C.; Cappellacci, L.; Petrelli, R.; Cianfaglione, K.; Morshedloo, M. R.; Tapondjou, L. A.; Rakotosaona, R.; Maggi, F.; Benelli, G. Effectiveness of Eight Essential Oils against Two Key Stored-Product Beetles, *Prostephanus truncatus* (Horn) and *Trogoderma granarium* Everts. *J. Food Chem. Toxicol.* **2020,** *139.* https://doi.org/10.1016/j.fct.2020.111255.

Khalil, A. A.; Rahman, U.; Khan, M.; Sahar, A.; Mehmoodac, T.; Khana, M. Essential Oil Eugenol: Sources, Extraction Techniques and Nutraceutical Perspectives. *J. R. Soc. Chem.* **2017,** *7.* doi: 10.1039/c7ra04803c

Khare, P.; Srivastava, S.; Nigam, N.; Singh, A. K.; Singh, S. Impact of Essential Oils of *E. citriodora, O. basilicum* and *M. arvensis* on Three Different Weeds and Soil Microbial Activities. *J. Environ. Technol. Innov.* **2019,** *14.* https://doi.org/10.1016/j.eti.2019.100343.

Kiani, S.; Minaei, S.; Ghasemi, M. Instrumental Approaches and Innovative Systems for Saffron Quality Assessment. *J. Food Eng.* **2018,** *216,* 1–10. doi: 10.1016/j.jfoodeng.2017.06.022

Langhasovaa, L.; Hanusovac, V.; Rezeka, J.; Stohanslovab, B.; Ambrozb, M.; Kralovac, V.; Vaneka, T.; Dong, J.; Li, Z.; Yange, J.; Skalovaba, L. Essential Oil from Myrica Rubra Leaves Inhibits Cancer Cell Proliferationand Induces Apoptosis in Several Human Intestinal Lines. *J. Indust. Crops Prod.* **2014,** *59,* 20–26. http://dx.doi.org/10.1016/j.indcrop.2014.04.018

León, G.; Osorio, M.; Martínez, S. R. Comparison of Two Methods for Extraction of Essential Oil from *Citrus sinesis L. Revfarmacia* **2015,** *49,* 742–750.

Leyva, M.; French, L.; Quintana, F.; Montada, D.; Castex, M.; Hernandez, A.; Marquetti, M. Melaleuca Quinquenervia (Cav.) S. T. Blake (Myrtales: Myrtaceae): Natural Alternative for Mosquito Control. *Asian Pacific J. Trop. Med.* **2016,** *9,* 979–984. doi: 10.1016/j.apjtm. 2016.07.034

Lin, L.; Changzhu, L. Antibacterial Mechanism of Oregano Essential Oil. *Indust. Crops Products* **2019,** *139,* 1–9. doi: 10.1016/j.indcrop.2019.111498

Madrid, T.; López, A.; Parra J. E. Effect of Inclusion Of Essential Oil of Oregano (*Lippia origanoides*) on Lipid Profile in Broiler Meat. *VITAE* **2018,** *25,* 75–82. doi: doi: 10.17533/ udea.vitae.v25n2a03

Mandal, S.; Mandal, V.; Kumar-Das, A. *Essentials of Botanical Extraction: Principles and Applications*; S. A., Ed., Vol. 1; Academic Press Inc, 2015; p 220. https://doi.org/10.1016/ C2014-0-02889-4

Manouchehria, R.; Saharkhiza, M. J.; Karamia, A.; Niakousarib, M. Extraction of Essential Oils from Damask Rose Using Green and Conventional Techniques: Microwave and Ohmic Assisted Hydrodistillation versus Hydrodistillation. *Sustain. Chem. Pharm.* **2018,** *8,* 76–81. doi: 10.1016/j.scp.2018.03.002

Mark, R.; Lyu, X.; Lee, J.; Parra, R.; Ning, Wei. Sustainable Production of Natural Phenolics for Functional Food Applications. *J. Funct. Foods* **2019,** *57,* 233–254. doi:10.1016/j. jff.2019.04.008

Maurya, A. K.; Mohanty, S.; Pal, A.; Chanotiya, C. S.; Bawankule, D. U. The Essential Oil from Citrus limetta Risso Peels Alleviates Skin Inflammation: In-Vitro and in-Vivo Study. *J. Ethnopharmacol.* **2018**, *212*, 86–94. doi:10.1016/j.jep.2017.10.018

Megawati; Fardhyantia, D. F.; Sediawan, W. B.; Hisyam, A. Kinetics of Mace (*Myristicae arillus*) Essential Oil Extraction Using Microwave Assisted Hydrodistillation: Effect of Microwave Power. *Indust. Crops Prod.* **2019**, *131*, 315–322.

Mendoza, A.; Aranda, S.; Bermeo, J. R.; Gómez, A.; Pozos, A.; Sánchez, L The Essential Oil of *Allium sativum* as an Alternative Agent against *Candida* Isolated from Dental Prostheses. *Revista Iberoamericana de Micología* **2017**. doi: 10.1016/j.riam.2016.11.008

Mendoza, A.; Aranda, S.; Bermeo, J. R.; Gómez, A.; Pozos, A.; Sánchez, L The Essential Oil of *Allium sativum* as an Alternative Agent against Candida Isolated from Dental Prostheses. *Revista Iberoamericana de Micología* **2017**. doi: 10.1016/j.riam.2016.11.008

Mohammadpour, H.; Sadrameli, S. M.; Eslami, F.; Asoodeh, A. Optimization of Ultrasound-Assisted Extraction of Moringa Peregrina Oil with Response Surface Methodology and Comparison with Soxhlet Method. *Indust. Crops Prod.* **2019**, *131*, 106–116. doi: 10.1016/j.indcrop.2019.01.030

Murfadunnisa, S.; Srinivasan, P. V.; Ganesan, R.; Nathan, S. S.; Kim, T. J.; Ponsankar, A.; Kumar, S. D.; Chandramohan, D.; Krutmuang, P. Larvicidal and Enzyme Inhibition of Essential Oil from *Spheranthus amaranthroids* (Burm.) against Lepidopteran Pest *Spodoptera litura* (Fab.) and Their Impact on Non-Target Earthworms. *J. Biocatalys. Agric. Biotechnol.* **2019**, *21*. https://doi.org/10.1016/j.bcab.2019.101324.

Nadar, S. S.; Rao, P.; Rathod, V. K. Enzyme Assisted Extraction of Biomolecules as an Approach to Novel Extraction Technology: A Review. *Food Res. Int.* **2018**, *108*, 309–330. doi:10.1016/j.foodres.2018.03.006

Noureddine Halla, N.; Fernandes, I. P.; Heleno, S-A.; Costa, P.; Boucherit-Otmani, Z: Boucherit, K.; Rodrigues, A. E.; Ferreira, I.; Barreiro, M. F. Cosmetics Preservation: A Review on Present Strategies. *Molecules* **2018**, *23*. doi:10.3390/molecules23071571.

Oliveiraa, P. F.; Alvesa, J. M.; Damascenoa, J.; Machado, R. A.; Diasb, H. J.; Crotti, A. E.; Crispim, D. Cytotoxicity Screening of Essential Oils in Cancer Cell Lines. *J. Revista Brasileira de Farmacognosia* **2015**, *25*, 183–188. http://dx.doi.org/10.1016/j.bjp.2015.02.009

Oreopoulou, A.; Tsimogiannis, D.; Oreopoulou, V. Chapter 15 -Extraction of Polyphenols from Aromatic and Medicinal Plants: An Overview of the Methods and the Effect of Extraction Parameters. *Acad. Press* **2019**, *2*, 243–259. https://doi.org/10.1016/B978-0-12-813768-0.00025-6

Organización de las Naciones Unidas para la Alimentación y la Agricultura. http://www.fao.org/emergencies/tipos-de-peligros-y-de-emergencias/plagas-y-enfermedades-de-las-plantas/es/ (accessed on 12 July 2019).

Pavelac, R.; Petrellid, R.; Cappellaccid, L.; Santinid, G.; Fiorinie, D.; Sutf, S.; Dall'Acquaf, S.; Canalea, A. The essential oil from industrial hemp (*Cannabis sativa L.*) By-Products as an Effective Tool for Insect Pest Managment in Organic Crops. *Indust. Crops Products* **2018**, *122*, 308–315. doi: 10.1016/j.indcrop.2018.05.032

Peter, K. V. *Handbook of Herbs and Spices*; S. A., Ed., Vol. 1; Woodhead Publishing, 2012; p 607.

Pinho. S. C. Antifungal Activity of Nanoemulsions Encapsulating Oregano (*Origanum vulgare*) Essential Oil: In Vitro Study and Application in Minas Padrão Cheese. *Bjmicrobiol* **2018**, *49*, 929–935. doi: 10.1016/j.bjm.2018.05.004

Polmanna,G.; Badiab, V.; Frena,, M.;Teixeira, G., Elisandra Rigo, E.; Block, J. M.; Camino, M:M. Enzyme-Assisted Aqueous Extraction Combined with Experimental Designs Allow the Obtaining of a High-Quality and Yield Pecan Nut Oil. *Food Sci. Technol.* **2019**, *113*. doi: 10.1016/j.lwt.2019.108283

Prado, J. M. Vardanega, R.; Debien, I.; Almeida, M. A.; Gerschenson, L. N.; Sowbhagya, H. B.; Chemat, S. Conventional Extraction-Chapter 6. *J. Food Waste Recov.* **2015**. http://dx.doi.org/10.1016/B978-0-12-800351-0.00006-7

Priyadarshi, R.; Sauraj; Kumar, B.; Deeba, F.; Kulshreshtha, A.; Singh, Y. Chitosan Films Incorporated with Apricot (*Prunus armeniaca*) Kernel Essential Oil as an Active Food Packaging Material. *Food Hydrocoll.* **2018**, accepted manuscript. doi: 10.1016/j.foodhyd.2018.07.003

Rakmai, J.; Cheirsilp, B.; Torrado-Agrasar, A.; Simal-Gándara, J.; Mejuto, J. C. Encapsulation of Yarrow Essential Oil in Hydroxypropyl-Beta-Cyclodextrin: Physiochemical Characterization and Evaluation of Bio-Efficacies. *CyTA J. Food* **2017**, *15*, 409–417. doi: 10.1080/19476337.2017.1286523

Ramesh, B. Application of HPLC and ESI-MS Techniques in the Analysis of Phenolic Acids and Flavonoids from Green Leaf Vegetables (GLVs). *J. Pharm. Analys.* **2017**, *7*, 349–364 doi: 10.1016/j.jpha.2017.06.005

Ren, M. Targeting Open Market with Strategic Business Innovations: A Case Study of Growth Dynamics in Essential Oil and Aromatherapy Industry. *J. Open Innov. Technol. Mark. Complex* **2019**, *5*, 7. doi:10.3390/joitmc5010007

Richardson, A. Genoprotective Effects of Essential Oil Compounds against Oxidative and Methylated DNA Damage in Human Colon Cancer Cells. *J. Food Sci.* **2019**, *0*, 1–7. doi: 10.1111/1750-3841.14665

Rodríguez, A.; Suarez, S.; Palacio, D. Effects of Pesticides on Health and the Environment. *Revista Cubana de Higiene y Epidemiología* **2014**, 52, 372–387.

Romero, C.; Navarro, C.; Martin, J.; Valero, A. Peppermint (*Mentha piperita*) and Albendazole against Anisakiasis in an Animal Model. *Trop. Med. Int. Health* **2014**, *19*, 1430–1436. doi: 10.1111/tmi.12399

Santana-Méridasa, O.; Polissiouc, M.; Izquierdo-Meleroa, M. E.; Astrakac, K.; Tarantilisc, P. A.; Herraiz-Peñalvera, D.; Sánchez-Vioquea, R. Polyphenol Composition, Antioxidant and Bioplaguicide Activities of the Solid Residue from Hydrodistillation of *Rosmarinus officinalis* L. *J. Indust. Crops Prod.* **2014**, *59*, 125–134. http://dx.doi.org/10.1016/j.indcrop.2014.05.008

Santos, K. A.; Klein, E. J.; Da Silva, C.; Da Silva, E. A.; Cardozo, L. Extraction of Vetiver (*Chrysopogon zizanioides*) Root Oil by Supercritical CO_2, Pressurized-Liquid, and Ultrasound-Assisted Methods and Modeling of Supercritical Extraction Kinetics. *J. Supercrit. Fluids* **2019**, *150*, 30–39. doi:10.1016/j.supflu.2019.04.005

Santos. M. C.; Adenir, T.; Menezes, M.; Pinto, D. M.; Arrigoni, M. F.; Cruz, E. M.; Santos, T.; Pimentel, A.;Coelho, C.; Fitzgerald, A. Bioactivity of Essential Oil from *Lippia gracilis* Schauer against Two Major Coconut Pest Mites and Toxicity to a Non-Target Predator. *J. Crop Protect.* **2019**, *125*. https://doi.org/10.1016/j.cropro.2019.104913

Sarkic, A.; Stappen, I. Essential Oils and Their Single Compounds in Cosmetics—A Critical Review. *Cosmetics* **2018**, *5*, 2–21. doi:10.3390/cosmetics501001

Saucedo, S.; Torres, J. A.; Castro, C.; Rojas, R.; Sánchez, E. J.; Ngangyo, M.; Martinez, G. C. G. Moringa Plants: Bioactive Compounds and Promising Applications in Food Products. *Food Res. Int.* **2018**, *111*, 438–450 doi:10.1016/j.foodres.2018.05.062

Smigielski, K. B.; Majewska, M.; Kunicka, A.; Gruska, R.; Stanczyk, Ł. The Effect of Commercial Enzyme Preparation-Assisted Maceration on the Yield, Quality, and Bioactivity of Essential Oil from Waste Carrot Seeds (*Daucus carota L.*). *Grasas Aceites* **2014**, *65*. doi: 10.3989/gya.0467141.

Souza, M.; Mellos, I.; Mello, D.; Cardoso, C. M.; Guedes, E.; Alves, M. A. Efficacy of Lemongrass Essential Oil and Citral in Controlling Callosobruchusmaculatus (Coleoptera: Chrysomelidae), a Post-Harvest Cowpea Insect Pest. *J. Crop Protect.* **2019,** *119,* 191–196. https://doi.org/10.1016/j.cropro.2019.02.007

Srivilai, J.; Phimnuan, P.; Jaisabai, J.; Luangtoomma, N.; Waranuch, N.; Khorana, N.; Wisutiprot, W.; Scholfield, N.; Champachaisri, K.; Ingkaninan, K. Curcuma aureginosa Roxb. Essential Oil Slows Hair-Growth and Lightens Skin in Axillae; a Randomised Double Blinded Trial. *Phytomedicine* **2016,** accepted manuscript. doi: 10.1016/j.phymed.2016.12.007

Syamsuddin, Y.; Gao, X.; Grayling, P.; Wu, H. Steam Distillation of Mallee Leaf: Extraction of 1,8-Cineole and Changes in the Fuel Properties of Spent Biomass. *Fuel* **2014,** *133,* 341–349. doi:10.1016/j.fuel.2014.05.03

Tabana, A.; Saharkhiz. M. J.; Niakousaric, M. Sweet Bay (*Laurus nobilis* L.) Essential Oil and Its Chemical Composition, Antioxidant Activity and Leaf Micromorphology under Different Extraction Methods. *Sustain. Chem. Pharm.* **2018,** *9,* 12–18. doi:10.1016/j.scp.2018.05.00

Tandon, S.; Mittal, A. Insecticidal and Growth Inhibitory Activity of Essential Oils of *Boenninghausenia albiflora* and *Teucrium quadrifarium* against *Spilarctia obliqua*. *Biochem. Syst. Ecol.* **2018,** *81,* 70–73.

Tisserand, R.; Young, R. 2-Essential Oil Composition. In *Essential Oil Safety, Churchill Livingstone,* 2nd ed.; 2014; pp 5–22. doi:10.1016/B978-0-443-06241-4.00002-3

Torres, C.; Núñez, A.; Rodríguez, J.; Castillo, S.; Leos, C.; Báez, J. G. Chemical Composition, Antimicrobial, and Antioxidant Activities of Orange Essential Oil and Its Concentrated Oils. *CyTA J. Food* **2017,** *15,* 129–135. doi: 10.1080/19476337.2016.1220021.

Vaillant, D. I.; Romeu, C. R.; Ramírez, R. Fungicide activity of *Melaleuca quinquenervia L.* Essential Oil. *Rev. Protección Veg.* **2015,** *30,* 47.

Vichi, S. Chapter 66 - Extraction Techniques for the Analysis of Virgin Olive Oil Aroma. Chapter 66 - Extraction Techniques for the Analysis of Virgin Olive Oil Aroma. **2010;** pp 615–623. https://doi.org/10.1016/B978-0-12-374420-3.00066-8

Wang, Y.; Li, R.; Jiang, Z. T.; Tan, J.; Tang, S. H.; Li, T. T.; Liang, L. L.; He, H. J.; Liu, Y. M.; Li, J. T.; Zhang, X. C. Green and Solvent-Free Simultaneous Ultrasonic-Microwave Assisted Extraction of Essential Oil from White and Black Peppers. *Indust. Crops Prod.* **2018,** *114,* 164–172. doi:10.1016/j.indcrop.2018.02.002

Wen, Y.; He, L.; Peng, R.; Lin, Y.; Zhao, L.; Li, X.; Ye, L.; Yang, J. A Novel Strategy to Evaluate the Quality of Herbal Products Based on the Chemical Profiling, Efficacy Evaluation and Pharmacokinetics. *JPBA* **2018,** *161,* 326–335. doi: 10.1016/j.jpba.2018.08.047

Yousefi, M.; Nasrabadi, M. R.; Pourminrtazavi, S. M.; Wysokowski, M.; Jesionowski, T.; Ehrlich, H.; Mirsadeghi, S. Supercritical Fluid Extraction of Essential Oils. *Trends Analyt. Chem.* **2019,** *118,* 182–193. doi: 10.1016/j.trac.2019.05.038

Zandi, N.; Hojjati, M.; Carbonell, A. Bioactivity of *Lantana camara L.* Essential Oil Against *Callosobruchus maculatus* (Fabricius). *Chilean J. Agric. Res.* **2012,** *72,* 502–506.

Zhang, D.; Zhao, S. Supercritical CO_2 Extraction of Eucalyptus Leaves Oil and Comparison with Soxhlet Extraction and Hydro-Distillation Methods. *Sep. Purif. Technol.* **2014,** *133,* 443–451. doi: 10.1016/j.seppur.2014.07.018

Zhao, S.; Zhang, D. Supercritical CO_2 Extraction of Eucalyptus Leaves Oil and Comparison with Soxhlet Extraction and Hydro-Distillation Methods. *Sep. Purif. Technol.* **2014,** *133,* 443–451. doi:10.1016/j.seppur.2014.07.018

CHAPTER 6

Physical and Chemical Water Quality from a Hybrid System for Wastewater Reclamation for Agricultural Purposes

MILTON TORRES-CERÓN[1], JUAN ANTONIO VIDALES-CONTRERAS[2*],
HUMBERTO RODRIGUEZ-FUENTES[2],
ALEJANDRO ISABEL LUNA-MALDONADO[2],
DONAJI JOSEFINA GONZALEZ-MILLE[3], and JUAN NAPOLES-ARMENTA[2]

[1]*Department of Ecology and Conservation Biology, Texas A&M AgriLife, Texas A&M University, College Station 778432258, Texas, United States*

[2]*Facultad de Agronomía, Universidad Autónoma de Nuevo León. Av. Francisco Villa S/N, col. Ex Hacienda el Canadá, Gral. Escobedo, Nuevo León 66050, México*

[3]*Facultad de Medicina-Centro de Investigación Aplicada al Ambiente y la Salud (CIAAS)-CIACyT, Universidad Autónoma de San Luis Potosí, Av. Sierra Leona, San Luis Potosí 78210, México*

*Corresponding author. E-mail: juan.vidalescn@uanl.edu.mx

ABSTRACT

In recent years, scarcity and quality deterioration of freshwater have threatened food security and public health. As a result, wastewater reuse has been an additional source of water to satisfy the increasing demand for crop irrigation in arid and semiarid lands. However, in addition to pathogen dissemination, reclaimed water reuse could lead to organic and inorganic soil pollution. For the current research work, a hybrid system was constructed to treat municipal wastewater, based on anaerobic reactors and small duckweed phytoremediation basins. The observed experimental parameters for irrigation water quality were analyzed by ANOVA and Tukey statistical tests. According to its high salinity, the treated wastewater effluents were nonrecommendable for low-salinity crop irrigation in soils with poor drainage. Since Riverside's classification showed that most of the water samples are in C2S1 and C3S1 categories. These results were confirmed by the Wilcox water quality index that showed more than 30% of Na content.

6.1 INTRODUCTION

6.1.1 THE PROBLEM OF WATER POLLUTION IN THE FRAMEWORK OF AGRICULTURAL USE

During the 20th century, the environmental impact on aquatic ecosystems of pollutant accumulation increased dramatically. Additionally, during the first decade of the 21st Century, the global average of the soil water content has been deceased apparently by a consistent increase of global warming leading to unsustainable food production for future generations because of water scarcity (Li et al., 2019; Qadir et al., 2010; Prăvălie, 2016; Leong, 2016; Fito and Van Hulle, 2020). As a consequence, wastewater reuse has been an important source for crop irrigation mostly in arid and semiarid regions (Fito and Van Hulle, 2020; Tran et al., 2016; Tabatabaei et al., 2020). This practice has resulted in the accumulation of organic and inorganic pollutants, changes in pH, soil salinity, pollution of the groundwater, and bioaccumulation of heavy metals and xenobiotics compounds in crops (Marofi et al., 2012; Iurciuc and Dima, 2013; Schwitzguébel, 2001; Mizyed, 2013).

However, even when apparently water availability for agricultural purposes may be increased, there are potential adverse effects for soil and water health due to the intrinsic chemical and microbiological composition of raw wastewater. Some of the negative consequences of crop irrigation with wastewater are as follows: (1) accumulation of metals in soils of urban and peri-urban areas, (2) accumulation of metals and xenobiotics in plants, leading to a bioaccumulative effect along food webs, (3) toxic effects of wastewater on plants and beneficial microbiota, (4) changes on soil salinity and pH, and (5) concentration increase of pathogens in both soils and plants (Iurciuc and Dima, 2013; Poustie et al., 2020; Chen et al., 2013; Becerra-Castro et al., 2015; Gupta et al., 2010; Qadir et al., 2010; Hanjra et al., 2012).

In Mexico, more than 70% of total freshwater is used for crop irrigation and more than 260,000 hectares are irrigated with wastewater (Marofi et al., 2012; Rojas-Valencia et al., 2011). Thus, inefficient planning of agriculture wastewater treatment and population growth (Tortajada and Castelán, 2003) have been linked to pollution of aquatic ecosystems. For instance, in Nuevo

Leon, the infiltration of wastewater, leachates from livestock farms, landfills, and industrial effluents are among the main sources of aquatic pollution (Dávila Pórcel et al., 2011). Consequently, in arid and semiarid regions, soil salinization frequently is an adverse situation for agriculture productivity as a result of evaporation and use of water with low quality for crop irrigation (Silva-García et al., 2006).

6.1.2   *WATER RECLAMATION BY HYBRID SYSTEMS AS A POTENTIAL SOLUTION TO WATER SCARCITY FOR AGRICULTURAL PURPOSES*

Due to the above scenario, a change of paradigm is being attested in which wastewater is considered a renewable resource, rather than just a waste resulting from human activities (Fito and Van Hulle, 2020; Zhang et al., 2021; Janeiro et al., 2020; Daigger, 2009; Esposito et al., 2005). In this shift of paradigm, wastewater reclamation has arisen as a practical and immediate way to bring the polluted water to a state where water resources fulfill the requirements for being reused for beneficial purposes (such as irrigation) (United States Environmental Protection Agency; Metzler and Russelmann, 1968). For that reason, water reclamation has become an acceptable solution to supply the increasing demand of water for human use and consumption, including agricultural irrigation, in order to replace traditional sources of freshwater (Zhang et al., 2021; Shingare et al., 2019; Lin et al., 2020).

Nevertheless, the costs associated with conventional wastewater treatment processes are expensive and they are highly and directly related to wastewater quality after treatment (Fito and Van Hulle, 2020; Hunter et al., 2019; Englande et al., 2015). As a consequence, research on new, efficient, cost-effective, and sustainable technologies for wastewater reclamation is a frequent topic in the scientific literature (Hassan and Dahlan, 2014; Lavee, 2011; Ibrahim et al., 2019). Among those emergent technologies are the hybrid systems that combine conventional technologies (such as aerobic and anaerobic reactors) with eco-remediation (usually, phytoremediation) modules (Chyan et al., 2013; Saeed et al., 2014; Ezechi et al., 2019) for low investment and operative costs, diminishing the environmental impact (Schwitzguébel, 2001; Schröder et al., 2007; Loupasaki and Diamadopoulos, 2013; Nasr et al., 2009; Pilon-Smits, 2005).

Regarding conventional technologies of wastewater treatment, in the present chapter, we are going to focus on anaerobic baffled reactors (ABR's) which are counted among the most efficient anaerobic treatment methods. In these systems, acidogenesis and acetogenesis are separated within them for low production of sludge and high remotion of organic matter (Hassan and

Dahlan, 2014; Bodkhe, 2009; Barber and Stuckey, 1999). The general structure of ABR's is a sequence of vertical baffles that let the influent flow from below to up continuously once the water enters into the reactor (Sarathai et al., 2010). As a result of this water flow, bacterial communities are in constant and smooth movement inside of reactor compartments, increasing the contact between wastewater and beneficial microorganisms (Wang et al., 2004).

Eco-remediation for wastewater treatment refers to the use of plants, microorganisms associated with plants, and agricultural tools to enhance pollutant removal, and optimizing the hydraulic retention time in the system (Schröder et al., 2007; Ning et al., 2014; Dzantor, 2007; Zhai, 2011). These types of emergent technologies offer the advantages of being noninvasive and environmentally friendly with a low cost for construction and operation so it is easy to incorporate them in systems that include conventional technologies (Zhang et al., 2007; Yang et al., 2007). In fact, plant-based eco-remediation (phytoremediation) is a common emergent technology used in wastewater treatment. Its pollutant removal efficiency depends on the kind of contaminant (or contaminants) present in the influent, physiology of hydrophyte species, and the metabolic route required to remove or transform wastewater pollutants (Zhai, 2011; Juwarkar et al., 2010; Rezania et al., 2016).

Once the treatment process is finished, evaluation of reclaimed water usually includes measurements of physical–chemical parameters, such as turbidity, pH, electric conductivity (EC), the concentration of organic constituents (greases, oils, biological, and chemical oxygen demand), microbial indicators, nutrients, and suspended solids (APHA/AWWA/WEF, 2005). However, regarding its use for agricultural purposes, the main water quality indicators are pH, CE, salinity, sodicity, and hardness (Hanjra et al., 2012; Silva-García et al., 2006; McGeorge, 1954; Vyas and Jethoo, 2015). Additionally, since effluent toxicity is a major concern issue, toxicity assays are performed to determine how harmful could be the use of reclaimed wastewater for the biotic component of soil and aquatic ecosystems (Englande et al., 2015; Maltby and Calow, 1989; Rizzo et al., 2011), however, the use of toxicological assays is beyond the scope of the present research manuscript.

In the present study, the construction and performance evaluation of a hybrid system for wastewater reclamation is presented. For this research, we hypostatized that a combination of conventional wastewater treatment (anaerobic baffled reactors, ABR) and eco-remediation technologies would reclaim municipal wastewater to meet a suitable quality for agricultural purposes. Hence, the aims of the actual research work were to (1) construct and operate a hybrid system for wastewater reclamation and (2) to analyze physical and chemical parameters related to water quality for crop irrigation.

6.2 MATERIALS AND METHODS

6.2.1 CONSTRUCTION OF HYBRID SYSTEM FOR WASTEWATER RECLAMATION

The designed hybrid system included both ABR's and constructed wetland basins to polish wastewater effluents from the reactors. The system was operated for 237 days under outdoor conditions to incorporate the environmental effects on wastewater reclamation process (Nasr et al., 2009). The experimental work was conducted at the School of Agronomy Experimental Station, Autonomous University of Nuevo Leon (UANL) located in Marin, Nuevo Leon, Mexico at 395326.90 W and 2862109.31 N UTM.

The wastewater reclamation system consisted of two 60-L plexiglass ABR with a series of vertical baffles that divide the reactors into eleven identical compartments. Waste rubber tire (WTR) with 86% porosity was added to one ABR as a substrate for enhancing microbial attachment and solid retention. After ABR treatment, the effluent went through three posttreatment units to polish the anaerobically treated wastewater. The first one was a duckweed (*Lemna minor*) phytoremediation basin with 3.89 kg of WRT; the second pond only was vegetated with duckweed, and the third one with 3.89 kg of WRT and without floating treatment plants. The hydraulic residence time (HRT) for each anaerobic phytoremediation system was 37 h. Figure 6.1 shows a schematic representation of both wastewater treatment units.

6.2.2 WATER SAMPLING AND ANALYTICAL METHODS

Each treatment system was sampled from September 2014 to April 2015 (n = 8) at (1) wastewater influent, (2) effluent of ABR with WRT and sites 4, 6, and 8 at the polishing ponds, (3) outlet of ABR without WTR and sites 5, 7, and 9. Water samples were collected in sterile polypropylene bottles and placed into an icebox to be transported to the laboratory for analysis.

The analyzed physical and chemical parameters were pH, EC, and Ca, Mg, Na, HCO_3, and Cl concentrations. Effective salinity (ES), potential salinity (PS), residual sodium carbonate (RSC), sodium adsorption ratio (SAR), and hardness were also analyzed (McGeorge, 1954; Ali, 2010). These parameters were represented on both USA Salinity Laboratory and Wilcox diagrams (McGeorge, 1954) constructed with Diagrammes 6.48 software developed by the Université d'Avignon (http://www.lha.univ-avignon.fr/).

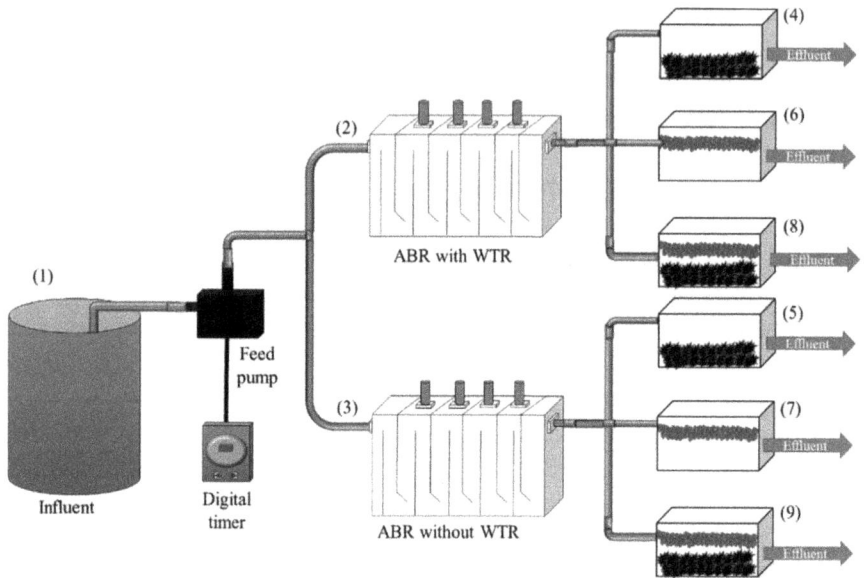

FIGURE 6.1 Schematic representation of the hybrid system for wastewater treatment showing the experimental design and the nine sampling locations as follows: (1) feed tank, (2) effluent of ABR with WRT and sites 4. 6, and 8 are the polishing ponds, and (3) outlet from the ABR without WTR and sites 5, 7, and 9 are its polishing ponds.

The statistical analysis was conducted using the Statistical Package for Social Science 17.0 (SPSS Inc., Chicago ILL). Tests to determine the significant differences between sampling periods and monitoring sites were conducted by ANOVA tests; whereas, those parameters that showed significant differences between samples were analyzed by Tukey ($p < 0.05$), and the relationship between physical and chemical water quality parameters is indicated by Pearson correlation.

6.3   RESULTS AND DISCUSSION

6.3.1   *PHYSICAL AND CHEMICAL WATER QUALITY*

The Pearson correlation values between physical and chemical water quality parameters are given in Table 6.1. In general, such values were lower than 0.58; for example, pH showed a significant correlation ($p < 0.05$) of -0.3 with either Mg or Ca concentrations. Whereas, the highest positive and significant correlation was found for EC, ES, and RSC.

TABLE 6.1 Pearson Correlation Coefficient between Physical-Chemical Parameters of Water Quality.

	pH	EC	Ca	Mg	Ni	HCO$_3$	Cl	ES	PS
EC	0.240*								
Ca	-0.324*	0.180							
Mg	-0.315*	0.101	0.315*						
Na	-0.240	0.441*	0.075	-0.163					
HCO$_3$	-0.231	0.567*	0.588*	0.455*	0.402*				
Cl	0.192	0.202	-0.065	0.206	-0.140	0.135			
ES	0.363*	0.676*	-0.125	0.247*	0.447*	0.406*	0.515*		
PS	0.225	0.260*	-0.053	0.188	0.017	0.150	0.966*	0.546*	
RSC	0.233*	0.492*	-0.264*	-0.264*	0.442*	0.481*	0.296*	0.432*	0.309*

*Signifficant correlation ($p = 0.05$)

For pH, a significant difference ($p < 0.05$) between treatments was observed. It is important to notice that the pH for raw wastewater was similar to the reported pH for urban wastewater in other studies (Belmont et al., 2004). In fact, collected water samples from the nine locations showed the strongest alkaline condition during the experimental period (Fig. 6.2). Probably, alkaline variations along the system could be a result of the different metabolic microbiological processes (Krishna et al., 2007). The behavior of the alkaline parameter was similar to the observed one in systems for wastewater that use WRT as substrate (Chyan et al., 2013). On the other hand, there was a negative correlation between pH and HCO$_3$ (-0.231). The presence of bicarbonates is commonly associated with microorganisms, which use these compounds to regulate water pH (Nzengy'a and Wishitemi, 2001). This chemical parameter was below the limit recommended by Mexican standards for treated wastewater discharges (SEMARNAP, 1996). However, the effluents with the best pH value, according to water quality irrigation criteria (Sundaray et al., 2009), were those from vegetated polishing treatments.

Tukey test did not show a significant difference between treatments for EC observations ($p < 0.05$). However, effluents from polishing treatments without plants showed the lowest EC values (Fig. 6.2). These variations of EC through the treatment trends could be an interaction between water minerals and plants (Nzengy'a and Wishitemi, 2001; Collins et al., 2013). According to the classification of McGeorge (1954), influent and effluent wastewater of the experimental treatment system are classified as highly saline for irrigation. Thus, the raw wastewater and the treated effluents could be used for halophytic plants or crops with a high tolerance to salinity (McGeorge, 1954).

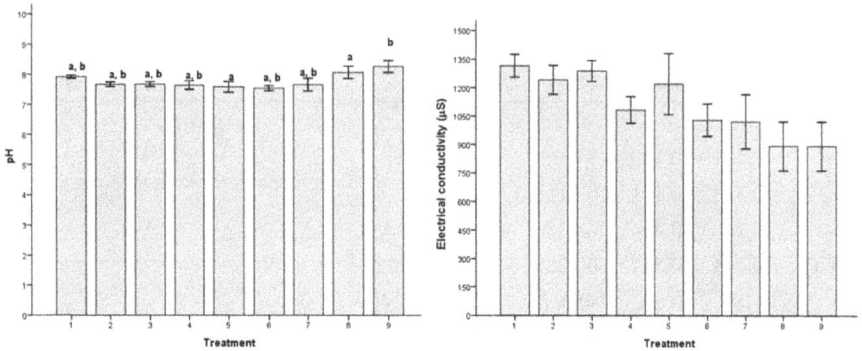

FIGURE 6.2   Mean value for pH and EC in the collected water samples. The groups with a significant difference ($p < 0.5$) are labeled with a different letter. Error bars indicate standard deviation.

A significant statistical difference was not found among observed Mg concentrations ($p < 0.05$) at sampling locations. However, a significant difference in Ca concentrations between influent and effluent wastewater was found. The highest removal was observed on the posttreatment with WTR and *L. minor.* In contrast, there was no significant difference in Na concentrations when influent and effluent locations were statistically contrasted. Concentrations of Mg, Ca, and Na were similar to the observed in a constructed wetland system for industrial wastewater (Hegazy et al., 2011). Even when the variation of these elements is usually related to EC. In the current study, only Na concentrations influenced significantly on that parameter. Figure 6.3 summarizes the concentrations of these elements.

Bicarbonates reached a concentration of 933 mg/L in the system during the experimental period (Fig. 6.3). The presence of these compounds indicates that biological processes have an important role in water chemical reactions inside the treatment units of the hybrid system (Moiseenko et al., 2013). This could be linked to the observed negative correlation between HCO_3 and pH probably associated with microbial homeostasis. This phenomenon is caused by microorganisms that produce HCO_3 for buffering water pH (Nzengy'a and Wishitemi, 2001; Collins et al., 2013).

Concentrations of Cl− were ranged between 130 and 192 mg/L (Fig. 6.3) with no significant differences among treatments ($p < 0.05$). The presence of chloride in water is associated with NaCl in the human diet, and its concentration has a direct relation with salinity parameters (APHA/AWWA/WEF, 2005).

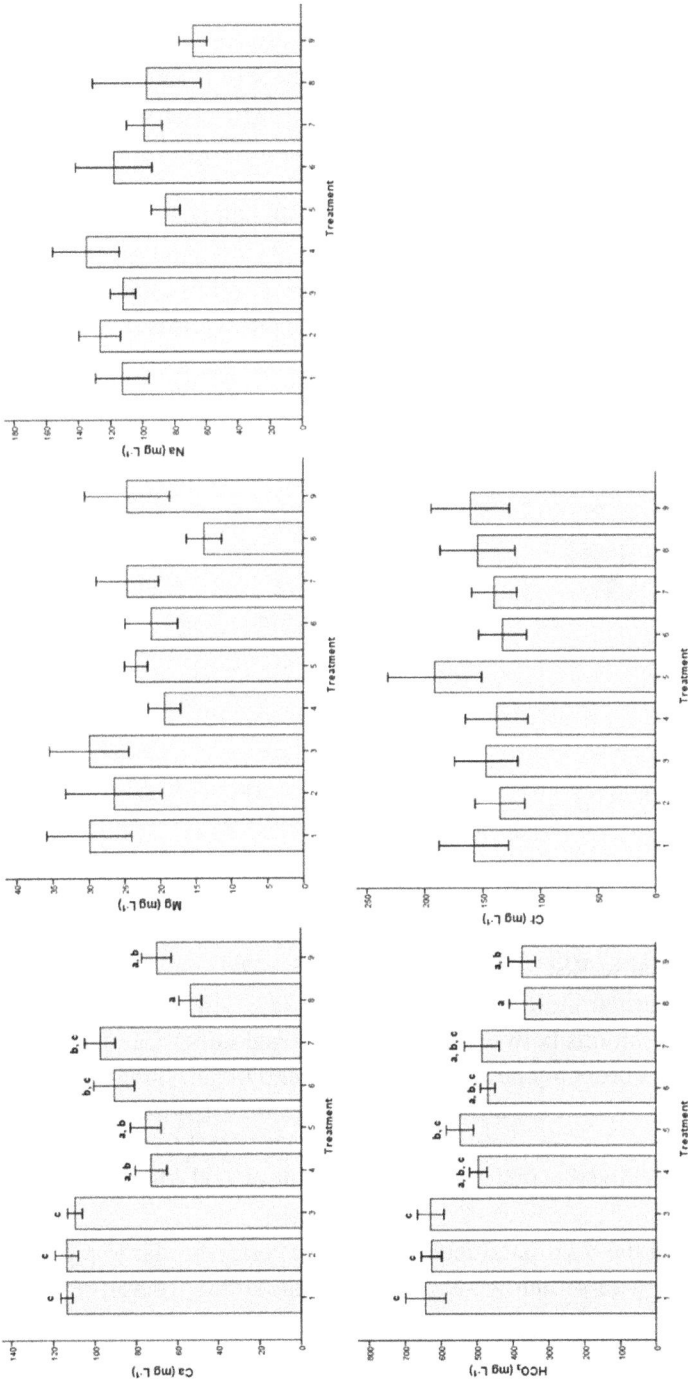

FIGURE 6.3 Mean values for inorganic compounds in the water samples. The groups with a significant difference ($p < 0.5$) are labeled with a different letter. Error bars indicate standard deviation.

The ANOVA conducted for ES and PS did not show significant differences between treatments ($p < 0.05$). According to Silva-Garcia et al. (2006), the observed water quality at the outlets of the system based on ES would have a potential effect on soil salinity when the wastewater treated by the hybrid system is used for crop irrigation. Whereas, based on PS, the reclaimed water from the hybrid systems would be recommended for crop irrigation because of low possibilities for soil salinization (Silva-García et al., 2006; McGeorge, 1954; Collins et al., 2013). ES and PS values probably are due to Cl⁻ concentration in water samples (Collins et al., 2013). This is supported by the high correlation between PS and chloride. Figure 6.4 shows the mean concentration of ES and PS.

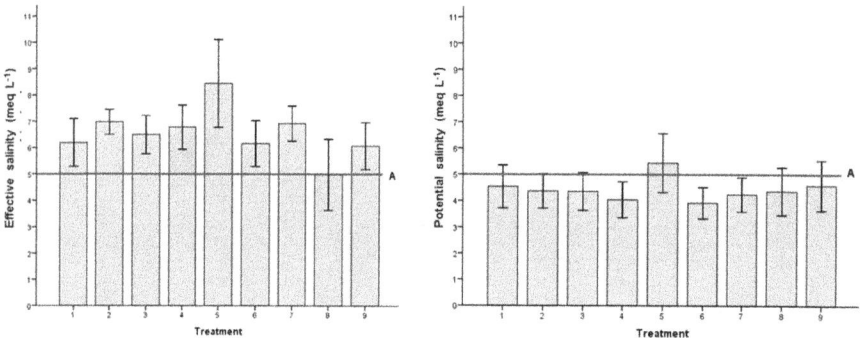

FIGURE 6.4 Mean values for ES and PS in the water samples. Salinity suitable for agricultural purposes is limited by the A-green line in both plots. Error bars indicate standard deviation.

There were no significant differences between treatments for RSC ($p < 0.05$). Based on the observed values, the effluents are not recommended for irrigation because RSC values are greater than 2.5 meq/L and also greater than the suggested limit of 1.25 meq/L (Fig. 6.5). In this study, there were negative correlations between this parameter and either Ca or Mg observed concentrations probably because of calcium and magnesium precipitation in the system (Collins et al., 2013). According to the USA Salinity Laboratory, most water samples are classified as C1S1, C2S1, and C3S2, representing a risk for soil salinization (McGeorge, 1954; Collins et al., 2013). In Figure 6.6, the USA Salinity Laboratory Diagram is shown.

Regarding the SAR parameter, their values were similar to reported ones for irrigation with groundwater (Silva-García, 2006). Therefore, according to the collected water samples, the treater wastewater effluent has a low risk of soil alkalization or salinization if that effluent is used for irrigation

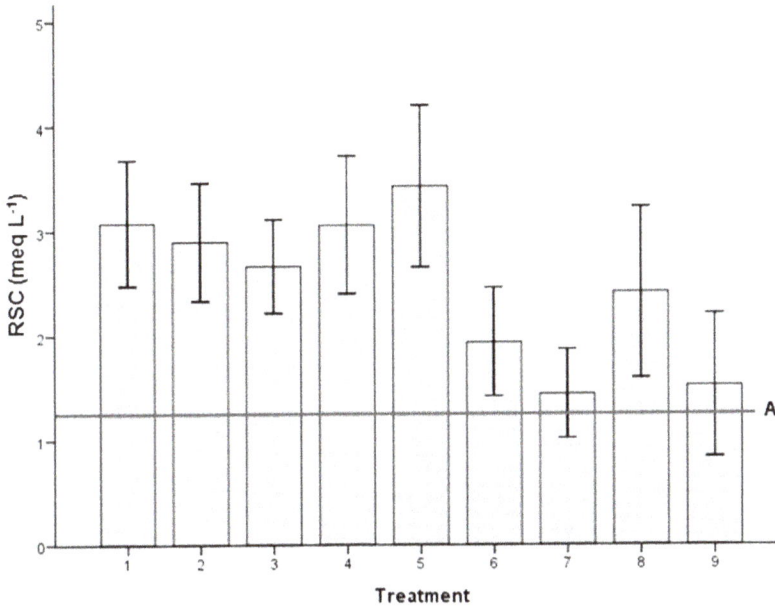

FIGURE 6.5 Mean value for RSC in the water samples. The A-line shows the RSC value suitable for agricultural purposes. Error bars indicate standard deviation.

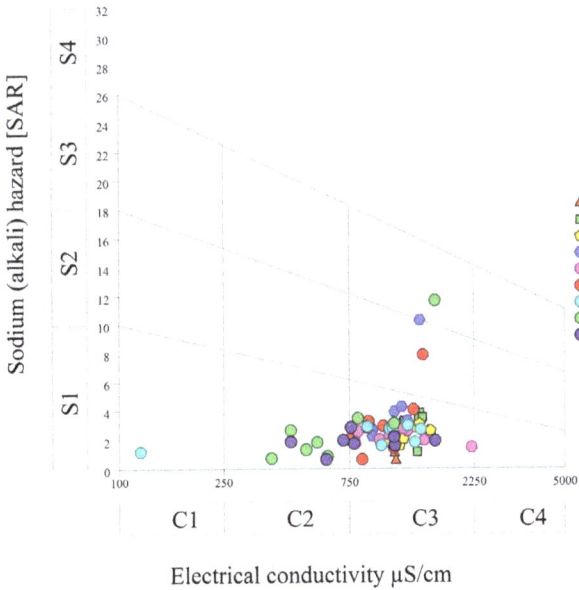

FIGURE 6.6 Classification of water samples from the hybrid wastewater treatment systems. US Salinity Laboratory Diagram.

(Silva-García, 2006; McGeorge, 1954). In contrast to SAR values, the Wilcox index for water quality showed that the treated wastewater in the hybrid system can be considered good to excellent for crop irrigation, as is shown in Figure 6.7.

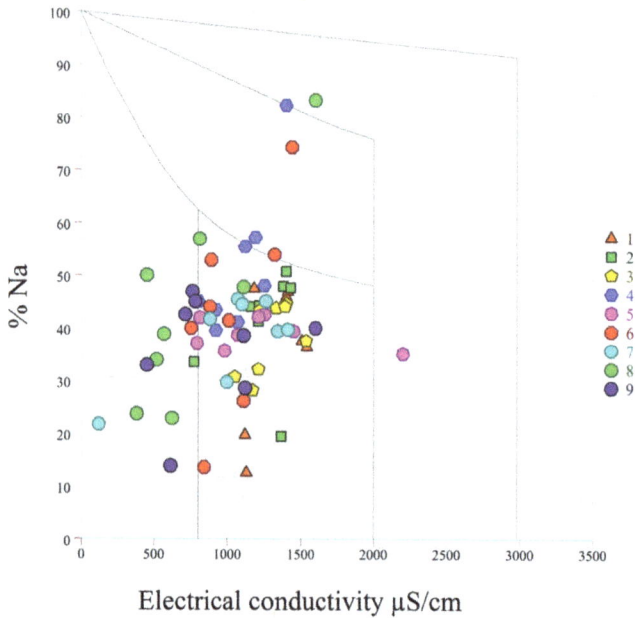

FIGURE 6.7   Wilcox Diagram for the classification of the water samples from the hybrid wastewater treatment system.

Finally, water samples were classified as hard and very hard (Fig. 6.8), whereas, the ANOVA test showed a significant difference between treatments ($p < 0.05$). In fact, the effluents from the polishing basins have a hardness significantly lowest than the influent. It is important to mention that the observed concentrations in the treatment system are a frequent condition in Nuevo Leon where Ca concentration is frequently associated with a high hardness water condition as a consequence of mineralization of calcareous rocks (Dávila Pórcel et al., 2011).

6.4   CONCLUSIONS

The effluents of the hybrid system for the treatment of municipal wastewater were suitable for low-salinity crop irrigation purposes. In general, the system was able to diminish the hardness of the influent. Whereas, the addition of WTR as a substrate influenced positively the wastewater reclamation processes showing that the behavior of physical and chemical parameters through the system was similar to the observed in natural and semi-natural wastewater systems.

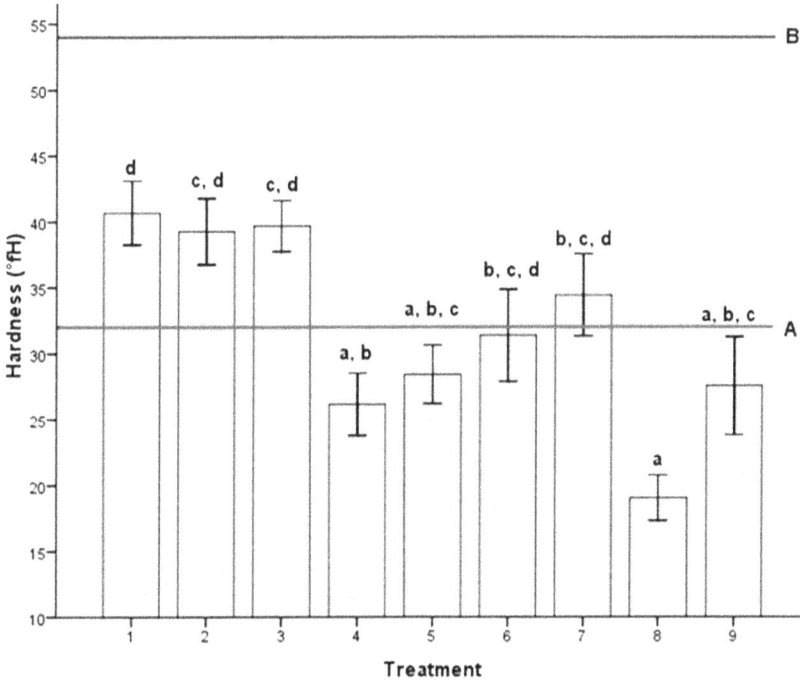

FIGURE 6.8 The A limit is the mean value for the hardness of the collected water samples. The groups with a significant difference ($p < 0.5$) are labeled with a different letter. Hard water (a) and very hard water (b). Error bars indicate standard deviation.

KEYWORDS

- **anaerobic baffled reactors**
- **wastewater reclamation**
- **water quality**
- **hybrid system**

REFERENCES

Ali, M. H. Water: An Element of Irrigation. In *Fundamentals of Irrigation and On-farm Water Management: Volume 1*; Ali, M. H., Ed., Vol. 1; Springer Science+Business Media, LLC 2010: New York, 2010; pp 1–560. https://doi.org/10.1007/978-1-4419-6335-2.

APHA/AWWA/WEF. *APHA: Standard Methods for the Examination of Water and Wastewater*, 21st ed.; American Public Health Association/American Water Works Association/Water Environment Federation: Washington, DC, 2005.

Barber, W. P.; Stuckey, D. C. The Use of the Anaerobic Baffled Reactor (ABR) for Wastewater Treatment: A Review. *Water Res.* **1999,** *33* (7), 1559–1578. https://doi.org/10.1016/S0043-1354 (98)00371-6.

Becerra-Castro, C.; Lopes, A. R.; Vaz-Moreira, I.; Silva, E. F.; Manaia, C. M.; Nunes, O. C. Wastewater Reuse in Irrigation: A Microbiological Perspective on Implications in Soil Fertility and Human and Environmental Health. *Environ. Int.* **2015,** *75,* 117–135. https://doi.org/10.1016/j.envint.2014.11.001.

Belmont, M. a.; Cantellano, E.; Thompson, S.; Williamson, M.; Sánchez, A.; Metcalfe, C. D. Treatment of Domestic Wastewater in a Pilot-Scale Natural Treatment System in Central Mexico. *Ecol. Eng.* **2004,** *23* (4–5), 299–311. https://doi.org/10.1016/j.ecoleng.2004.11.003.

Bodkhe, S. Y. A Modified Anaerobic Baffled Reactor for Municipal Wastewater Treatment. *J. Environ. Manage.* **2009,** *90* (8), 2488–2493. https://doi.org/10.1016/j.jenvman.2009.01.007.

Chen, W.; Lu, S.; Pan, N.; Jiao, W. Impacts of Long-Term Reclaimed Water Irrigation on Soil Salinity Accumulation in Urban Green Land in Beijing. *Water Resour. Res.* **2013,** *49* (11), 7401–7410. https://doi.org/10.1002/wrcr.20550.

Chyan, J.-M. M.; Senoro, D.-B. B.; Lin, C.-J. J.; Chen, P.-J. J.; Chen, I.-M. M. A Novel Biofilm Carrier for Pollutant Removal in a Constructed Wetland Based on Waste Rubber Tire Chips. *Int. Biodeterior. Biodegrad.* **2013,** *85,* 638–645. https://doi.org/10.1016/j.ibiod.2013.04.010.

Collins, K. E.; Doscher, C.; Rennie, H. G.; Ross, J. G. The Effectiveness of Riparian 'Restoration' on Water Quality-A Case Study of Lowland Streams in Canterbury, New Zealand. *Restor. Ecol.* **2013,** *21* (1), 40–48. https://doi.org/10.1111/j.1526-100X.2011.00859.x.

Daigger, G. T. Evolving Urban Water and Residuals Management Paradigms: Water Reclamation and Reuse, Decentralization, and Resource Recovery. *Water Environ. Res.* **2009,** *81* (8), 809–823. https://doi.org/10.2175/106143009x425898.

Dávila Pórcel, R. A.; De León Gómez, H.; Schüth, C. Urban Impacts Analysis on Hydrochemical and Hydrogeological Evolution of Groundwater in Shallow Aquifer Linares, Mexico. *Environ. Earth Sci.* **2011,** *66* (7), 1871–1880. https://doi.org/10.1007/s12665-011-1411-3.

Dzantor, E. K. Phytoremediation: The State of Rhizosphere ' Engineering ' for Accelerated Rhizodegradation of Xenobiotic Contaminants. *J. Chem. Technol. Biotechnol.* **2007,** *232,* 228–232. https://doi.org/10.1002/jctb.

Englande, A. J.; Krenkel, P.; Shamas, J. Wastewater Treatment &Water Reclamation. In *Reference Module in Earth Systems and Environmental Sciences*; Elsevier, 2015. https://doi.org/10.1016/b978-0-12-409548-9.09508-7.

Esposito, K.; Tsuchihashi, R.; Anderson, J.; Selstrom, J. *The Role of Water Reclamation in Water Resources Management in the 21 St Century*; 2005.

Ezechi, E. H.; Kutty, S. R. M.; Muda, K.; Yaqub, A. A Comparative Evaluation of an Integrated Hybrid Bioreactor Treating Industrial Wastewater. *J. Water Process Eng.* **2019,** *31,* 1–9. https://doi.org/10.1016/j.jwpe.2019.100805.

Fito, J.; Van Hulle, S. W. H. Wastewater Reclamation and Reuse Potentials in Agriculture: Towards Environmental Sustainability. *Environ. Dev. Sustain.* **2020**. https://doi.org/10.1007/s10668-020-00732-y.

Gupta, S.; Satpati, S.; Nayek, S.; Garai, D. Effect of Wastewater Irrigation on Vegetables in Relation to Bioaccumulation of Heavy Metals and Biochemical Changes. *Environ. Monit. Assess.* **2010**, *165*, 169–177. https://doi.org/10.1007/s10661-009-0936-3.

Hanjra, M. A.; Blackwell, J.; Carr, G.; Zhang, F.; Jackson, T. M. Wastewater Irrigation and Environmental Health: Implications for Water Governance and Public Policy. *Int. J. Hyg. Environ. Health* **2012**, *215* (3), 255–269. https://doi.org/10.1016/j.ijheh.2011.10.003.

Hassan, S. R.; Dahlan, I. Anaerobic Wastewater Treatment Using Anaerobic Baffled Bioreactor: A Review. *Cent. Eur. J. Eng.* **2014**, *3* (3), 389–399. https://doi.org/10.2478/s13531-013-0107-8.

Hegazy, A. K. K.; Abdel-Ghani, N. T. T.; El-Chaghaby, G. A. A. Phytoremediation of Industrial Wastewater Potentiality by Typha Domingensis. *Int. J. Environ. Sci. Technol.* **2011**, *8* (3), 639–648. https://doi.org/10.1007/BF03326249.

Hunter, R. G.; Day, J. W.; Wiegman, A. R.; Lane, R. R. Municipal Wastewater Treatment Costs with an Emphasis on Assimilation Wetlands in the Louisiana Coastal Zone. *Ecol. Eng.* **2019**, *137*, 21–25. https://doi.org/10.1016/j.ecoleng.2018.09.020.

Ibrahim, Y.; Banat, F.; Naddeo, V.; Hasan, S. W. Numerical Modeling of an Integrated OMBR-NF Hybrid System for Simultaneous Wastewater Reclamation and Brine Management. *Euro-Mediterranean J. Environ. Integr.* **2019**, *4* (1). https://doi.org/10.1007/s41207-019-0112-2.

Iurciuc, C. E.; Dima, M. Wastewater for Irrigation in Agriculture: Some Effects of Effluent on Soil Quality and Canola (Brassica Napus Oleifera) Growth. *Environ. Eng. Manag. J.* **2013**, *12* (4), 801–806.

Janeiro, C. N.; Arsénio, A. M.; Brito, R. M. C. L.; van Lier, J. B. Use of (Partially) Treated Municipal Wastewater in Irrigated Agriculture; Potentials and Constraints for Sub-Saharan Africa. *Phys. Chem. Earth* **2020**, *118–119*, 1–12. https://doi.org/10.1016/j.pce.2020.102906.

Juwarkar, A. a.; Singh, S. K.; Mudhoo, A. A Comprehensive Overview of Elements in Bioremediation. *Rev. Environ. Sci. Bio/Technol.* **2010**, *9* (3), 215–288. https://doi.org/10.1007/s11157-010-9215-6.

Krishna, G. V. T. G.; Kumar, P.; Kumar, P. Complex Wastewater Treatment Using an Anaerobic Baffled Reactor. *Environ. Prog.* **2007**, *26* (4), 391–398. https://doi.org/10.1002/ep.10239.

Lavee, D. A Cost-Benefit Analysis of Alternative Wastewater Treatment Standards: A Case Study in Israel. *Water Environ. J.* **2011**, *25* (4), 504–512. https://doi.org/10.1111/j.1747-6593.2010.00246.x.

Leong, C. Resilience to Climate Change Events: The Paradox of Water (In)-Security. *Sustain. Cities Soc.* **2016**, *27*, 439–447. https://doi.org/10.1016/j.scs.2016.06.023.

Li, A.; Kroeze, C.; Kahil, T.; Ma, L.; Strokal, M. Water Pollution from Food Production: Lessons for Optimistic and Optimal Solutions. *Curr. Opin. Environ. Sustain.* Oct 1, **2019**, pp 88–94. https://doi.org/10.1016/j.cosust.2019.09.007.

Lin, X.; Xu, J.; Keller, A. A.; He, L.; Gu, Y.; Zheng, W.; Sun, D.; Lu, Z.; Huang, J.; Huang, X.; Li, G. Occurrence and Risk Assessment of Emerging Contaminants in a Water Reclamation and Ecological Reuse Project. *Sci. Total Environ.* **2020**, *744*. https://doi.org/10.1016/j.scitotenv.2020.140977.

Loupasaki, E.; Diamadopoulos, E. Attached Growth Systems for Wastewater Treatment in Small and Rural Communities: A Review. *J. Chem. Technol. Biotechnol.* **2013**, *88* (2), 190–204. https://doi.org/10.1002/jctb.3967.

Maltby, L.; Calow, P. The Application of Bioassays in the Resolution of Environmental Problems; Past, Present and Future. *Hydrobiologia* **1989,** *188–189* (1), 65–76. https://doi.org/10.1007/BF00027772.

Marofi, S.; Parsafar, N.; Rahim, G. H.; Dashti, F.; Marofi, H. The Effects of Wastewater Reuse on Potato Growth Properties under Greenhouse Lysimeteric Condition. *Int. J. Environ. Sci. Technol.* **2012,** *10* (1), 133–140. https://doi.org/10.1007/s13762-012-0108-9.

McGeorge, W. T. T. Diagnosis and Improvement of Saline and Alkaline Soils. *Soil Sci. Soc. Am. J.* **1954,** *18* (3), 348. https://doi.org/10.2136/sssaj1954.03615995001800030032x.

Metzler, D. F.; Russelmann, H. B. Wastewater Reclamation as a Water Resource. *J. Am. Water Works Assoc.* **1968,** *60* (1), 95–102. https://doi.org/10.1002/j.1551-8833.1968.tb03521.x.

Mizyed, N. R. Challenges to Treated Wastewater Reuse in Arid and Semi-Arid Areas. *Environ. Sci. Policy* **2013,** *25*, 186–195. https://doi.org/10.1016/j.envsci.2012.10.016.

Moiseenko, T. I.; Gashkina, N. a.; Dinu, M. I.; Kremleva, T. A.; Khoroshavin, V. Y. Aquatic Geochemistry of Small Lakes: Effects of Environment Changes. *Geochem. Int.* **2013,** *51* (13), 1031–1148. https://doi.org/10.1134/S0016702913130028.

Nasr, F. A.; Doma, H. S.; Nassar, H. F. Treatment of Domestic Wastewater Using an Anaerobic Baffled Reactor Followed by a Duckweed Pond for Agricultural Purposes. *Environmentalist* **2009,** *29* (3), 270–279. https://doi.org/10.1007/s10669-008-9188-y.

Ning, D.; Huang, Y.; Pan, R.; Wang, F.; Wang, H. Effect of Eco-Remediation Using Planted Floating Bed System on Nutrients and Heavy Metals in Urban River Water and Sediment: A Field Study in China. *Sci. Total Environ.* **2014,** *485–486* (1), 596–603. https://doi.org/10.1016/j.scitotenv.2014.03.103.

Nzengy'a, D. M.; Wishitemi, B. E. L. The Performance of Constructed Wetlands for, Wastewater Treatment: A Case Study of Splash Wetland in Nairobi Kenya. *Hydrol. Process.* **2001,** *15* (17), 3239–3247. https://doi.org/10.1002/hyp.185.

Piegorsch, W. W.; Bailer, A. J. *Analyzing Environmental Data*; John Wiley & Sons, Ltd: Chichester, UK, 2005.

Pilon-Smits, E. Phytoremediation. *Annu. Rev. Plant Biol.* **2005,** *56* (1), 15–39. https://doi.org/10.1146/annurev.arplant.56.032604.144214.

Poustie, A.; Yang, Y.; Verburg, P.; Pagilla, K.; Hanigan, D. Reclaimed Wastewater as a Viable Water Source for Agricultural Irrigation: A Review of Food Crop Growth Inhibition and Promotion in the Context of Environmental Change. *Science of the Total Environment.* Elsevier B.V. October 15, 2020. https://doi.org/10.1016/j.scitotenv.2020.139756.

Prăvălie, R. Drylands Extent and Environmental Issues. A Global Approach. *Earth-Sci. Rev.* **2016,** *161*, 259–278. https://doi.org/10.1016/j.earscirev.2016.08.003.

Qadir, M.; Bahri, A.; Sato, T.; Al-Karadsheh, E. Wastewater Production, Treatment, and Irrigation in Middle East and North Africa. *Irrig. Drain. Syst.* **2010,** *24* (1–2), 37–51. https://doi.org/10.1007/s10795-009-9081-y.

Qadir, M.; Wichelns, D.; Raschid-Sally, L.; McCornick, P. G.; Drechsel, P.; Bahri, a.; Minhas, P. S. The Challenges of Wastewater Irrigation in Developing Countries. *Agric. Water Manag.* **2010,** *97* (4), 561–568. https://doi.org/10.1016/j.agwat.2008.11.004.

Rezania, S.; Taib, S. M.; Md Din, M. F.; Dahalan, F. A.; Kamyab, H. Comprehensive Review on Phytotechnology: Heavy Metals Removal by Diverse Aquatic Plants Species from Wastewater. *J. Hazard. Mater.* Nov 15, **2016,** 587–599. https://doi.org/10.1016/j.jhazmat.2016.07.053.

Rizzo, L. Bioassays as a Tool for Evaluating Advanced Oxidation Processes in Water and Wastewater Treatment. *Water Res.* **2011,** *45* (15), 4311–4340. https://doi.org/10.1016/j.watres.2011.05.035.

Rojas-Valencia, M. N.; Velásquez, M. T. O. De; Franco, V. Urban Agriculture, Using Sustainable Practices That Involve the Reuse of Wastewater and Solid Waste. *Agric. Water Manag.* **2011,** *98* (9), 1388–1394. https://doi.org/10.1016/j.agwat.2011.04.005.

Saeed, T.; Al-Muyeed, A.; Afrin, R.; Rahman, H.; Sun, G. Pollutant Removal from Municipal Wastewater Employing Baffled Subsurface Flow and Integrated Surface Flow-Floating Treatment Wetlands. *J. Environ. Sci. (China)* **2014,** *26* (4), 726–736. https://doi.org/10.1016/ S1001-0742(13)60476-3.

Sarathai, Y.; Koottatep, T.; Morel, A. Hydraulic Characteristics of an Anaerobic Baffled Reactor as Onsite Wastewater Treatment System. *J. Environ. Sci.* **2010,** *22* (9), 1319–1326. https://doi.org/10.1016/S1001-0742(09)60257-6.

Schröder, P.; Navarro-Aviñó, J.; Azaizeh, H.; Goldhirsh, A. G.; DiGregorio, S.; Komives, T.; Langergraber, G.; Lenz, A.; Maestri, E.; Memon, A. R.; Ranalli, A.; Sebastiani, L.; Smrcek, S.; Vanek, T.; Vuilleumier, S.; Wissing, F. Using Phytoremediation Technologies to Upgrade Waste Water Treatment in Europe. *Environ. Sci. Pollut. Res. Int.* **2007,** *14* (7), 490–497.

Schwitzguébel, J.-P. Hype or Hope: The Potential of Phytoremediation as an Emerging Green Technology. *Remediat. J.* **2001,** *11* (4), 63–78. https://doi.org/10.1002/rem.1015.

SEMARNAP. Norma Oficial Mexicana NOM-001-ECOL-1996, Que Establece Los Límites Máximos Permisibles de Contaminantes En Las Descargas de Aguas Residuales En Aguas y Bienes Nacionales. *D. of. la Fed.* **1996.**

Shingare, R. P.; Thawale, P. R.; Raghunathan, K.; Mishra, A.; Kumar, S. Constructed Wetland for Wastewater Reuse: Role and Efficiency in Removing Enteric Pathogens. *J. Environ. Manage.* Sept 15, **2019,** 444–461. https://doi.org/10.1016/j.jenvman.2019.05.157.

Silva-García, J. T. T.; Ochoa-Estrada, S.; Cristóbal-Acevedo, D.; Estrada-Godoy, F. Calidad Química Del Agua Subterránea de La Ciénega de Chapala Como Factor de Degradación Del Suelo. *Terra Latinoam.* **2006,** *24* (4), 503–513.

Sundaray, S. K.; Nayak, B. B.; Bhatta, D. Environmental Studies on River Water Quality with Reference to Suitability for Agricultural Purposes: Mahanadi River Estuarine System, India—A Case Study. *Environ. Monit. Assess.* **2009,** *155* (1–4), 227–243. https://doi.org/ 10.1007/s10661-008-0431-2.

Tabatabaei, S. H.; Nourmahnad, N.; Kermani, S. G.; Tabatabaei, S. A.; Najafi, P.; Heidarpour, M. Urban Wastewater Reuse in Agriculture for Irrigation in Arid and Semi-Arid Regions—A Review. *Int. J. Recycl. Org. Waste Agric.* **2020,** *9* (2), 193–220. https://doi.org/10.30486/ ijrowa.2020.671672.

Tortajada, C.; Castelán, E. Water Management for a Megacity: Mexico City Metropolitan Area. *Ambio* **2003,** *32* (2), 124–129.

Tran, Q. K.; Schwabe, K. A.; Jassby, D. Wastewater Reuse for Agriculture: Development of a Regional Water Reuse Decision-Support Model (RWRM) for Cost-Effective Irrigation Sources. *Environ. Sci. Technol.* **2016,** *50* (17), 9390–9399. https://doi.org/10.1021/acs. est.6b02073.

United States Environmental Protection Agency. Basic Information about Water Reuse https:// www.epa.gov/waterreuse/basic-information-about-water-reuse.

Vyas, A.; Jethoo, A. S. Diversification in Measurement Methods for Determination of Irrigation Water Quality Parameters. *Aquat. Procedia* **2015,** *4*, 1220–1226. https://doi. org/10.1016/j.aqpro.2015.02.155.

Wang, J.; Huang, Y.; Zhao, X. Performance and Characteristics of an Anaerobic Baffled Reactor. *Bioresour. Technol.* **2004,** *93* (2), 205–208. https://doi.org/10.1016/j.biortech.2003.06.004.

Yang, Q.; Chen, Z.; Zhao, J.; Gu, B. Contaminant Removal of Domestic Wastewater by Constructed Wetlands: Effects of Plant Species. *J. Integr. Plant Biol.* **2007,** *49* (4), 437–446. https://doi.org/10.1111/j.1672-9072.2006.00389.x.

Zhai, G. Phytoremediation: Right Plants for Right Pollutants. *J. Bioremediation Biodegrad.* **2011,** *02* (03), 102. https://doi.org/10.4172/2155-6199.1000102e.

Zhang, Q. yu; Liu, L. sheng; Liu, Z. jin. Application of Safety and Reliability Analysis in Wastewater Reclamation System. *Process Saf. Environ. Prot.* **2021,** *146*, 338–349. https://doi.org/10.1016/j.psep.2020.09.010.

Zhang, X.; Liu, P.; Yang, Y.; Chen, W. Phytoremediation of Urban Wastewater by Model Wetlands with Ornamental Hydrophytes. *J. Environ. Sci.* **2007,** *19*, 902–909.

PART II

Plant Physiology Innovations

CHAPTER 7

Effect of Films Based on Pectin-Hydroponic Mucilage on the Quality and Shelf Life of Golden Apple

BRENDA LUNA-SOSA[1], DULCE CONCEPCIÓN GONZÁLEZ-SANDOVAL[1], C. G. GUILLERMO MARTÍNEZ-ÁVILA[1], JUANA ARANDA-RUÍZ[1], MAYRA TREVIÑO-GARZA[2], and ROMEO ROJAS[1*]

[1]*Universidad Autonoma de Nuevo Leon, Research Center and Development for Food Industries, School of Agronomy, 66050 General Escobedo, Nuevo León, México*

[2]*Universidad Autonoma de Nuevo Leon, Research Center and Development for Biological Sciences, School of Biological Sciences, Pedro de Alba, Niños Héroes, Ciudad Universitaria, San Nicolás de los Garza, N.L., México*

Corresponding author. E-mail: romeo.rojasmln@uanl.edu.mx

ABSTRACT

There is currently a strong demand for healthy and nutritiously enriched fruits. However, all these products are perishable to a different extent despite their ascorbic acid content, which serves to inhibit the enzymatic activity and preserve the natural state of the fruits. This has led to the development of technologies that prolong the shelf life of different products. The cause of food spoilage can be of chemical, physical, or microbiological origin. The apple market continues its upward trend and expands by the attributes that characterize it. However, the fruit suffers browning in a thermal environment, so it is important to explore, study, and preserve the product in environmental and safe conditions for consumption, maintaining its properties, and reducing costs for storage.

For this study, the Golden delicious apple was sanitized and covered by immersion with hydroponic nopal mucilage in order to assess its shelf life.

The stability attributes in the physicochemical parameters (pH, soluble solids, vitamin C, and ascorbic acid), microbiological (fungal and yeast inhibition), and quality (texture, flavor, color, and DR attributes) were evaluated under storage conditions (35 days at $25 \pm 1°C$). Physicochemical factors such as weight loss were registered between 135 g and 125 g as a minimum value on day 17. The color varied during ripeness of apples (68.41–70.24) in brightness. The firmness remained stable, registering a maximum value of 625.49 g in the application of the Vi mucilage at the end of the trial period. On the other hand, stability was also found in the microbiological determination (<35 CFU/g) during the 35 days of evaluation. Heartburn and vitamin C turned out to be higher for apples covered with Vi. Sensory acceptance was higher for apples covered with CF1 mucilage. Finally, the addition of edible films based on nopal mucilages does not affect the sensory quality of fruits after application, being suitable for consumption.

7.1 INTRODUCTION

Currently, in Mexico and throughout the world, food is wasted from initial production to fresh consumption by consumers, which occurs accidentally or intentionally, causing low availability of edible products for humanity. According to the above, when an excessive limit is reached by consumers, wasting generates significant losses in labor and supplies used for the food provision without taking into account that most of the discarded products can be in acceptable state for consumption while many people suffer from hunger every day. About one-third of the food produced in the world for human consumption (1.3 million t/year) is lost. The Food and Agriculture Organization of the United Nations (FAO) estimates that the generated food waste is of 95–115 kg/year per person in Europe and North America, while in sub-Saharan Africa and South Asia and southeast, this figure represents only 6–11 kg/year (FAO, 2012).

The stages where food waste develops include agricultural production (criteria for handling the crop from planting to harvesting of fruit), postharvest management (microbiological, chemical, mechanical, and environmental causes), distribution, and consumption (market for sale and purchase by consumers). The importance of reducing crop losses and maintaining the quality of fresh fruits for longer periods is a priority for all producers, which leads to developing new technologies for fruit conservation and thus these technologies meet the demand for healthier and more natural foods for human consumption (de Azeredo et al., 2018).

The edible packaging industry has had remarkable growth in recent years and is expected to have a major impact on the food market for the future. This growth is due to the demand of food consumers who meet higher standards of quality, knowledge, and technology in the area of edible films through the science and technology of material processing. The impact of sustainability and interest in the use of renewable resources has created interest for scientists to generate research in the area of "edible coatings and films" with the perspective of ensuring the safety and quality of food products. Due to the concern of replacing plastic packaging with a biodegradable one, there are several studies of edible films which, in addition to the characteristics granted, can be ingested with the food. It is important to keep in mind that the preparation of edible films depends both on the additives used and their manipulation to confer desirable mechanical and barrier properties to the matrix in order to preserve moisture and oxygen in the product to which it is intended to apply the film without modifying the organoleptic characteristics, such as taste, smell, texture, among others (Padmanabhan et al., 2018).

In Mexico, the *Golden Delicious* apple is one of the most consumed fruits. The national production of *Golden Delicious* apples is 375,055 t/year and the state of Coahuila produces around 10% of these. The apple growing region is located in the Sierra de Arteaga, mainly in places like Carbonera, Los Lirios, El Tunal, Jamé, San Antonio de Alazanas, and Huachichil. However, the production of Golden Delicious apple is strongly affected by its short shelf life in the postharvest because it is a perishable fruit of the climaterial type, which means that even under refrigeration conditions, the fruit can decompose (León de Zapata et al., 2018). The production of apple involves a series of stages, such as harvesting, transportation, and storage, in which certain parameters and care must be taken to avoid losses or damage to the product. Although at the harvest stage there may be problems due to improper handling of the fruit, most of the problems appear in the postharvest stage. Humidity and temperature are important factors that contribute to the appearance of diseases in apples, facilitating the presence of pests and microorganisms. One of the most frequent microorganisms in the apple and that decreases its useful life is the fungus of the genus *Penicillium*, mainly the species *Penicillium expansum*. The main apple varieties are early red one, top red, red delicious, starking, royal gala, grany smith, golden supreme, golden delicious, gray reineta of Canada and McIntosh (Gudkovsky et al., 2012; Varela et al., 2008; Juhnevica et al., 2014).

Despite various studies on the subject, there is still little information on the effect of edible films based on nopal–pectin mucilage on the quality and shelf life of apples. The application of a nonperceptible edible film based on nopal mucilage, pectin, and glycerol to *Golden Delicious* apples, could increase its resistance to fungi and yeasts without reducing water loss and prolongs the shelf life without altering the organoleptic properties.

7.2   MATERIALS AND METHODS

7.2.1   *BIOLOGICAL MATERIAL*

Nopal hydroponic mucilage of Copena F1 and Villanueva with a state of maturity of 45 days previously characterized, pectin (Sigma-Aldrich, Canada) of galacturonic acid of $\geq 74.0\%$ and glycerol $\geq 99.5\%$ (Sigma-Aldrich) were used in the making of the films. We worked with a batch of 250 Golden apples purchased at "Mercado de Abastos La Estrella," San Nicolás de los Garza, N.L., from Ciudad Juárez, Chihuahua. To have homogeneous batches, apples were selected by eliminating those that had mechanical damage and only those with a firm appearance, free of visible fungi and of a homogeneous color, as well as with characteristic smell were accepted.

7.2.2   *PREPARATION OF IMMERSION-BASED FILMS*

The films CF1sf and Vacsf were selected due to the properties they provided according to the results of the previous characterization performed on these treatments. Three film-forming solutions were prepared in sterile distilled water by mechanical stirring every 20 minutes, until complete homogenization. CF1sf and Vasf with the same concentration (2.5 g of mucilage plus 1.5 g of pectin and 0.5 g of glycerol in 100 mL of distilled water). The forming solutions were made in a 100-ml sterilized beaker, covered with parafilm paper and stored at room temperature. The formulations were subjected to the removal of excess bubbles through injection vacuum with a hose held in the 250 mL Erlenmeyer flasks. Five repetitions were performed for each treatment.

7.2.3   *APPLICATION OF FILM FORMING SOLUTIONS ON APPLES*

The selected apples were washed with running water and disinfected by immersion in a hypochlorite solution (250 mg/kg) for 1 min, washed with distilled water and dried with paper towels. All materials (surfaces and utensils) in contact with the fruit were disinfected prior to use. Three different treatments were used: (1) apple without film (Ctrl), (2) apple covered with Villanueva mucilage without fiber (Visf), and (3) apple covered with Copena F1 mucilage without fiber (CF1sf). The whole apples were immersed for 10 min in clean $20 \times 15 \times 7''$ plastic trays containing the film formulating solution. The fruits were left to drain to remove the excess of the coating until drying in aluminum gratings at room temperature (20–25°C) in a laminar flow hood (biobase). The fruit without film-forming solution was used as Ctrl. Finally, control and covered apples were stored on a stainless steel table wrapped in foil at room temperature of 25°C with a relative humidity (RH) of 59% for 35 days. The respective analyses were performed for each of the treatments and parameters (physicochemical every 7 days), the sensory and microbiological analyses were performed at 0, 17, and 35 days in a period of 35 days.

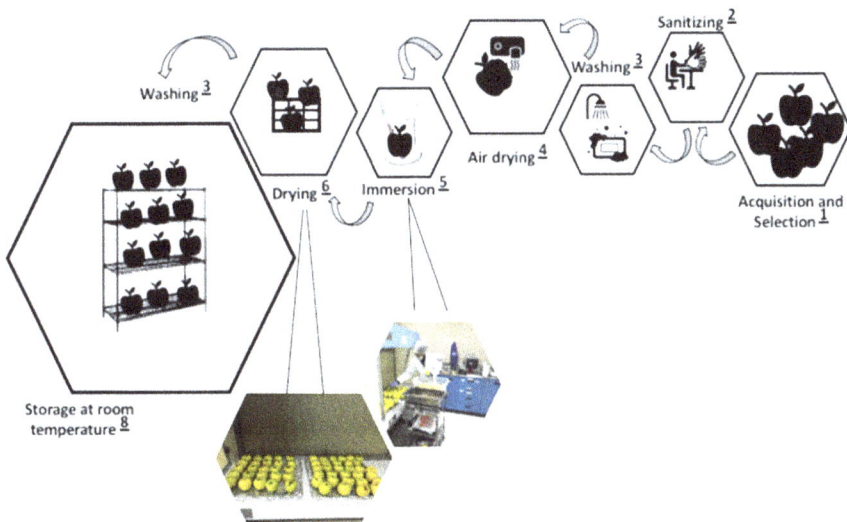

FIGURE 7.1   Golden apple processing for conservation at 25 ± 1°C.

7.2.4   *APPLE MICROBIOLOGICAL ANALYSIS*

7.2.4.1   *SAMPLE PREPARATION AND DILUTION*

Ten grams of apple were weighed in sterile bags (Nasco Whiril-Pak 18 oz) with 90 mL bag of sterile peptonated water (0.85% NaCl and 0.10%

peptone, prepared in the dilution bottles) and homogenized for 1 min (this is the 10–1 dilution). Subsequently, the serial dilutions necessary for the microbial analysis were performed (10–2–10–6) to make the 10–2 dilution, and 1 mL of the 10–1 dilution was taken and placed in 9 mL of sterile saline (0.85% NaCl, prepared in dilution tubes) and so on (Treviño-Garza et al., 2015).

7.2.4.2   SOWING ON PLATE FOR FUNGAL AND YEAST GROWTH

One milliliter of each dilution was seeded in sterile Petri dishes and between 15 and 20 mL of Papa dextrose Agar (PDA acidified with 10% sterile tartaric acid; 1.4 mL of tartaric acid per 100 mL of medium) was added. Plates were homogenized and incubated at $24 \pm 2°C$ for 5 days. The results were expressed as colony forming units per gram of fruit by CFU/g (Treviño-Garza et al., 2015).

7.2.4.3   APPLE SENSORY EVALUATION

Sensory analyses were performed to evaluate the acceptance of the products. The sensory parameters of color, smell, taste, texture, and acceptance were evaluated by untrained panelists ($n = 40$) during a period of 35 days of storage (days 0, 7, 17, and 35). The panelists were students and teachers (20–60 years old) of our institution who like apples. Fresh fruits cut with and without coating were presented randomly to the panelists and the variables were evaluated with qualitative values in a range of 1–5 (bad to excellent). Samples were considered acceptable when they received scores greater than or equal to 2.5. This criterion was used to determine the shelf life of the product (Treviño-Garza et al., 2015).

7.2.4.4   DECAY RATE (DR)

Panelists evaluated the absence or presence of fungi in detail, both in fruits coated with the different formulations and in the control. The symptoms of deterioration caused by the presence of fungi in the fruits were evaluated visually according to the following scale: 1 = not damaged (0%), 2 = light damage (0–25%), 3 = moderate damage (25–50%), 4 = severe damage (50–75%), and 5 = completely damaged (75–100%) (Treviño-Garza et al., 2015).

7.2.5 PHYSICOCHEMICAL ANALYSIS OF APPLES WITH THE FILM-FORMING SOLUTION

7.2.5.1 WEIGHT LOSS, °BRIX, COLOR, FIRMNESS, TITRATABLE ACIDITY, AND VITAMIN C

Individual weights ($n = 10$) were registered based on the AOAC method (AOAC, 1984). The difference between the initial and final weight of the fruits was considered as the total weight loss and the results were expressed as percentages. For the determination of the total soluble solids content (SST), the sample (1.5 g) of $n = 5$ was ground for each treatment (León de Zapata et al., 2018) and one drop was placed in the digital refractometer (0–23°Brix). The CIELAB profile of the color in the apple was analyzed directly in four angles (quadrant) with a Minolta fruit colorimeter ($n = 3$) (Gheribi et al, 2018). For firmness, a penetrometer was used in $n = 5$ apples with the execution of a standardized 2-mm needle disc for pomaceous (León de Zapata et al., 2018). The results were expressed in kg. A total of 5 g of apple ($n = 5$) with five repetitions of each treatment were ground for the digital pH analysis previously calibrated with buffers 4 and 7. Titratable acidity was determined based on the AOAC 942.15 method by titration with a NaOH solution 0.1 N using phenolphthalein as an indicator (pink turn). The result was expressed in citric acid %. Vitamin C was performed by titration. To 10 g of apple, 4 mL of acetic acid, 10 mL of potassium iodide and 1 mL of starch solution were added, then a titration with iodine at 0.01 N was performed (Tavarez, 2005). The % of vitamin C was determined by the following equation:

$$\text{Vitamin C} = \frac{\left(mL\ de\ Yodo\right)\left(N\right)\left(0.089\right)}{\text{(Sample weight in g)}} \tag{7.1}$$

N = normality of iodine (0.01 N); Meq = milliequivalents of iodine (0.1 N = 0.089) and V = volume of iodine spent.

7.2.5.2 STATISTIC ANALYSIS

The results of the physicochemical and microbiological analyses were subjected to an analysis of variance (ANOVA) and Tukey test. A completely randomized design with $n = 10$ was used for the different analyses and an apple for the batch of 250 apples was taken as an experimental unit. The results that were in percentage were transformed to normalize the data and then the ANOVA was applied using the following formula: $P' = \text{arcsen}\sqrt{P}//P' = Sen^{-1}\sqrt{P}$. The results of the sensory analysis were submitted to a Kruskal–Wallis test in Excel (2010) and in the IBM SPSS Statistics program to perform the analysis of variance (ANOVA).

7.3 RESULTS AND DISCUSSION

7.3.1 MOLDS AND YEAST COUNT

According to the homogeneity groups indicated with letters in the graphs, there were significant differences ($P < 0.05$) between treatments (Table 7.1). In the case of molds and yeasts, they ranged from 13.50 and 24.79 CFU/g for day 21. For day 35, the values were 11.45, 12.28, and 34.88 CFU/g. The microbiological stability between treatments was due to the barrier protection generated in apples and the antimicrobial activity of mucilages during storage. Nopal-based mucilage films maintain high fruit color values. As far as we know, there are no studies of microbiological evaluations in whole apples covered with Visf and CF1sf maintained in environmental conditions.

TABLE 7.1   Effects of Edible Films on the Microbiological Analysis (CFU/g) of Whole Apples During Day 0, 21 and 35 Stored at Room Temperature. The Controls were Molds and Yeasts.

Storage (d) molds and yeasts (CFU/g)	Ctrl	Visf	CF1sf
0	ND	ND	ND
21	24.79 ± 1.20[b]	13.50 ± 0.71[a]	ND
35	34.88 ± 0.93[b]	12.28 ± 0.66[a]	11.45 ± 0.73[a]

[a,b]Means within a row which do not have a common superscript letter, are significantly different ($P < 0.05$).

ND= No detected.

Previous studies have mixed mucilages together with waxes obtained from plant extracts, antimicrobial agents, and polysaccharides that are considered nonbiological-based. León de Zapata et al. (2018) evaluated shelf life in apples with nanocubers of bioactive compounds under industrial conditions and increased significantly. Ma et al. (2019) used methylcyclopropene (1-MCP) in apples to know their shelf life and antioxidant power.

Shin et al. (2017) reported values of 4–6 log CFU/g of fungi and yeasts in apples covered with antioxidant and microbial agents from plants. Noshad and Rahmati (2019) evaluated the effects of the mucilage by incorporating it into apples cut and stored at 4°C for 10 days and found values of 2.2–3 log CFU/g of microorganisms (*Plantago major*, *Plantago psyllium* and *D. Sophia*). These values were higher than our values, since the microbial activity was different from ours in the development of fungi and yeasts. This was possibly observed due to the diffusion of phenolic compounds into the cytoplasmic membrane, the interrupted proton movement force, and the electric current and the coalescent cell content (Noshad and Rahmati, 2019).

7.3.2 SENSORY EVALUATION

7.3.2.1 COLOR, ODOR, FLAVOR, TEXTURE, DR, AND GENERAL ACCEPTANCE

According to the sensory analysis evaluated, no significant differences were found between the sensory variables ($P < 0.05$) on all days of storage for the odor factor, texture, appearance, and decay index (DR). These variables have an impact on the typical attributes of the apple, and this influences its acceptance or rejection (Fig. 7.2).

During the storage of the apples, the highest value for the color variable was 4.25 for CF1 on day 21, followed by Ctrl 3.62 and Visf of 3.92. On day 35 for the taste of apples, Ctrl and CF1sf were statistically equal ($P < 0.05$) with values 3.77 and 3.90. However, Visf was 3.15.

There were no significant differences observed during fruit storage. For the initial time, the data remained stable (Fig. 7.1) and presented positive evaluations, since the fruits retained their natural state. The odor described the intensity of breathing of the apples as their development increased with the time. The color generated a transition from yellow to a hydrated tone due to the chlorophyll effect of the polymers, specifically those of the mucilages. When chlorophyll breaks down, it generates framed dyes (carmine, purple and opaque coffee). In addition, it increases the production of reddish and yellow colors typical of ripening.

The flavor and aroma changes were characteristic for day 21 in apples, which provided the carbohydrates the typical sweetness in decreasing the initial sour taste. Temperature is a parameter that influences the shelf life in fruits rich in vitamin E, since it provided changes in the aroma. The softening behavior and decay index were attributed to the added pectin and protopectin

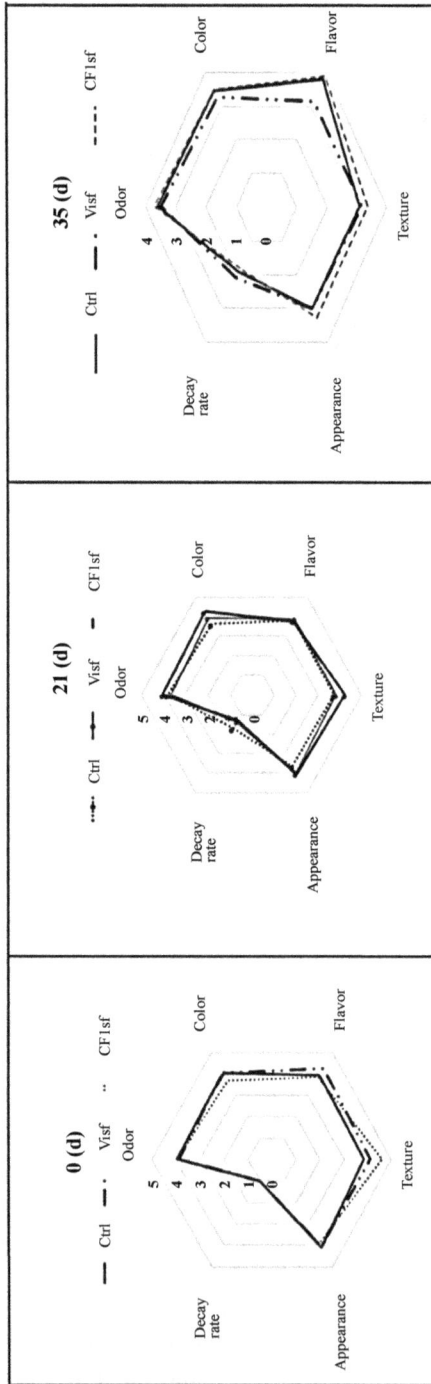

FIGURE 7.2 Effect of mucilage films on sensory quality parameters: (a) odor, (b) color, (c) flavor, (d) texture, (e) appearance, and (f) decay rate of apples stored for 0, 21, and 35 days at room temperature $25 \pm 1°C$. Scale 1–5 (unacceptable to excellent) ($n = 5$).

that traps water forming a kind of mesh, which gives the immature apple its particular texture. With maturation, this substance decreases generating soluble pectins.

The application of these materials reduced the development of fungi and yeasts in a period of 35 days. Therefore, the overall sensory effectiveness presented the following order: CF1sf, Ctrl, and Visf.

7.3.2.2 DECAY RATE (DR)

All treatments had positive scores before visible damage caused by fungi and yeasts compared with Ctrl. The DR remained constant during the first 17 days of storage, however, after 21, the treatments were stable, but the damage caused by senescence was increased in Ctrl. On days 28 and 35, the fungal damage was severe in Ctrl, indicating that the mucilages helped to delay the damage caused by microorganisms. León de Zapata et al. (2018) estimated the shelf life of the cut apple in frigo conservation in 56 days, however, candelilla wax was used to inhibit the passage of microorganisms.

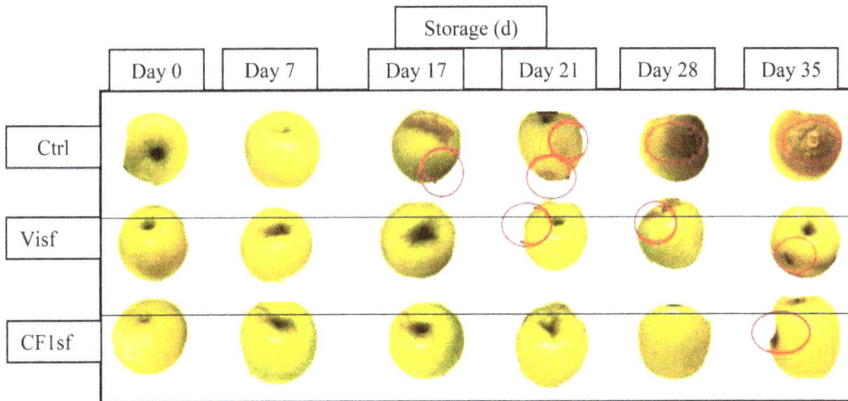

FIGURE 7.3 Effect of mucilage films on decay rate and appearance after 0, 21, and 35 days at $25 \pm 1°C$.

7.3.2.3 WEIGHTLOSS

Apples covered with films showed significant differences ($P < 0.05$) from each other. The fundamental value of this parameter is the mass/volume ratio of apples specifically. Figure 7.4 shows the behavior of moisture stability in fruits. The weight loss of Ctrl was 2.59% for day "0," 4.73% for day "17," and 9.99% for day "35." The weight losses registered for the average Ctrl during apples storaging was 5.88%. On the other hand, apples covered with mucilages had 5.30–8.25%, and these values were similar compared with Ctrl. However, they were statistically equal.

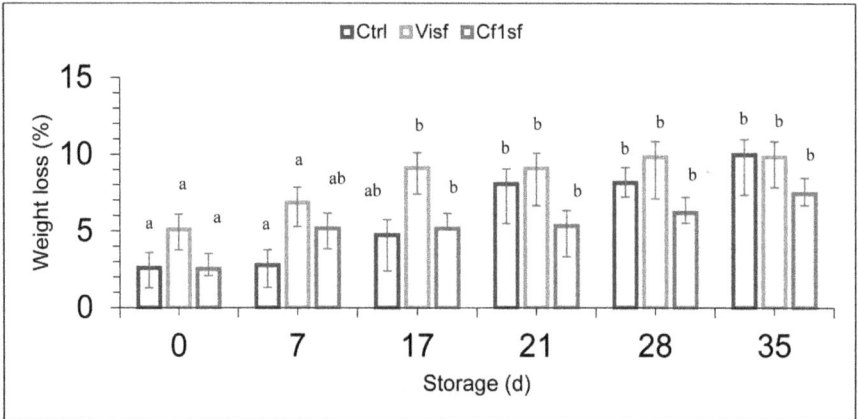

FIGURE 7.4   Effect of mucilage films on weight loss after 0, 21, and 35 days at 25 ± 1°C.

The treatments were more efficient on days 17, 28, and 35, since their values varied minimally over time, achieving stability despite the slight variation on day 21. This indicates that the stable behavior observed could be due to the content of sugars and organic acids present in the mucilage. Also, the rest of the compounds could be a source of substrate for the breathing of the apples during the production of ethylene (Fan et al., 2018). Another peculiar characteristic is that of the mucilages that were able to retain and keep the flow of water on the apples stable.

7.3.2.4   TOTAL SOLUBLE SOLIDS

This parameter indicated the state of maturity and the total soluble solids content stored in apples. There were slight variations between the treatments observed in Figure 7.5. All the initial TSS values ranged from 18°Brix, being equal. These results were similar to those found in previous studies (Nelson and Nicolás, 2011). Significant differences were observed from day 7, 28, and 35 during storage time ($P < 0.05$). The average°Brix for the final period was 18.40° for Ctrl, 13.71° for Visf, and 13.71° for CF1sf. Our values were above the values evaluated by Capistrán-Carabarín et al. (2017), being those of 10.2 and 8.4°Brix. The mucilages act as a protective matrix to the interaction with gases and give little release of O_2 that delays the maturity in the fruits. In addition, the stability of TSS is generated by the hydrolysis of starches and organic acids that are directly transformed into simple sugars as storage time passes.

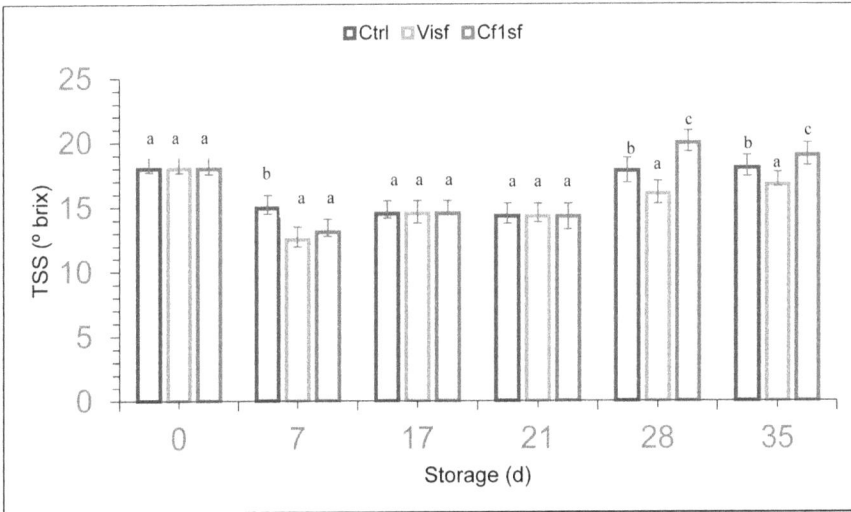

FIGURE 7.5 Effect of mucilage films on TSS after 0, 21, and 35 days at 25 ± 1°C.

7.3.2.5 COLOR AND FIRMNESS LOSS

The color analysis for apples covered with mucilage presented significant differences ($P < 0.05$). Through storage, it was observed in Table 7.2 that the values were maintained regarding the Visf and CF1sf treatments. Therefore, the control was the one that changed the most during the ripening of the apples. The initial value of L* for the initial time in the Ctrl was 72.08, and this is the highest. However, Visf ranged from 70.76 and finally CF1sf with 67.85. By day 35, the variable L* was statistically the same between treatments (68.41, 69.19, and 70.24). For the a* coordinate, there was a decrease in the values as of day 28 (3.26, 4, and 4.87) and a slight increase for day 35 (4.13, 5.31, and 8.97). The luminosity and the coordinate are essential parameters to observe the darkening changes of the fruits. The oxidation in apples appeared slightly from day 28 and 35.

TABLE 7.2 Effect of Mucilage Films on Color (*L*, *a**, *c* and *b*) for 0, 7, 17, 21, 28, 35 Days at 25 ± 1°C.

Storage (d)	Treatment	L	a*	c	b
0	Ctrl	72.08 ± 1.41[b]	6.56 ± 1.90[a]	49.12 ± 1.03[a]	48.12 ± 1.47[a]
	Visf	70.76 ± 2.67[ab]	5.86 ± 2.30[a]	51.90 ± 7.78[ab]	52.08 ± 7.35[a]
	CF1sf	67.85 ± 2.45[a]	9.26 ± 2.56[a]	57.40 ± 3.09[b]	54.33 ± 4.06[a]
7	Ctrl	68.88 ± 2.70[a]	1.70 ± 4.46[a]	48.78 ± 2.10[a]	47.78 ± 3.52[a]
	Visf	68.84 ± 3.97[a]	2.90 ± 2.70[a]	46.52 ± 4.45[a]	46.18 ± 4.51[a]
	CF1sf	66.95 ± 3.07[a]	−068 ± 1.42[a]	43.70 ± 4.54[a]	43.80 ± 4.47[a]
	Ctrl	71.28 ± 1.07[a]	4.82 ± 1.70[a]	52.18 ± 1.75[a]	51.94 ± 1.67[a]
17	Visf	69.54 ± 6.11[a]	4.76 ± 2.18[a]	48.46 ± 8.57[a]	48.16 ± 8.83[a]
	CF1sf	69.88 ± 1.71[a]	6.32 ± 3.00[a]	49.18 ± 4.63[a]	48.66 ± 4.97[a]
21	Ctrl	70.24 ± 1.77[a]	18.28 ± 20.09[a]	55.42 ± 3.00[b]	55.62 ± 3.16[b]
	Visf	69.20 ± 3.65[a]	6.84 ± 3.18[a]	50.64 ± 1.48[b]	51.18 ± 1.64[ab]
	CF1sf	69.12 ± 3.30[a]	8.68 ± 3.81[a]	50.46 ± 2.94[a]	49.58 ± 3.34[a]
28	Ctrl	70.25 ± 3.23[a]	3.26 ± 2.50[a]	50.48 ± 1.45[a]	47.95 ± 2.34[a]
	Visf	68.13 ± 3.27[a]	4.87 ± 2.35[a]	47.49 ± 4.63[a]	49.13 ± 3.17[a]
	CF1sf	65.63 ± 2.79[a]	4.00 ± 1.80[a]	46.44 ± 1.94[a]	49.06 ± 3.07[a]
35	Ctrl	70.24 ± 0.90[a]	4.13 ± 2.29[a]	53.80 ± 1.46[b]	49.86 ± 1.99[a]
	Visf	69.19 ± 2.88[a]	5.31 ± 1.17[ab]	49.69 ± 4.86[ab]	47.17 ± 5.02[a]
	CF1sf	68.41 ± 1.43[a]	8.97 ± 2.92[b]	47.81 ± 2.88[a]	46.23± 2.05[a]

[a,b]Means within a row which do not have a common superscript letter, are significantly different ($P < 0.05$).

Different letters indicate a significant difference ($P \leq 0.05$) between treatments ($n = 3$) between columns.

The chroma or saturation coordinates were maintained in the treatments. Visf started with 51.90 and finished with 49.64. CF1sf started with 57.40 and by day 35, it was 47.81. The Ctrl was higher at day 35 (53.80) compared with the initial one (49.12). The above values describe the positive effect that the mucilage films had on the physical conservation of apples due to the slight transformation of phenols to quinones that the fruits had during storage.

The loss of firmness in the fruits is due to the precursor processes, such as perspiration and respiration of these fruits before a damage is generated in some tissue of the product that is caused by microorganisms, in apples, manly fungi. As shown in Figure 7.6, the values remained significantly equal ($P < 0.05$).

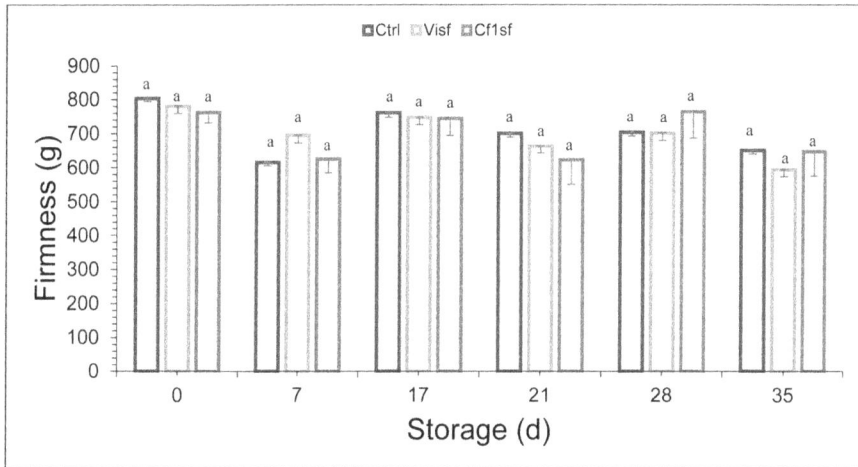

FIGURE 7.6 Effect of mucilage films on firmness after 0, 21, and 35 days at 25 ± 1°C.

Covered apples have values of at least 592.92 g to 873.82 g maximum. Attributes were determined in the Golden apple and the identified firmness values of 8555.41–9146.85 g-force during 10 days of storage, being higher than our values. On the other hand, León de Zapata et al. (2018) detected the firmness values of 3100 and 4800 g-force in apples covered with candelilla wax. The loss of weight and firmness occurs mainly by the process of perspiration after the harvest of the fruit (Gol et al., 2013). However, due to the loss of water during the fruit perspiration process, the total soluble solids content increases (Saucedo-Pompa et al., 2007; Ochoa-Reyes et al., 2013) and the resulting metabolic activity results in the conversion of organic acids into sugars during the ripening process (Togrul and Arslan, 2004). However, the interaction in the storage of apples was the stability to softening, such as changes in firmness and breathing speed.

7.3.2.6 pH, TITRATABLE ACIDITY AND VITAMIN C

The pH results were recorded at 5.01 for Ctrl, 4.98 for the Visf, and 2.95 for the CF1sf at the initial time (Table 7.3). Subsequently, they remained stable during the days 7–35. The trends of the values during the apple storage were observed as slightly acidic, being Visf the largest with 5.33, followed by CF1sf with 4.95, and the minimum value of 3.05 was observed for the Ctrl. Our values were higher than those reported by Azeredo (2004) with a maximum value of 3.8. The differences in the analysis between treatments are due to the chemical nature of the mucilages during their extraction and incorporation in apples that were positively increased. As the pH value of fruits decreases (they are more acidic), microorganisms have more difficult conditions to survive and grow. The above turns out to be beneficial to improve the shelf life of apples.

TABLE 7.3 Effect of Mucilage Films on pH for 0, 7, 17, 21, 28, 35 Days at 25 ± 1°C.

Storage (days)	Treatment	pH
0	Ctrl	2.95 ± 0.18[a]
	Visf	4.98 ± 0.37[b]
	CF1sf	5.01 ± 0.16[b]
7	Ctrl	3.51 ± 0.11[a]
	Visf	5.92 ± 0.14[c]
	CF1sf	4.68 ± 0.22[b]
17	Ctrl	3.23 ± 0.11[a]
	Visf	5.45 ± 0.16[c]
	CF1sf	4.85 ± 0.15[b]
21	Ctrl	3.37 ± 0.10[a]
	Visf	5.69 ± 0.11[c]
	CF1sf	4.76 ± 0.18[b]
28	Ctrl	3.16 ± 0.13[a]
	Visf	5.33 ± 0.21[c]
	CF1sf	4.89 ± 0.15[b]
35	Ctrl	3.05 ± 0.15[b]
	Visf	5.16 ± 0.29[c]
	CF1sf	0.46 ± 0.15[a]

[a,b]Means within a row which do not have a common superscript letter, are significantly different ($P < 0.05$).

Total acidity was evaluated in apples and reported in percentages to determine the presence of citric acid in fruits (Fig. 7.7). The application of films with mucilages in apples significantly affected ($P < 0.05$) the values of this parameter, which were maintained from day 17 between 0.120 and 0.193%, and for the final day, the results of 0.05 were obtained for the Ctrl and 0.11% for Visf treatments and 0.54% for CF1sf. The treatments turned out to be effective on this variable for the transformation of acids, such as citric, malic, and tartaric acids present during the production of ethylene in apples throughout the evaluation time. Mucilages are rich in malic acid, which could have intervened positively on the application in apples. However, previous authors reported higher values (0.99%, 1.14%, and 1.31%) compared with our values.

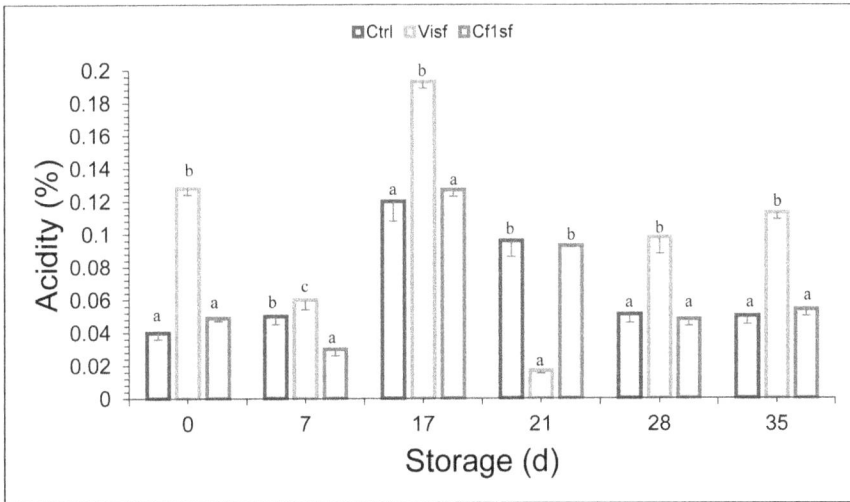

FIGURE 7.7   Effect of mucilage films on acidity after 0, 21, and 35 days at 25 ± 1°C.

The results of the ascorbic acid content of the covered apples and the Ctrl are shown in Figure 7.8. The values ranged from 0.50 to 0.54 mg 10 g^{-1} of apple on the initial day and 0.35, 0.39, and 0.40 mg 10 g^{-1} of apple on the last day. The treatments kept the behavior of ascorbic acid. Ascorbic acid is used to prevent fruit browning. The phenolases present in apples oxidize orthodiphenols and flavonoids present in them, giving dark-colored orthoquinones. Therefore, the incorporation of mucilages helps to maintain a stable concentration of ascorbic acid and others in apples. Figueroa et al. (2016) establish values of 0.07, 0.12, 0.14, 0.18, and 0.23 mg 10 g^{-1} of ascorbic acid in apples treated with ultraviolet light, these values are slightly lower compared with ours. The concentration of ascorbic acid that the mucilage naturally has ranges from 0.007 to 0.0076 mg/g according to report by Guerrero (2014) where it is mentioned that the ascorbic content is related to many components, such as folic acid, choline, and niacin.

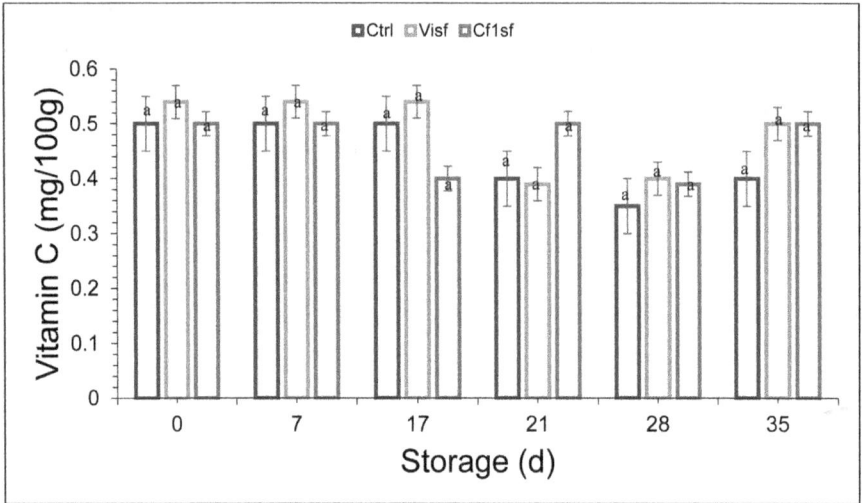

FIGURE 7.8   Effect of mucilage films on vitamin C after 0, 21, and 35 days at $25 \pm 1°C$.

7.4   CONCLUSIONS

Films with hydroponic nopal mucilages keep the physicochemical properties of apples stable at a room temperature of 24–26°C in contrast to Ctrl films. Polymer formulations proved to be efficient through immersion to evaluate the shelf life in Golden delicious products. The application of these materials reduced the development of fungi and yeasts in a period of 35 days. Therefore, the overall sensory effectiveness is presented in the following order: CF1sf, Ctrl, and Visf.

The physical appearance of apples covered with films formulated with mucilage generated greater brilliance compared with those without the polysaccharide layer. Therefore, the fruits had a better attraction to the evaluator.

The change of the DR was more visible in the Ctrl films than those containing mucilages due to the microbial reduction given by the effect of the films over time. This is due to the minimal electrostatic interaction induced by the materials in the fruits.

The application of films with mucilages in apples showed that there was a stability in the conservation of their condition due to the presented adjustment.

Total soluble solids and color coordinates (L^* and a^*) remained similar at the beginning and end of apple storage with the application of films with and without mucilages.

The apples remained acidic with the application of films with mucilages. The films with CF1sf and Ctrl mucilages were not so affected by the acidity parameter.

Vitamin C and firmness remained between treatments during storage (0, 7, 17, 21, and 35 days).

ACKNOWLEDGMENTS

Authors thank to Sectorial fund for research, development and forest technological innovation CONAFOR-CONACYT projects 2018-B-S-65769 and 2018-2-B-S-131466.

KEYWORDS

- **shelf life**
- **apple**
- **villanueva**
- **copena F1 mucilages**

REFERENCES

AOAC. *Official Methods of Analysis*, 14th ed., 1984.

Azeredo, H. M. Fundamentos de Estabilidad de Alimentos. EMBRAPA. Fortaleza, Brasil. 195 p. cactus (*Opuntia elatior* Mill.) Sobre la Calidad de Frutos de Piña Mínimamente Procesados. *Bioagro* **2004,** *29* (2), 129–136.

Capistrán-Carabarín, A.; Aquino-Bolaños, E. N.; Chávez-Servia, J. L.; Velásquez-Melgarejo, V.; Vera-Guzmán, A.; Viveros-Contreras, R.; Guzmán, V. Cambios en Los Parámetros Fisicoquímicos en Tomate de Arbol (*Solanum betaceum*) Durante su Almacenamiento Poscosecha a 5°C. *Avan. Cie. Méx.* **2017,** 1020–1027.

de Azeredo, H. M. C.; Rosa, M. F.; Figueirêdo, M. C. B. Lignocellulosic-Based Nanostructures and Their Use in Food Packaging. In *Micro and Nanotechnologies, Nanomaterials for Food Packaging*; Cerqueira, M. Á. P. R., Lagaron, J. M., Castro, L. M. P., de Oliveira Soares Vicente, A. A. M. (Eds.); Elsevier: Amsterdam, 2018, Chapter 3; pp 47–69. ISBN: 9780323512718.

Fan, F.; Huber, J. D.; Su, Z.; Hu, M.; Gao, Z.; Li, M.; Shi, X.; Zhang, Z. Effect of Postharvest Spray of Apple Polyphenols on the Quality of Fresh-Cut Red Pitaya Fruit During Shelf Life. *Food Chem.* **2018,** *243*, 19–25.

FAO. *Plataforma técnica sobre la medición y la reducción de las pérdidas y el desperdicio de alimentos*. Düsseldorf, Alemania, 2012.

Gheribi, R.; Puchot, L.; Verge, P.; Jaoued-Grayaa, N.; Mohamed, M.; Habibi, Y.; Khwaldia, K. Development of Plasticized Edible Films from *Opuntia ficus-indica* Mucilage: A Comparative Study of Various Polyol Plasticizers. *Carbohydr. Polym.* **2018**, *190*, 204–211.

Gudkovsky A. V.; Kozhina V. L.; Balakirev A. E.; Nazarov Y. B. *Современная система хранения плодов и ягод. Перспективы развития технологий хранения и переработки плодов и ягод в современных экономических условиях*, 2012; pp 75–83.

Juhnevica-Radenkova, K.; Skudra, L.; Skrivele, M.; Radenkovs, V.; Seglina, D. Impact of the Degree of Maturity on Apple Quality During the Shelf Life. 2014.

León de Zapata, A. M.; Ventura-Sobrevilla, M. J.; Salinas-Jasso, A. T.; Flores-Gallegos, C. A.; Rodríguez-Herrera, R.; Pastrana-Castro, L.; Rua-Rodríguez, L. M.; Aguilar, N. C. Changes of the Shelf Life of Candelilla Wax/Tarbush Bioactive Based-Nanocoated Apples at Industrial Level Conditions. *Sci. Hortic.* **2018**, *231*, 43–48.

Ma, Y.; Ban, Q.; Shi, J.; Dong, T.; Jiang, C.-Z.; Wang, Q. 1-Methylcyclopropene (1-MCP), Storage Time, and Shelf Life and Temperature Affect Phenolic Compounds and Antioxidant Activity of 'Jonagold' Apple. *Postharvest Biol. Technol.* **2019**, *150*, 71–79.

Nelson, P.; Nicolás, M.. *Desarrollo de manzana trozada minimamente procesada y determinación de vida útil*. Tesis Licenciatura. Universidad de Chile, 2011.

Noshad, M.; Rahmati, M. Effects of Natural Mucilage as an Edible Coating on Quality Improvement of Freshly Cut Apples. *Nutr. Food Sci. Res.* **2019**, 6 (2), 21–27.

Ochoa-Reyes, E.; Martínez-Vazquez, G.; Saucedo-Pompa, S.; Montañez, J.; Rojas-Molina, R.; León de Zapata, M. A. Improvement of Shelf Life Quality of Green Bell Peppers Using Edible Coating Formulations. *J. Microbiol. Biotechnol. Food Sci.* **2013**, *2*, 2448–2451.

Padmanabhan, C. S.; Cruz-Romero, C. M.; Kerry, P. J.; Morris, A. M. Food Packaging: Surface Engineering and Commercialization. In *Micro and Nanotechnologies, Nanomaterials for Food Packaging*; Cerqueira, M. Â. P. R., Lagaron, J. M., Castro, L. M. P., de Oliveira Soares Vicente, A. A. M. (Eds.); Elsevier: Amsterdam, 2018; Chapter 11, pp 301–328. ISBN: 9780323512718.

Saucedo-Pompa, S.; Jasso-Cantu, D.; Ventura-Sobrevilla, J.; Sáenz-Galindo, A.; Rodríguez-Herrera, R.; Aguilar, C. N. Effect of Candelilla Wax with Natural Antioxidants on the Shelf Life Quality of Fresh-Cut Fruits. *J. Food Qual.* **2007**, *30*, 823–836.

Shin, S.-H.; Chang, Y.; Lacroix, M.; Han, J.; Solís, R. E. C.; Rivera, G. M.; Tamayo, J. C.; Sauri E. D. *Elaboración de un recubrimiento comestible a base de quitina para retrasar el oscurecimiento y deshidratación de manzanas cortadas en rebanadas*. División de Estudios de Posgrado Instituto Tecnológico de Mérida: Yucatán, México, 2008.

Tavarez, J. A. R. *Guía de laboratorio. Determinación de vitamina C*. Área Ciencias Fisiológicas, PUCMM, 2005.

Togrul, H.; Arslan, N. Extending Shelf-Life of Peach and Pear by Using CMC from Sugar Beet Pulp Cellulose as a Hydrophilic Polymer in Emulsions. *Food Hydrocoll.* **2004**, *18*, 215–226.

Treviño-Garza, M. Z.; Árevalo, N.-K.; Galán, W.-L. J.; Alanís, G.-M. G.; Heredia, R.-N. L.; Flores G.-M. S. *Evaluación y comparación de recubrimientos comestibles a base de mucílagos, quitosán y pululano en la calidad y vida de anaquel de la piña fresca cortada*. Tesis Doctoral. Universidad Autónoma de Nuevo León, México, 2016.

Varela, P.; Salvador, A.; Fiszman, S. Shelf-Life Estimation of 'Fuji' Apples: The Behaviour of Recently Harvested Fruit During Storage at Ambient Conditions. *Postharvest Biol. Technol.* **2008**, *50*, 64–69.

Vargas-Rodríguez, L.; Arroyo Figueroa, G.; Herrera Méndez, C. H.; Pérez Nieto, A.; García Vieyra, M. I.; Rodríguez Núñez, J. R. Propiedades Físicas del Mucílago de Nopal. *Acta Universitaria*, **2016,** *26* (NE-1), 8–11.

Velázquez-Moreira, A.; Guerrero Beltrán, J. A. Algunas Investigaciones Recientes en Recubrimientos Comestibles Aplicados en Alimentos. *Temas Selectos de Ingeniería de Alimentos*; Universidad de las Américas Puebla; San Andrés: Cholula, Puebla, México, 2014, 8 (2), pp 5–12.

CHAPTER 8

Vegetable Products from the Huasteca Potosina: A Healthy and Sustainable Option

ABIGAIL REYES-MUNGUÍA[1], MARÍA LUISA CARRILLO- INUNGARAY[1], PEDRO AGUILAR-ZÁRATE[2*], RAFAEL GERMÁN CAMPOS-MONTIEL[3], DIANA JAQUELINE PIMENTEL-GONZÁLEZ[3], JORGE ENRIQUE WONG-PAZ[2], and DIANA BEATRÍZ MUÑIZ-MÁRQUEZ[2]

[1]*Autonomous University of San Luis Potosi, Multidisciplinary Academic Unit, Huasteca Zone, Romualdo del Campo No. 501, Fracc. Rafael Curiel C.P 79060, Ciudad Valles, S. L. P, Mexico*

[2]*Departamento de Ingenierías, Tecnológico Nacional de México, Campus Ciudad Valles, Carretera al Ingenio Plan de Ayala Km. 2, Colonia Vista Hermosa, Ciudad Valles, San Luis Potosí, C.P 79010, Mexico*

[3]*Autonomous University of the State of Hidalgo, Institute of Agricultural Sciences, College Ranch, Av. Universidad Km 1 Ex-Hacienda de Aquetzalpa A.P. 32, C.P 43600 Tulancingo, Hidalgo, Mexico*

Corresponding author. E-mail: pedro.aguilar@tecvalles.mx

ABSTRACT

Objective: To value the nutritional potential of edible plants from the Huasteca Potosina and to promote their consumption and sustainable use. **Methodology:** Food samples were collected in local markets and domestic yards in the Huasteca Potosina, throughout the year 2018. The content of protein, fat, carbohydrates, and vitamin C were determined using official techniques. The samples were analyzed in triplicate. The macronutrient content of some ethnic foods was compared with conventional foods. **Results:** The major components of the edible plants from the Huasteca were carbohydrates, followed by proteins and in lesser quantity by lipids. The foods with the highest protein content were suyo (*Ipomea dumosa*), moringa (*Moringa oleifera*) and chamal seeds (*Dion edule*). **Conclusion:** The nutritional quality of many of the regional plant products is similar to that of the conventional foods. Promoting their consumption could be a strategy to achieve food security in this region, which will affect positively on the maintenance of the population's health.

8.1 INTRODUCTION

According to the Food and Agriculture Organization (FAO, 2009), agricultural production, and in particular crop production, will increase substantially in order to meet the needs of a population projected to grow by 40% in 2050. To ensure that all those suffering from hunger and malnutrition have access to nutritious food, edible plants can be used as food resources. Edible plants could be the basis of subsistence since they supply basic nutrition needs, help to solve problems, such as hunger and poverty, and are therefore the biological basis for achieving food security (FAO, 2012).

The exchange of knowledge and experiences of population about the consumption of edible vegetables has led to the development of ethnobotany as an empirical science. It aims to understand what people know about plants, and how humans use plants to reproduce their social and cultural life. In addition, the plants are used for medicinal purposes, and also food is one of the most important, especially for rural and indigenous populations who consume them as part of their diet (Cilia-López et al., 2015), and of which there are no reports on their nutritional quality.

In the Huasteca Potosina, a wide varieties of plants are grown and are consumed by the indigenous population. Because they grow in the domestic yards, near the houses, or have been planted by the inhabitants for their consumption. Some of these foods are produced permanently or seasonally. However, those who consume them do not know the nutritional composition of those plants.

The Huasteca Potosina is located in the northeast of Mexico, between the Sierra Madre Oriental and the coastal plains of the Gulf of Mexico in the extreme coordinates of 98°20' in the municipality of Ébano (extreme east), 99°32'–22°12' in Tamasopo (extreme west), 99°32'–22°44' in El Naranjo (extreme north), and 98°49'–21°10' in Tamazunchale (extreme south), also including the municipalities of Aquismón, Tanlajás, Tampacán, Ciudad Valles, Huehuetlán, San Antonio, and Tancanhuitz in the state of San Luis

Potosí (Gallardo, 2004). Because of its location, this region has different climates such as warm humid tropical climate with an accumulated annual rainfall of 1800–2000 mm and an average annual temperature of 21°C. The Huasteca Potosina is warm subhumid tropical in the transitional lands between humid and arid tropical regions with rainfall between 600 and 1500 mm, average temperatures above 20°C and a dry period of 5–9 months, which causes the vegetation to be made up of low and medium altitude forests with great biotic diversity (Lárraga, 2015).

The Huasteca has 250,000 indigenous inhabitants of Nahuatl, Tenek, and Pame origin who have a diverse knowledge about the use of plants, such as which are used for food, medicinal, and ceremonial purposes among others. However, very little is known about the products and preparations used in traditional food. Traditional ingredients are not always of local origin. They are a set of contributions of products and preparations incorporated into local diets coming from migratory processes or productive transformations (Consejo Nacional de la Cultura y las Artes, 2014). This region still has rural characteristics. The agricultural and agro-industrial productive sector predominates, and whose warm and humid climate favors the growth of a wide diversity of flora including plants that the inhabitants use as food.

The traditional cuisine in the Huasteca region originates from the culinary traditions of its inhabitants. The regional raw materials are mixed with the Spanish influence during the colonization of Mexico. Although, the culinary tradition has resisted the onslaught of time. The population, strongly tied to its origins, has seen its food consumption decisions affected by the modernization processes that have transformed many of the traditions of consumption and taste with respect to food. These changes have affected the diet of the population. Hence, it is important in valuing the nutritional and biological potential of the edible plants of this region that are not yet supported with scientific studies. Tastes and food preferences are influenced by factors, such as social class, ethnicity, religion, age, education, and health, which shape the aspects of taste at the social level beyond the individual, thus manifesting a cultural identity (Ayuso Peraza and Castillo León, 2017; Nunes, 2007).

Currently, the type of food has been affected by several factors including the context of globalization, greater access to food products and scarcity of time (Meléndez and Cañez, 2009). This has caused the rural and indigenous population to change their varied diet to one that is poor in nutritional quality, and a monotonous lifestyle, which is reflected in high rates of malnutrition (Popkin et al., 2012). A few years ago, the inhabitants of rural communities would go to the *tianguis* (fresh market) to offer their products and bring the products of others. Now, fewer and fewer products are sold and more are bought. Among the products still offered in the *tianguis* are vegetable products. Research on these products is important for the benefit of the community. Worldwide, the trend is the consumption of natural foods with nutrients and compounds with healthy biological properties and minimally processed. These are characteristics of edible plants which are a viable option due to their availability and nutritional balance.

Cilia et al. (2015) identified the local species of edible plants that contribute to the population's diet in an indigenous community of the Huasteca. They found 54 species of edible plants, of which, the most mentioned were zarabando (*Vigna unguiculata*), suyo (*Ipomea dumosa*), banana (*Musa paradisiaca*), and pemoche (*Erythrina americana*). The culinary preparations in the Huasteca have become widespread throughout the population and are now an attraction for tourists. Dishes are dominated by ingredients, such as nixtamalized corn, meat, vegetables, and lard (Table 8.1).

TABLE 8.1 Culinary Preparations that Predominates in the Gastronomic Offer of the Huasteca Potosina.

Food	Ingredients
Bocoles	Corn dough, lard, and different stews
Bolín	Corn dough, chicken meat, cascabel chili, ancho chili, pork lard, and salt
Coffee	Roasted and milled coffee, piloncillo, and cinnamon
Chochas	Chochas, garlic, tomatoes, onion and serrano pepper
Enchiladas Huastecas	Corn dough, milled pepper, cheese, meat, beans, and avocado
Entomatadas	Corn dough, tomatoes, cheese, meat, beans, avocado
Menudo	Cow intestines, cascabel chili, onion, and coriander
Palmito guisado	Palmito, garlic, onion and tomatoes
Pemoches	Pemoches, onion, garlic, tomatoes, sesame, and red chili
Pemoles	Nixtamalized maize and piloncillo
Sopes	Corn dough, beans, different stews, lettuce, tomato, onion, and cheese
Tamales huastecos	Corn dough, chili, pork meat, chicken, and banana leaves
Yuca en dulce	Yuca and piloncillo
Zacahuil	Masa de maíz, carne de puerco o pollo, chile cascabel, manteca de cerdo y sal

When foods are combined without taking into account the appropriate proportion, there is a risk of having an unbalanced diet. So, it is necessary to disseminate the information generated in the scientific field to train the population and show to the new generations the knowledge of the potential of plant resources in the region as an option for the diet.

The nutritional profile of leaves, fruits, and edible roots that are produced in the Huasteca Potosina has not been evaluated yet. The study of the foods in the region constitutes a great work for the scientific community. On the other hand, the food producers of the Huasteca Potosina require information for complementing their activities allowing the sustainable use of the natural resources of the region (Arredondo Gómez et al., 2012).

Therefore, the objective of this work was to assess the nutritional potential of edible plants from the Huasteca Potosina in order to promote their consumption and sustainable use by comparing the nutrient content with conventional foods.

8.2 METHODOLOGY

8.2.1 SAMPLES

During the year 2018, samples of ethnic foods from the Huasteca Potosina were collected in domestic yards, and others were acquired in the local market of Ciudad Valles, San Luis Potosí. The samples were kept at 10°C until analysis.

8.2.2 CHEMICAL ANALYSIS

The collected samples were analyzed according to the methods described in AOAC (2012). Moisture (Method 925.09) in drying oven at 110°C, ash (Method 923.03) by muffle incineration at $550 \pm 1°C$, crude protein (Method 977.14) by the Kjeldalh method, lipids in Soxhlet equipment (Method 920.29). Carbohydrates were calculated by the difference from the other macronutrients and within these, fiber was determined by the enzymatic method (Sigma-Aldrich), and sugars were determined by the Fehling–Soxhlet technique. Each determination was carried out in triplicate. The energy contribution of each plant was calculated as described by Ledesma et al. (2010) and was reported in kJ and kcal as indicated in NOM-051-SSA-2010: 1 g protein, 1 g fat, and 1 g carbohydrate contribute 4, 9, and 4 kcal, respectively, and 4 kcal are equivalent to 17 kJ.

8.2.3 DATA ANALYSIS

Each determination was performed in triplicate and the results were expressed as means ± standard deviation. The ANOVA test and the Tukey test ($P < 0.05$) were carried out using the Statistica 7 software (Statsoft, Tulsa, OK).

8.3 RESULTS AND DISCUSSION

By visiting the *tianguis* of the region, it was possible to identify the times when food is produced, finding that another reason to promote its consumption, is its availability at different times of the year. Although food quality is not only determined by the content of macronutrients, it is common to make these determinations in order to know their nutritional value (Abbasi and Xinbo, 2015). Table 8.2 depicts the results of the chemical composition of plant products growing in the Huasteca.

The main components of the edible plants were carbohydrates, followed by proteins and in smaller quantities by lipids. Among the foods with the highest protein content, moringa and chamal seeds were important. Moringa leaves (*Moringa oleifera* Lam) are rich in polyphenols and their extraction is of great importance, since they can enrich food products with the consequent impact on human health. Cabrera-Carrión et al. (2017) reported that *M. oleifera* contains secondary metabolites that can be safely used as functional ingredients.

The seeds of chamal (*Dioon edule*), come from the family of *Cycadales*. The palm known as chamal grows exclusively in the Huasteca. It is the oldest plant in Mexico and stands out among some of the longest living plants on the planet, as its origin dates back to more than 250 million years and each one could last up to 2500 years (NOM-059-SEMARNAT-2010, 2010). The inhabitants of the Huasteca practice agriculture based on corn, beans, and sugar cane, and also collect the seeds of *D. edule*. The harvest of these seeds plays an important role in their diet and they use it as a substitute for corn. With the dough obtained from boiled and crushed seeds of *D. edule*, they prepare tortillas, gorditas, and tamales. Chamal seeds can be a good substitute for corn during seasons when it is not grown, since it can also undergo a process of nixtamalization for consumption. These results set the tone for continuing the study of this palm species and promoting its propagation.

In recent years, foods of natural origin have aroused great interest because they are a source of various bioactive metabolites that give plants biological properties. In addition to the basic nutritional value, the ethnic foods of the Huasteca Potosina contain compounds with beneficial biological properties to

TABLE 8.2 Chemical Composition of the Plants Used as Main Ingredients in the Preparation of Ethnic Foods in the Huasteca Potosina.

Common name	Moisture	Ashes	Protein	Total fat	Saturated fat	Unsaturated fat	Total Carbohydrates	Fiber	Sugar	Ascorbic acid (mg)	Energy content (kJ)(kcal)
Curcuma (*Curcuma longa*)	78.7 (±1.3)	1.2 (±0.1)	4.1 (±0.7)	0.0 (±0.1)	0.0 (±0.3)	0.0 (±0.1)	9.6 (±1.2)	1.2 (±0.1)	8.4 (±0.8)	5.0 (±0.6)	233 (54)
Coffee (*Coffea arabica*)	5.45 (±0.3)	5.70 (±0.8)	0.0 (±0.2)	0.0 (±0.1)	0.0 (±0.1)	0.0 (±0.1)	0.0 (±0.0)	0.0 (±0.0)	1.0 (±0.1)	0.0 (±0.0)	17 (4)
Sugar cane (*Saccharum officinarum*)	76.0 (±1.8)	0.74 (±0.1)	0.5 (±0.1)	0.0 (±0.1)	0.0 (±0.1)	0.0 (±0.1)	17.2 (±0.5)	1.9 (±0.1)	15.3 (±0.9)	8.0 (±1.0)	110 (26)
Capulin (*Prunus salicifolia*)	77.2 (±1.0)	0.5 (±0.1)	1.5 (±0.3)	0.0 (±0.1)	0.0 (±0.2)	0.0 (±0.1)	16.8 (±1.9)	0.6 (±0.1)	14.2 (±2.1)	13.0 (±0.6)	291 (68)
Star fruit (*Averrhoa carambola*)	90.3 (±1.7)	0.4 (±0.1)	0.0 (±0.1)	0.0 (±0.1)	0.0 (±0.1)	0.0 (±0.1)	9.3 (±0.2)	2.8 (±0.8)	6.5 (±0.7)	34.4 (±1.9)	111 (26)
Cilantrón (*Coriandrum sp.*)	89.5 (±1.9)	0.9 (±0.2)	2.5 (±0.5)	0.0 (±0.1)	0.0 (±0.1)	0.0 (±0.1)	6.2 (±0.3)	4.2 (±0.1)	2.0 (±0.1)	11 (±1.3)	148 (35)
Plum (*Prunus domestica*)	82.5 (±1.6)	0.2 (±0.1)	1.0 (±0.1)	0.6 (±0.2)	0.0 (±0.1)	0.6 (±0.1)	13.0 (±1.5)	1.7 (±0.6)	11.3 (±1.4)	10.0 (±1.1)	233 (55)
Corozo (*Acrocomia aculeata*)	75.59 (±1.2)	1.76 (±0.3)	13.7 (±1.1)	2.6 (±0.2)	2.1 (±0.1)	0.5 (±0.1)	6.8 (±0.2)	6.8 (±0.1)	0.0 (±0.0)	0.0 (±0.0)	448 (105)
Chalauite (*Inga spuria*)	75 (±1.5)	0.3 (±0.1)	1.0 (±0.5)	0.5 (±0.1)	0.0 (±0.0)	0.5 (±0.1)	17.9 (±1.0)	3.0 (±0.5)	4.7 (±0.4)	1.6 (±0.5)	340 (80)
Chaya (*Cnidoscolus aconitifolius*)	76.8 (±1.5)	1.0 (±0.1)	8.2 (±1.4)	1.0 (±0.1)	0.0 (±0.2)	0.8 (±0.2)	6.7 (±0.7)	2.5 (±0.2)	4.2 (±0.3)	235 (±0.0)	147 (35)

TABLE 8.2 *(Continued)*

Common name	Moisture	Ashes	Protein	Total fat	Saturated fat	Unsaturated fat	Total Carbohydrates	Fiber	Sugar	Ascorbic acid (mg)	Energy content (kJ)(kcal)
Squash (*Sechium edule*)	87.1± (1.4)	0.3 (±0.1)	1.0 (±0.1)	0.3 (±0.0)	0.06 (±0.0)	0.2 (±0.1)	6.6 (±0.6)	1.9 (±0.1)	4.7 (±0.4)	12 (±0.1)	141 (33)
Chicozapote (*Manilkara zapote*)	76.0 (±1.1)	0.3 (±0.1)	1.0 (±0.1)	0.0 (±0.0)	0.0 (±0.0)	0.0 (±0.0)	18.0 (±0.6)	1.6 (±0.0)	18.4 (±0.4)	12 (±0.1)	323 (76)
Chile piquing (*Capsicum annuum*)	88.0 (±0.8)	0.5 (±0.0)	1.0 (±0.8)	0.3 (±0.1)	0.0 (±0.0)	0.2 (±0.1)	9.9 (±1.3)	7.9 (±1.1)	2.0 (±0.1)	36 (±1.8)	198 (46)
Chochas (*Yucca filifera*)	90.2 (±1.31)	0.2 (±0.1)	1.0 (±0.1)	0.5 (±0.1)	0.2 (±0.1)	0.3 (±0.1)	8.2 (±0.1)	2.0 (±0.1)	6.2 (±0.1)	1.0 (±0.1)	176 (41)
Chote (*Parmentiera edulis*)	85.0 (±1.7)	1.0 (±0.3)	0.5 (±1.2)	0.0 (±0.1)	0.0 (±0.1)	0.0 (±0.1)	12.6 (±1.6)	2.0 (±1.1)	10.0 (±0.5)	10.0 (±0.1)	222 (52)
Flor de ortiga (*Urtica dioica*)	82.0 (±0.9)	3.6 (±0.8)	2.0 (±0.3)	0.0 (±0.0)	0.0 (±0.1)	0.0 (±0.0)	5.3 (±0.4)	1.6 (±0.5)	3.7 (±1.2)	11.0 (±1.5)	94 (22)
Ayocote beans (*Phaseolus coccineus*)	7.5 (±0.3)	3.6 (±0.1)	19.0 (±1.1)	4.2 (±0.1)	0.4 (±0.1)	3.9 (±0.2)	67.1 (±0.1)	4.3 (±0.1)	62.8 (±0.9)	0.0 (±0.0)	1631 (383)
Guava (*Psidium guajava*)	83 (±1.5)	1.6 (±0.3)	3.0 (±0.2)	1.5 (±0.1)	0.0 (±0.1)	1.5 (±0.2)	12.0 (±1.9)	5.0 (±1.9)	7.8 (±0.1)	236 (±1.9)	312 (74)
Avocado leaves (*Persea americana*)	80.3 (±1.6)	2.0 (±0.9)	1.0 (±0.3)	2.5 (±0.2)	0.5 (±0.2)	2.0 (±0.6)	14.0 (±0.6)	1.5 (±0.3)	12.5 (±0.0)	0.0 (±0.0)	350 (83)
Hoja santa (*Piper auritum*)	80.4 (±1.0)	0.4 (±0.1)	4.2 (±1.0)	1.8 (±0.1)	0.0 (±0.0)	1.8 (±0.2)	8.2 (±0.9)	2.5 (±0.3)	6.0 (±0.5)	49.0 (±1.0)	280 (66)

TABLE 8.2 (Continued)

Common name	Moisture	Ashes	Protein	Total fat	Saturated fat	Unsaturated fat	Total Carbohydrates	Fiber	Sugar	Ascorbic acid (mg)	Energy content (kJ)(kcal)
Litchi (*Litchi sinensis*)	84.0 (±2.0)	0.2 (±0.1)	0.8 (±0.1)	0.0 (±0.0)	0.0 (±0.0)	0.0 (±0.1)	15 (±0.8)	0.4 (±0.1)	14.0 (±0.9)	37 (±0.1)	268 (63)
Sweet lemon (*Citrus limetta*)	89.0 (±0.2)	0.5 (±0.1)	0.9 (±0.1)	0.2 (±0.1)	0.0 (±0.1)	0.2 (±0.1)	9.7 (±0.1)	1.2 (±0.1)	8.5 (±0.1)	56.0 (±0.1)	188 (44)
Lemon (*Citrus aurantifolia*)	86.7 (±1.1)	0.6 (±0.1)	1.0 (±0.1)	1.0 (±0.1)	0.0 (±0.1)	1.0 (±0.1)	12.2 (±0.1)	2.5 (±1.1)	9.7 (±2.1)	77.0 (±1.3)	263 (62)
Mango (*Mangifera indica*)	87.2 (±1.5)	0.3 (±0.1)	0.8 (±0.1)	0.0 (±0.1)	0.0 (±0.1)	0.5 (±0.1)	12.1 (±0.1)	1.56 (±0.1)	10.0 (±0.1)	39 (±1.0)	219 (52)
Passion fruit (*Passiflora edulis*)	86.13 (±0.9)	0.4 (±0.3)	2.4	1.0 (±0.1)	0.0 (±0.1)	1.0 (±0.1)	9.1 (±0.1)	1.5 (±0.1)	7.5 (±0.1)	23.00	234 (55)
Honey (*Apis mellifera*)	17.2 (±1.0)	1.0 (±0.2)	0.0 (±0.0)	0.0 (±0.0)	0.0 (±0.0)	0.5 (±0.1)	82.4 (±1.5)	0.0 (±0.0)	82.4 (±1.9)	0.5 (±0.1)	1401 (330)
Moringa (*Moringa oleifera*)	80.6 (±0.3)	0.2 (±0.6)	9.5 (±1.4)	1.3 (±0.2)	0.5 (±0.1)	1.2 (±0.1)	8.4 (±0.5)	2.0 (±1.1)	6.0.0 (±0.6)	50 (±0.1)	354 (83)
Orange (*Citrus sinensis*)	84.93 (±2.0)	0.55 (±0.1)	0.8 (±0.1)	0.2 (±0.1)	0.0 (±0.1)	0.02 (±0.1)	10.0 (±0.1)	1.0 (±0.1)	8.9 (±0.1)	50.0 (±0.1)	191 (45)
Loquat (*Eriobotrya japónica*)	88.0 (±1.5)	0.2 (±0.1)	0.5 (±0.2)	0.2 (±0.1)	0.0 (±0.0)	0.2 (±0.1)	12.0 (±0.3)	0.5 (±0.1)	11.7 (±1.1)	1.0 (±0.1)	220 (52)
Noni, mulberry (*Morinda citrifolia*)	85.3 (±0.8)	1.0 (±0.3)	0.0 (±0.1)	0.0 (±0.1)	0.0 (±0.1)	0.0 (±0.1)	0.0 (±0.1)	2.0 (±0.1)	0.0 (±0.1)	0.0 (±0.1)	34 (8)

TABLE 8.2   *(Continued)*

Common name	Moisture	Ashes	Protein	Total fat	Saturated fat	Unsaturated fat	Total Carbohydrates	Fiber	Sugar	Ascorbic acid (mg)	Energy content (kJ)(kcal)
Palmito (*Chamaerops humilis*)	85.93 (±1.2)	0.9 (±0.1)	2.7 (±0.5)	0.3 (±0.3)	0.0 (±0.3)	0.2 (±0.3)	19.0 (±0.3)	2.0 (±0.3)	17.0 (±0.3)	1.0 (±0.3)	380 (90)
Papaloquelite (*Porophyllum ruderale*)	93.0 (±1.3)	0.5 (±0.1)	1.9 (±0.5)	0.3 (±0.1)	0.0 (±0.0)	0.2 (±0.1)	3.0 (±0.6)	1.0 (±0.1)	1.9 (±0.2)	19 (±0.3)	95 (22)
Pemoche (*Erythrina collaroides*)	75 (±0.5)	0.4 (±0.1)	2.3 (±0.4)	0.5 (±0.1)	0.2 (±0.1)	0.3 (±0.0)	20.5 (±1.3)	2.5 (±0.3)	17.5 (±0.5)	0.0 (±0.0)	116 (27)
Piloncillo (*Saccharum officinarum*)	7.4 (±0.3)	1.0 (±0.1)	0.0 (±0.0)	0.0 (±0.0)	0.0 (±0.0)	0.0 (±0.0)	90 (±0.0)	0.0 (±0.0)	90 (±0.0)	2.0 (±0.0)	1530 (360)
Pitahaya (*Hylocereus undatus*)	84.4 (±1.0)	0.4 (±0.1)	1.5 (±0.2)	0.6 (±0.1)	0.0 (±0.0)	0.6 (±0.2)	10.4 (±1.1)	2.0 (±0.5)	8.3 (±0.6)	16 (±0.1)	157 (37)
Banana (*Musa paradisiaca*)	71.7 (±1.2)	1.0 (±0.2)	1.2 (±0.1)	0.5 (±0.2)	0.0 (±0.0)	0.5 (±0.1)	23.4 (±1.1)	2.5 (±0.5)	21.0 (±1.0)	90 (±0.8)	437 (103)
Puam (*Muntingia calabura*)	78.9 (±0.4)	0.5 (±0.1)	1.0 (±0.2)	0.0 (±0.0)	0.0 (±0.0)	0.0 (±0.0)	24.5 (±1.1)	2.0 (±0.2)	22.2 (±0.9)	15 (±0.4)	433 (102)
Quelite (*Amaranthus hybrids*)	83.5 (±0.9)	2.0 (±0.5)	4.0 (±0.2)	0.5 (±0.1)	0.0 (±0.0)	0.5 (±0.1)	5.3 (±0.3)	1.3 (±0.1)	3.9 (±0.4)	42.0 (±0.2)	177 (42)
Sarabando bean (*Vigna unguiculata*)	7.1 (±0.2)	3.59 (±0.1)	20.9 (±0.4)	3.3 (±0.2)	0.4 (±0.1)	2.8 (±0.3)	62.0 (±0.9)	4.0 (±0.4)	57 (±0.5)	0.0 (±0.0)	1536 (361)
Chamal seeds (*Dion edule*)	14.7 (±0.2)	0.6 (±0.1)	20.65 (±3.67)	3.21 (±0.22)	0.2 (±0.0)	0.5 (±0.1)	60.69 (±0.21)	1.5 (±0.2)	8.0 (±0.8)	5.0 (±0.3)	239 (56)

TABLE 8.2 (Continued)

Common name	Moisture	Ashes	Protein	Total fat	Saturated fat	Unsaturated fat	Total Carbohydrates	Fiber	Sugar	Ascorbic acid (mg)	Energy content (kJ)(kcal)
Suyo (*Ipomea dumosa*)	84.7 (±1.1)	1.6 (±0.8)	4.2 (±0.5)	1.2 (±0.1)	0.0 (±0.0)	1.0 (±0.3)	8.3 (±0.9)	2.4 (±0.3)	5.8 (±0.3)	0.2 (±0.1)	255 (60)
Tamarindo (*Tamarindus indica*)	29.3 (±1.1)	0.8 (±0.1)	2.0 (±0.3)	0.0 (±0.1)	0.0 (±0.0)	0.0 (±0.0)	65 (±2.6)	6.6 (±1.2)	58.5 (±2.2)	5.0 (±0.8)	1139 (268)
Tomatoes (*Solanum lycopersicum*)	90.6 (±1.6)	2.66 (±0.9)	3.5 (±0.9)	0.0 (±0.1)	0.0 (±0.0)	0.0 (±0.0)	7.6 (±0.8)	1.3 (±0.1)	0.3 (±0.1)	46 (±2.1)	189 (44)
Grapefruit (*Citrus paradisi*)	88.0 (±1.0)	0.3 (±0.1)	0.5 (±0.2)	0.0 (±0.2)	0.0 (±0.1)	0.0 (±0.0)	9.0 (±1.1)	2.5 (±0.2)	6.5 (±0.7)	38.0 (±1.8)	162 (38)
Vanilla (*Vainilla plantifolia*)	98 (±0.6)	0.0 (±0.1)	0.0 (±0.1)	0.0 (±0.1)	0.0 (±0.1)	0.0 (±0.1)	2.0 (±0.3)	0.0 (±0.0)	2.0 (±0.4)	0.0 (±0.1)	34 (8)
Cassava (*Manihot esculenta*)	61.4 (±0.5)	1.0 (±0.3)	1.4 (±0.4)	0.0 (±0.1)	0.0 (±0.3)	0.0 (±0.1)	35.0 (±1.6)	1.2 (±0.1)	33.0 (±1.1)	21.0 (±1.1)	155 (36)

The values are the average of three replicas ± Standard Deviation. *Source:* Own elaboration.

Note: Data are expresed as percentage composition.

maintain human health. The potential of some fruits that grow in the Huasteca Potosina as a natural source of bioactive compounds has been proven. Zhang et al. (2015) identified phenolic compounds, carotenoids, and organosulfur compounds in lychee, tomato, and coffee. *Manilkara zapota* contains alkaloids, sterols, insaturations, sesquiterpenolactones, and saponins in chicozapote. Morales et al. (2012) reported that this fruit contains organic acids and vitamins, which justifies its use to make food products from them. The differences in the concentration of active compounds among edible plants are associated with extrinsic and intrinsic factors that directly influence the chemical composition of the plants, such as the age of the plant, climate, temperature, light, humidity, altitude, and even factors of biological origin (Cuéllar, 2001). All bioactive compounds have a beneficial function in the body. The phytosterols are naturally found in small amounts in many natural foods, such as fruits, vegetables, vegetable oils, nuts, and cereals. Although these compounds have a similar structure to cholesterol, they cannot be absorbed by the human body. However, when ingested in the context of a normal diet, phytosterols interfere with the absorption of cholesterol in the human gut, causing a decrease in cholesterol concentrations in the blood (Muñoz et al., 2011). The semitropical climate of the Huasteca Potosina allows for the growth of exotic fruits, such as pitahaya, star fruit, jackfruit, litchi, and passion fruit. These fruits have compounds with antioxidant properties, such as vitamin C, flavonoids, and fiber (Chavez-Villasana et al., 2014). In addition to their pleasant appearance and taste, the content of bioactive compounds in these fruits makes their consumption even more recommendable.

Regarding the content of macronutrients in ethnic and conventional foods, some of the foods from the Huasteca Potosina have nutrients in greater quantities than the conventional foods (Table 8.3).

TABLE 8.3   Proximal Composition (%) of Some Ethnic and Conventional Foods.

Nutrients	Conventional foods			Nonconventional foods		
	Milk	Maize	Spinach	Moringa	Suyo	Chamal
Protein	3.1 ± 0.9^a	9.4 ± 1.2^b	2.9 ± 0.2^a	9.5 ± 1.4^b	4.2 ± 0.5^a	3.3 ± 0.7^a
Fat	3.8 ± 1.0^b	4.7 ± 0.8^b	0.4 ± 0.1^a	1.3 ± 0.2^a	1.2 ± 0.1^a	0.7 ± 0.3^b
Carbohydrates	4.7 ± 0.8^a	74.3 ± 0.6^c	3.9 ± 0.9^a	8.4 ± 0.5^a	8.3 ± 0.9^b	9.2 ± 0.4^b
Moisture	87.9 ± 1.1^a	0.8 ± 0.3^b	88.1 ± 1.1^a	80.6 ± 0.3^a	84.7 ± 1.1^a	14.7 ± 0.2^b
Minerals (mg)	0.5 ± 0.1^a	1.1 ± 0.2^b	0.9 ± 0.2^b	0.2 ± 0.6^a	1.6 ± 0.8^b	0.6 ± 0.09^a
Energetic content	278 kJ $(65 \text{ kcal})^a$	1603.87 $(377.38)^c$	131 kJ $(31 \text{ kcal})^b$	354 kJ $(83 \text{ kcal})^a$	255 kJ $(60 \text{ kcal})^a$	1505.56 $(354.25)^c$

The values are the average of three replicas ± Standard Deviation. Different letters between columns indicate significant difference ($P < 0.05$).

Source: Own elaboration.

The protein content in moringa seeds is similar to corn. While chamal seeds and suyo leaves have similar protein content to milk and spinach. The consumption of nonconventional plants are food alternatives that can contribute to improving the food security of the population.

Chamal seeds have more protein (20.65 ± 3.67%) than corn (9.42 ± 1.23%) but have less lipids (3.21 ±0.22%) and carbohydrates (60.69 ± 0.21%) than corn and beans. The ash content was lower than corn and beans, whose values are 1.25–1.40%, showing a low mineral content in the sample. However, chamal seeds supply nutrients and energy, similar to those grains.

Cárdenas et al. (2000) report that *Phaseolus vulgaris* cultivars are low in fat, high in carbohydrates and medium in protein. The nutritional value of legumes is due especially to their contribution of proteins (20%–35%). In the case of cereals, their main contributions are carbohydrates (24–68%). The seeds of chamal constitute a good alternative of proteins for the diet like bean and corn, besides having a minor caloric contribution. These values are similar to those of other varieties of palms (*Astrocaryum mexicanum*, *Chamaedorea alternans*, and *Chamaedorea tepejilote*) (Centurión-Hidalgo et al., 2009). In rural communities, consumption of maize and beans is the main source of energy. In addition, chamal seeds undergo a nixtamalization process similar to that of corn, so the nutritional content can be encrusted (Cabrera, 1992). Nixtamalization is a selective process of the proteins. For example, during the cooking of corn, the zein (a protein deficient in lysine and tryptophan) decreases its solubility, while glutelin which has a higher nutritional value increases the solubility, and with it the availability of the essential amino acids. Hence, after nixtamalization, there is an increase of 2.8 times lysine, 8 times tryptophan, and the isoleucine to leucine ratio increases by 1.8 times (Waliszeski et al., 2003). In addition, calcium plays a very important role during nixtamalization. The treatment of seeds with hydrated lime facilitates the removal of the pericarp during cooking and resting, controls microbial activity, improves taste, aroma, color, and improves shelf life and nutritional value (Waliszeski et al., 2003). It is important to mention that in the nixtamalization process for chamal seeds, the shell of the seeds is removed to extract the gametophyte. This process allows the elimination of toxic substances such as the nonprotein amino acids derived from alanin, that is, the β-N-methylamino-L-alanine (L-BMAA), a compound that at high concentrations is neurotoxic to mammals and birds (Brener et al., 2003; Schneider et al., 2002). Considering that corn and beans are two grains used in most of the Mexican diet Rosado, (2001), and when the comparison in terms of protein content of *D. edule* seeds was made with these two legumes, chamal seeds can be considered as a nonconventional source for the production of foods similar to those made from corn. Comparing the nutrient content of some edible plants and foods with some conventional foods is an option for consuming nutritious low-cost foods.

8.4   CONCLUSIONS

The nutritional information that has been generated from the scientific research has allowed to recognize the potential of edible plants consumed in the Huasteca Potosina. Due to the quality of these foods, it is of the utmost importance to disseminate the knowledge to the population. The plant species consumed in the Huasteca Potosina have the potential to become resources for food and contribute to self-consumption agriculture, not only for ethnic groups, but also for the general population.

KEYWORDS

- **ethnic foods**
- **proximal composition**
- **energy content**
- ***Ipomea dumosa***
- **macronutrients**

REFERENCES

Abbasi, A. M.; Xinbo Guo, X. Proximate Composition, Phenolic Contents and in Vitro Antioxidant Properties of *Pimpinella stewartii* (a Wild Medicinal Food). *J. Food Nutr. Res.* **2015,** *3* (5), 330–336.

Association of Official Analytical Chemist (AOAC). *Official Methods of Analysis*, 16th ed.; Association of Official Analytical Chemist: Washington, D.C, 2012; vol. 1.

Arredondo Gómez, A.; Ávila Ayala, R.; Muñoz Gutiérrez, L. *Diagnóstico del Viverismo en la Huasteca Potosina*; Instituto Nacional de Investigación Forestal, Agrícolas y Pecuarias (INIFAP): México, 2012.

Ayuso Peraza, G.; Castillo León, M. T. Globalización y Nostalgia. *Camb. Aliment. Fam. Yucat.* **2017,** *27* (50), 1–16.

Brener, E. D.; Stevenson, D. W.; Twigg, R. W. Cycads: Evolutionary Innovations and the Role of Plant-Derived Neurotoxins. *Trends Plant Sci.* **2003,** *8* (9), 446–452.

Cabrera, L. *Diccionario de Aztequismos*, 5th ed.; Colofón: México, 2001.

Cabrera-Carrión, J. L.; Jaramillo-Jaramillo, C.; Dután-Torres, F.; Cun-Carrión, J.; García, P. A.; Rojas de Astudillo, L. Variación del Contenido de Alcaloides, Fenoles, Flavonoides y Taninos en *Moringa oleifera* Lam. en Función de su Edad y Altura. *Bioagro* **2017,** *29* (1), 53–60.

Chávez-Villasana, A.; Ledesma-Solano, J.A.; Mendoza-Martínez, E.; Calvo-Carrillo, C.; Castro-González, M. I.; Ávila-Curiel, A.; Sánchez-Castillo, C. P.; Pérez-Gil Romo, F. *Tablas de Uso Práctico de los Alimentos de Mayor Consumo "Miriam Muñoz"*, 3rd ed. McGraw Hill: México, 2014.

Centurión-Hidalgo, D.; Alor-Chávez, M. J.; Espinosa-Moreno, J.; Gómez-García, E.; Solano, M. L.; Poot-Matu, J. E. Contenido Nutricional de Inflorescencias de Palmas en la Sierra del Estado de Tabasco. *Univ. Cien.* **2009,** *25* (3), 193–199.

Cilia López, V.G.; Aradillas, C.; Díaz-Barriga, F. Las Plantas Comestibles de una Comunidad Indígena de la Huasteca Potosina, San Luis Potosí. *Entreciencias* **2015,** *3* (7), 143–152.

Consejo Nacional de la Cultura y las Artes. *Arca del gusto: catálogo alimentario patrimonial.* https://www.cultura.gob.cl/wp-content/uploads/2014/12/arca-del-gusto.pdf (Accesado Feb 24, 2020).

Cuéllar, A.; Miranda, M. *Farmacognosia y Productos Naturales*; La Habana: Félix Varela, 2001.

FAO. *Organización de la Naciones Unidas para la Agricultura y la Alimentación: Los recursos fitogenéticos para la alimentación y la agricultura en el mundo*; FAO, 2009 http://www.fao.org/agriculture/seed/sow2/en/ (Accesado Feb. 24, 2020).

FAO. *Organización de la Naciones Unidas para la Agricultura y la Alimentación. Segundo plan de acción mundial para los recursos fitogenéticos para la alimentación y la agricultura*; FAO, 2012. http://www.fao.org/3/i2624s/i2624s00.pdf (Accesado Feb. 24, 2020).

Gallardo, A. P. *Huastecos de San Luis Potosí. Pueblos indígenas del México contemporáneo.* Comisión Nacional para el Desarrollo de los Pueblos Indígenas y Programa de las Naciones Unidas para el Desarrollo, 2004.

Lárraga-Lara, R. *La vivienda tradicional en la Huasteca Potosina en componentes de sostenibilidad de la vivienda tradicional en el ámbito rural de la región huasteca de San Luis Potosí: Hacia una arquitectura rural sostenible*. Universidad Autónoma de San Luis Potosí. México, 2015, pp 14–36.

Ledesma Solano, J. A.; Chávez Villasana, A.; Pérez-Gil Romo, F.; Mendoza Martínez, E.; Calvo Carrillo, C. *Composición de alimentos. Miriam Muñoz de Chávez. Valor nutritivo de los alimentos de mayor consumo*, 2nd ed.; McGraw Hill: México, 2010; p 15.

Meléndez, J.; Cañez, G. La cocina tradicional regional como un elemento de identidad y desarrollo local. El caso de San Pedro El Saucito Sonora, México. *Estud. Soc.* **2009,** *17*, 181–204.

Morales, P.; Ramírez-Moreno, E.; Sánchez-Mata, M.; Carvalho A. (*Opuntia joconostle* F.A.C. Weber ex Diguet and *Opuntia matudae* Scheinvar) of High Consumption in México. *Food Res. Int.* **2012,** *46*, 279–285.

Moura, R. M.; De Lira, R. V.; Farías, I.; Menezes, M.; Santana, A. A. D. Fungal Rots of the Spineless Cactus in the State of Pernambuco. *Fitopatol. Bras.* **1998,** *23*, 2.

Muñoz, A.; Alvarado, C.; Encina, C. Fitoesteroles y Fitoestanoles: Propiedades Saludables. *Rev. Horiz. Méd.* **2011,** *11*, 2.

Norma Oficial Mexicana (NOM-059-SEMARNAT-2010). *Protección ambiental—Especies nativas de México de flora y fauna silvestres—Categorías de riesgo y especificaciones para su inclusión, exclusión o cambio—Lista de especies en riesgo*, 2010. https://dof.gob.mx/nota_detalle_popup.php?codigo=5173091 (Accesado Feb. 25).

Nunes, C. Somos Los Que Comemos. Identidad Cultural, Hábitos Alimenticios y Turismo. *Estud. Perspect. Tur.* **2007,** *16*, 234–242.

Popkin, B.; Adair, L. S.; Wen, S. How and Then: The Global Nutrition Transition: The Pandemic of Obesity in Developing Countries. *Nutr. Rev.* **2012,** *70* (1), 3–21.

Rosado, L. J. Dietary Fiber in Mexico: Recommendations and Actual Consumption Patterns. In *Handbook of Dietary Fiber*; Cho, S. S.; Dreher, L. M., Eds.; Marcel Dekker Inc: New York, 2001; p 68.

Schneider, D.; Wink, M.; Sporer, F.; Lounibos, P. Cycads: Their Evolution, Toxins, Herbivores and Insect Pollinators. *Naturwissenschaften* **2002,** *89* (7), 281–294.

Villareal-Ibarra, E. C. *Estudio Etnofarmacológico de Especies Vegetales con Actividad Hipoglicemienate en la Comunidad de Malpasito en Huimanguillo, Tabasco*. Tesis de Doctorado. Instituto de Enseñanza e Investigación en Ciencias Agrícolas. Tabasco, México. 2014.

Waliszeski, K. N.; Estrada, Y.; Pardio, V. Recovery of Lysine and Tryptophan from Fortified Nixtamalized Corn Flour and Tortillas. *Int. J. Food Sci. Technol.* **2003,** *38* (1), 73–75.

Zhang, Y.-J.; Gan, R.-Y.; Li, S.; Zhou, Y.; Li, A.-N.; Xu, D.-P.; Li, H.-B. Antioxidant Phytochemicals for the Prevention and Treatment of Chronic Diseases. *Molecules* **2015,** *20*, 21138–21156.

CHAPTER 9

Strategies for the Recovery of Phytochemicals from Turmeric (*Curcuma longa* L.) Cultivated in the Huasteca Potosina

FERNANDA ANDRADE-DAMIÁN, MARIELA R. MICHEL,
NAOMI GABRIELA ÁLVAREZ-DÍAZ, DIANA BEATRÍZ MUÑIZ-MÁRQUEZ,
JORGE ENRIQUE WONG-PAZ, FABIOLA VEANA-HERNÁNDEZ, and
PEDRO AGUILAR-ZÁRATE*

*Departamento de Ingenierías. Tecnológico Nacional de México,
Campus Ciudad Valles, Carretera al Ingenio Plan de Ayala Km. 2,
Colonia Vista Hermosa, Ciudad Valles, San Luis Potosí*

Corresponding author. E-mail: pedro.aguilar@tecvalles.mx

ABSTRACT

Curcuma longa L. also known in some places as Indian saffron or curcuma is a rhizome from southeast India. The rhizome grows on places with hot and wet weather at 30°C. Due to the phytochemical composition, the rhizome is used for its medicinal value in pharmaceutical industry, and by its pigments, it is used in cosmetic and food industries. The Huasteca Potosina is a region that possesses the favorable climatic conditions for the cultivation of curcuma. Because of this reason, researchers are in search of alternatives to valorize the rhizome and to promote the cultivation of the plant. Curcumin is one of the compounds present in high amounts in the rhizome. The rhizome also contains other curcuminoids. The industries that use these compounds from this rhizome must use huge amounts of organic solvents for the extraction. The yield and purity obtained by the conventional methods are low and the costs for the recovery are elevated. The present manuscript reviews the main compounds found in curcuma, the extraction and concentration methods, as well as the main application of phytochemicals.

9.1 INTRODUCTION

The turmeric (*Curcuma longa L*) is a plant belonging to the family *Zingiberaceae*. It is originated from Southeast Asia. It is known worldwide as an aromatic spice, used in Asian cuisine to add a touch of color and spice to dishes (Akram et al., 2010). The phytochemical compounds present in its characteristic orange rhizome include curcumin and curcuminoids. These compounds provide the turmeric plant important medicinal properties. This plant is cultivated on a small scale in the Huasteca Potosina, where the appropriate environmental conditions for the development of the crop are met. However, despite the knowledge of the wide applications of the plant's pigments, the crop has not been exploited. One of the main challenges faced by industries applying turmeric pigments is the recovery of these pigments, as due to the low polarity of curcumin and curcuminoids, the use of organic solvents is necessary for their recovery.

Curcumin or diferuloylmethane with the chemical formula (1,7-bis(4-hydroxy-3-methoxyphenyl)-1,6-heptadiene-3,5-dione) (Fig. 9.1) and other curcuminoids constitute the main phytochemicals in *Curcuma longa L.* (Zorofchian Moghadamtousi et al., 2014).

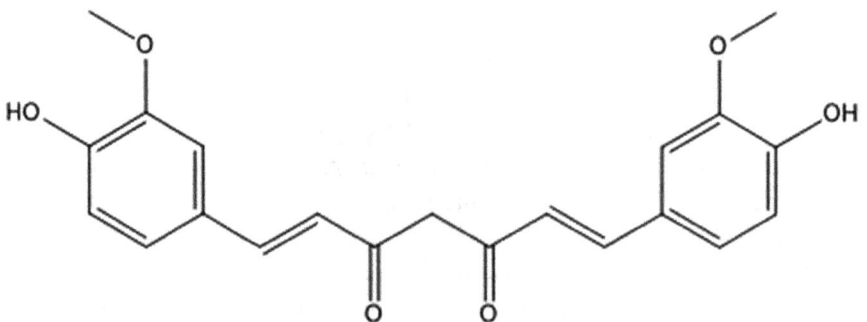

FIGURE 9.1 Chemical structure of curcumin.

The rhizome of turmeric has been widely used for centuries in traditional medicine for the treatment of inflammations and other diseases. Its medicinal properties have been attributed primarily to the constituent curcuminoid present in the turmeric rhizomes (Maheshwari et al., 2006).

The potential health benefits provided by the Indian saffron rhizomes have led to interest in their use for incorporation into food products, as they are considered nutraceutical agents. However, factors, such as the high melting point, low water solubility, and low oral bioavailability of curcumin make its incorporation into many functional foods difficult. In addition, curcumin is highly susceptible to chemical degradation when exposed to aqueous environments, particularly at pH values close to neutral, which also limits its bioavailability (Zou et al., 2015).

Several methods have been reported for the isolation of curcumin and other curcuminoids from *C. longa* rhizomes. Mainly, solvent extraction (Janaki and Bose, 1967; Sastry, 1970), hot and cold percolation (Krishnamurthy et al., 1976), use of alkaline solutions (Stransky, 1979), and salt insolubilization (Tønnesen et al., 1989) have been reported. Recovery of curcuminoids has also been demonstrated using microwave-assisted, ultrasound-assisted, and supercritical carbon dioxide-assisted extractions (Wakte et al., 2011). The process of extracting turmeric pigments using solid-state fermentation (SSF) is an alternative tool for releasing curcumin and curcuminoids from rhizomes (Andrade-Damián et al., 2019).

In this chapter, the different strategies that can be applied to the rhizome of *Curcuma longa L.* in order to obtain curcumin and curcuminoids from *C. longa* with important industrial applications are revised. In this sense, it is intended to promote and valorize the cultivation of turmeric in the Huasteca Potosina.

9.1.1 OVERVIEW OF TURMERIC

Turmeric is a plant with a long history in Asian cultures. It is believed to have originated in western India, and its first use dates back to 2500 years. Its name is derived from the ancient Arabic Kurkum plant, better known as saffron, in fact, turmeric is known as Indian saffron (Saiz de Cos and Perez-Urria, 2014). Therefore, in the Huasteca Potosina, it is commonly known as saffron. A distinction must be made between *Crocus sativus* saffron and *Curcuma longa* saffron or turmeric.

It was originally used to dye fabrics and then introduced as a condiment. Its habitat is in South Asia and the Pacific Islands (India, Indonesia including Hawaii), and in the Americas, it is traditionally cultivated in southern Mexico and the Andes. It is adapted to arid areas. It can be found from Polynesia and Micronesia to Southeast Asia, mainly in India (especially in Andhra Pradesh and Tamil Nadu with 600,000 tons annually), Indonesia, Vietnam, and the

Philippines. In Latin America, *C. longa* is cultivated extensively. In Mexico, it was introduced for ornamental purposes and is known as camotillo, raicilla, or azafrán (Martínez, 1970).

It is a tropical plant that belongs to the Zingiberaceae family. This crop in particular requires special climatic conditions. The rhizome of the plant is orange (Fig. 9.2) and is the mostly used part in the food industry, in medicine, and natural cosmetics (Neerja Pant et al., 2013). *C. longa* adapts excellently to regions with high relative humidity and temperature, as well as in regions with constant rainfall throughout the year, such as tropical and subtropical regions (Tahani et al., 2016). The soil should be slightly acidic (pH 6) and should have good drainage to avoid waterlogging and damage to the crop. The plant has the ability to grow at sea level and up to 1500 m of altitude, requires sun exposure to produce more rhizomes.

FIGURE 9.2   Rhizome of the *Curcuma longa* plant cultivated in the Huasteca Potosina.
Source: Own source.

The Huasteca Potosina is one of the regions where it is possible to cultivate turmeric since it meets the climate, soil, humidity, and rainfall characteristics for the development of the crop in small and large quantities. Generally, in this region, the cultivation of turmeric is carried out in the backyard of the houses. The population of this region uses it mainly as a condiment and coloring for food, as well as natural medicine for the treatment of inflammations and wound healing. There are no data on turmeric cultivation in the Huasteca Potosina region, however, the rhizomes can be found for sale in the local market and in supermarkets (Fig. 9.3), which indicates the widespread use of this product by the population.

FIGURE 9.3   Commercialization of the turmeric rhizome in the municipal market of Ciudad Valles.

Source: Own source.

9.1.2 BIOACTIVE COMPOUNDS FROM TURMERIC

Natural bioactive compounds include a wide diversity of structures and functionalities that provide an excellent set of molecules for the production of nutraceutical products, functional foods, food additives, among others of high economic value (Gil-Chávez et al., 2015).

Natural colors and their formulations find wide applications from food to pharmaceutical products and from dyes to cosmetic products. Important natural food colors are the yellow, orange, and red pigments that are widely distributed in plants and animals. The increasing demand (consumption) for natural food colors has led not only to their recovery from natural sources but also to their synthesis. The value of global consumption of natural food colors is estimated to be more than 1 billion US dollars (Lashkari, 1999).

Commercial preparations obtained from the extracts of turmeric rhizomes contain three constituents, with curcumin being the majority (75%), as well as the isomers demethoxycurcumin, and bis-demethoxycurcumin (Jayaprakasha et al., 2002). Therefore, they are the phytochemicals sought after in the extraction processes. Since they are very expensive and have a low extractable content, their recovery needs to be very efficient and fast.

The commercial extraction of various active biological compounds (extracts and essences) of natural origin from plants (leaves, flowers, fruits, seeds, roots, and tubers) is done by means of organic solvents. The extracts and essences are mainly used as colorants (anthocyanins, betacyanins, anthocyanins, carotenoids, flavonoids, chlorophyll, xanthophylls, carmine acid and kermisic acid), flavors, aromas and additives, among others. Extraction by organic solvents hase the disadvantages that the solvents are not completely eliminated (resulting in contamination of the final product) and the high operating temperature during the process causes the denaturation of the extracted compounds (Elizalde, 2008; Martínez, 2008).

9.1.3   PHYTOCHEMICAL EXTRACTION METHODS

9.1.3.1   SOXHLET EXTRACTION

Several authors have conducted studies on curcuminoid extraction with organic solvents using Soxhlet equipment, and have found that despite being a method that requires long extraction times and large amounts of solvent, it is an efficient, simple method, commonly used as a reference against which the extraction performance of other methods is compared (Wakte et al., 2011; Mandal et al., 2008). Braga et al. (2003) compared the curcumin extractions using various techniques, such as supercritical fluids, low-pressure solvent extraction, Soxhlet, and hydrodistillation, and found that the highest yields were obtained by the Soxhlet method.

A large amount of organic and inorganic solvents has been reported in the literature for Soxhlet type extractions. However, many of these solvents cannot be used in obtaining curcuminoids from *Curcuma longa*, especially if these are to be used as a food-coloring additive. The most commonly used solvents in this type of extraction are acetone, ethanol, and hexane (Wakte et al., 2011). The performance and efficiency of the extraction depend directly on the residence time of the process and the solute/solvent ratio, which varies considerably according to the type of solvent.

The extraction of curcuminoids and oleoresin by the Soxhlet method is performed in different steps. Initially, the sample is placed in a porous cartridge or thimble, which consists of a cylindrical container that is supported at the base of the extraction equipment, the condensed solvent vapors fall and gradually fill the volume where the thimble is, leaching the sample, until it exceeds the overflow level, then the leachate is conducted by a siphon that discharges it into the extraction ball. This operation is repeated until the extraction of curcuminoids is completed (Luque de Castro and García-Ayuso, 1998).

Despite the benefits of Soxhlet extraction, it is important, especially at the industrial level, to develop, adapt, and implement other extraction technologies that have technical, operational, and environmental advantages.

9.1.3.2   MICROWAVE EXTRACTION

This extraction technology is based on microwave irradiation of the medium to be extracted. Microwaves are electromagnetic waves composed of a magnetic field and an electric field that oscillate perpendicularly to each other with frequencies between 0.3 and 300 GHz. Microwaves have the ability to penetrate certain materials and interact with polar components to vibrate their molecules generating heat, which is selective and directed at certain materials that can be heated based on their dielectric constants (Chan et al., 2011).

The heating given by the energy of the microwaves acts directly on the molecules as a result of collisions between particles and is produced by two mechanisms, such as ionic conduction and dipole rotation. Ionic conduction is the electrophoretic migration of charge-carrying components, ions and electrons, under the influence of the electric field produced by microwaves and dipole rotation is the movement of polar molecules trying to align with the electric field generating collisions (Wang et al., 2016).

During the microwave-assisted extraction of natural components, the energy produced by the microwaves is absorbed by some substances in the plant material, mainly water and polar substances. Therefore, the internal temperature of the plant cells increases dramatically, generating an overheating that evaporates the liquids present and breaks the cell walls and/or plasma membranes, releasing the components to be extracted, which favors the transfer of mass to the solvents, thus allowing the efficient extraction (Zhang et al., 2011).

Microwave-assisted extraction technology has been applied for the extraction of curcumin. Several studies have been reported in which the optimal conditions for extraction and the effects of different process parameters on the extraction rate are sought (Li et al., 2014; Mandal et al., 2008; Wakte et al., 2011). The microwave energy has been used for drying the turmeric powder in order to improve its quality by suppressing the enzymatic browning via inhibition of polyphenol oxidase (Hirun et al., 2014). Also, the starch modification in turmeric by curing the rhizome by microwave was explored. Higher retention of curcumin and lower hardness in turmeric rhizomes using microwave-assisted curing (Hmar et al., 2017) was obtained. However, the microwave irradiation of turmeric dry powder resulted in low curcumin recovery (40%), while irradiating the water-soaked turmeric powder, the recovery yield of curcumin increased to reach 90.47% (Wakte et al., 2011). The depletion of the use of organic solvents in the extraction of curcumin and curcuminoids has made researchers to explore the use of microwave-assisted extraction combined with other methods. For example, Liang et al. (2017) used ionic liquid-based microwave-assisted extraction yielding 1.77% of curcuminoids recovery. The use of microwave-assisted extraction is a highly efficient and pollution-free method that generally requires short periods of extraction and less energy consumption.

9.1.3.3 SOLID-STATE FERMENTATION-ASSISTED EXTRACTION

The extraction of bioactive compounds by fermentation is also an interesting alternative that can be explored. This technology makes it possible to obtain extracts with high biological activity and eliminates the toxicity associated with organic solvents. The application of SSF represents a real alternative for the recovery of phytochemicals (Martins et al., 2011). SSF is defined as the process of fermentation in the absence or near absence of free water. However, substrates can possess sufficient moisture to promote the growth and metabolism of microorganisms (Aguilar et al., 2008). The solid substrates used in FMS are mainly natural products obtained from agriculture or residues from the agro-food industry (Mussatto et al., 2011). SSF is an excellent alternative for the use of residues from agriculture and agro-industry, because they can be used to extract and/or produce compounds, such as enzymes, pigments, organic acids, and flavors.

There is little information related to the application of SSF as a tool for the extraction of curcumin or curcuminoids from turmeric. This could be due to the high antimicrobial activity of turmeric phytochemicals. However, there are reports evidencing the use of *Aspergilli* and *Trichoderma* strains capable of growing and releasing curcumin from turmeric rhizomes. The strains *Aspergillus niger* GH1 and PSH were grown on pulverized turmeric rhizome and were capable of releasing 58.42 and 58.21 mg of curcumin per gram of dry material, respectively (Andrade-Damián et al., 2019). Mohamed et al. (2016) evaluated the influence of SSF on the solubility of phenolic compounds from commercial turmeric rhizomes using six species of *Trichoderma* spp. They found that SSF enhanced and improved the total phenolic content, antioxidant activity, and antibacterial activity of the phytochemicals extracted from turmeric.

9.1.3.4   ULTRASOUND-ASSISTED EXTRACTION

This technology is widely used because of the benefits it presents for food and in industries. It is friendly to the environment, and some call it green technology. One of the advantages of this type of extraction is that it reduces the use of solvents and improves the quality of the product. It does not involve expensive process. Therefore, it is favorable to use in small or large scales and the results are in a matter of minutes compared with other types of extractions (Chemat et al., 2011). It consists of a water bath that uses electrical energy to emit high or low frequency waves that are measured by two groups, such as low and high intensity. The differences between these groups include (1) the application of low intensity waves is to obtain information from the medium without emitting any modification, (2) the application of high intensity waves on the other hand is to produce a permanent change in the environment (Robles-Ozuna and Ochoa-Martinez, 2012).

The ultrasound has the function to produce gaseous cavities, they are microbubbles in the liquid medium that are emitted by the pressure of high frequency, when hitting, these microbubbles expand their waves in the medium, these waves inactivate the bacteria and this effect is called cavitation, and depending on the used frequency, the changes generated in the medium include physical, chemical, and biochemical (Robles-Ozuna and Ochoa-Martinez, 2012; Soleno Wilches, 2015).

Therefore, ultrasound-assisted extraction is considered by Chemat et al. (2011) as an emerging technology that is capable of accelerating heat transfer. Thanks to the fact that ultrasound waves are capable of modifying the environment in both physical and chemical state, the cavitation effect that it has facilitates the release of the compounds to be extracted.

The use of ultrasound-assisted extraction is cost-effective, but yields lower efficiency in the extraction of curcumin compared with traditional extraction methods (Soxhlet) (Sahne et al., 2016). For that reason, efforts have been focused in modifying the ultrasound operating parameters. Operating parameters, such as type of solvent, extraction time, extraction temperature, solid to solvent ratio, particle size, and ultrasonic power have been investigated in detail for the approach of ultrasound-assisted extraction (Shirsath et al., 2017). Xu et al. (2017) reported the use of ultrasonic extraction combined with ammonium sulfate/ethanol aqueous two-phase system for the extraction of curcumin. Optimal extraction conditions (material–solvent ratio 3.29:100, ultrasound intensity 33.63 W/cm^2, time 17 min) yielded in the recovery of 46.91 mg/g. Authors mentioned that this method maintained the extraction yields in scale-up efforts. The simplest technology of extraction by ultrasound is the ultrasonic bath. It has been applied in the extraction of pigments from *Curcuma longa* using a smaller amount of solvents, the components to extract from the rhizome are in a more natural state compared with other types of extractions. The process was combined with column chromatography in order to enhance the recovery and to purify the components (Torres-Rodríguez et al., 2014).

9.1.3.5   CHROMATOGRAPHICAL STRATEGIES

The use of chromatography in the extraction of *Curcuma longa* rhizome is used to separate phytochemical compounds. Thanks to the separation, the compounds have been used for different types of investigations.

Different curcuminoids were purified from the rhizome of the plant, first, using thin layer chromatography to verify the presence of these curcuminoids and later followed by column chromatography for their separation (García Ariza et al., 2017). Semi-preparative HPLC is a method that allows the recovery of high-purity compounds. This method is commonly used after an efficient extraction process. Xu et al. (2017) carried out the purification of curcumin after optimizing the extraction process. This method allowed us to obtain curcumin with 85.58% of purity. The complete methods of extraction, isolation, identification, and purification of curcuminoids could be carried out by column chromatography, followed by purity analysis using high-performance liquid chromatography (HPLC). The latter has proven to be a sensitive, precise, and an accurate method in the detection and quantification of curcuminoids in *Curcuma longa* rhizome extract (Revathy et al., 2011; Paramapojn and Gritsanapan, 2009). Schiborr et al. (2010) determined by means of high-performance liquid chromatography that it is possible to quantify curcumin in rat brain tissue, and phenolic compounds in plasma. On the other hand, it was determined that the same method is useful for the quantification of heptanoids and other phytochemical compounds from *Curcuma longa* (Gupta et al., 1999). Jiang et al. (2006) stated that liquid chromatography–electrospray ionization–tandem mass spectrometry (LC–ESI–MS/MS) is a fast and efficient method in the identification of diarylheptanoids present in fresh *Curcuma longa* rhizome extract.

9.1.4 APPLICATION OF CURCUMA LONGA L PIGMENTS

Curcuma longa has a quite wide field of application and has been used for different purposes since ancient times. This rhizome is mainly used in the pharmaceutical industry for the great medicinal value found in different researches thanks to the curcuminoids it possesses, used in the cosmetologically industry in application as dye, and in the food industry as natural coloring and spice for food (Tahani et al., 2016).

9.1.5 APPLICATION IN THE FOOD INDUSTRY

The acceptance of a food is often given by the influence of the color that is provided by the ingredients of the food. Nowadays, the preference for natural pigments has increased due to the strong association with the positive health effects they can provide and the fact that synthetic pigments are associated with toxic effects. Pigments obtained from plant extracts have been used as food colorants. Among these pigments, those obtained from turmeric (*Curcuma longa*) stand out. Curcuminoids are compounds present in turmeric with orange to yellow coloration (Fig. 9.4), which are synthesized in the rhizomes of the plant (Bello-Pérez and Jiménez-Aparicio, 2000; Jayaprakasha et al., 2002).

The turmeric is used in the food industry thanks to the natural coloring that it possesses. The coloring that characterizes the turmeric is due to the curcumin, a phenolic compound that is used as a colorant in the industry (called E-100). This is a natural colorant used to dye and to aromatize different foods like butter, cheeses, mustard, and popcorn. It is the main component of curry (Saiz de Cos and Pérez-Urria, 2014) in food items. Another application in the industry is that it is used as a pH indicator in a range of 8–9 for the detection of boron.

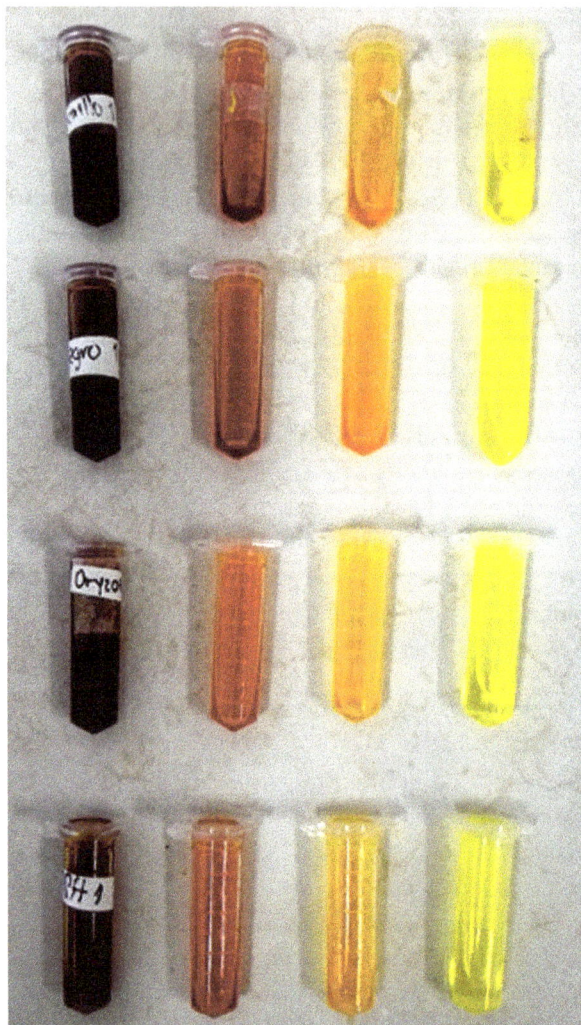

FIGURE 9.4   Different shades obtained by the *Curcuma longa* pigments.
Source: Own source.

9.1.6   MEDICINAL USE OF CURCUMA LONGA

Since ancient times in India and other cultures, this rhizome has been used to treat digestive problems, due to its main active component curcumin. Turmeric is considered a drug for many diseases, including digestive problems, cancer, inflammation, hepatitis, and other relevant (Priyadarsini, 2014).

Turmeric contains two ingredients called curcumin and turmerin. They have anti-inflammatory and wound healing properties when applied as a powder and paste to wounds and ulcers or taken with water, milk, or honey. Mucilage is a substance related to starch with therapeutic action because of its slippery texture. This helps to give calming effects to inflammation in the inner lining of the digestive system (Fan et al., 2015; Kant et al., 2014; Kuo et al., 2009).

The recommended form in which this rhizome can be consumed is as a crushed powder or in an infusion. Scientific research has shown that curcumin is a potential agent against arthritis, cancer prevention, and anti-inflammatory drugs (Ríos et al., 2009).

Curcumin has been commonly consumed orally using turmeric rhizome powder. However, there are studies reporting the pharmacokinetic effect of curcumin injected intravenously in mice. The authors were able to identify by chromatographic techniques the residual curcumin in plasma, as well as to increase 162 times the half-life of the compound in the bloodstream (Schiborr et al., 2010).

9.2 CONCLUSION

The use of turmeric is not yet widespread in Mexico. In the Huasteca Potosina, it has been used due to the easy adaptation of the plant to the climatic conditions of the region. It is mainly used as a pigment and as a seasoning agent in the preparation of dishes. However, turmeric rhizome has phytochemicals which are very importantto the food, pharmaceutical, and cosmetic industries. It is important to propose strategies for the use and exploitation of turmeric, and in this way to value and promote the use of this crop.

KEYWORDS

- **curcumin**
- **curcuminoids**
- **pigments**
- **ultrasound extraction**
- **extraction by fermentation**

REFERENCES

Aguilar, C. N.; Gutiérrez-Sánchez, G.; Prado-Barragán, L. A.; Rodríguez-Herrera, R.; Martínez-Hernandez, J. L.; Contreras-Esquivel, J. C. Perspectives of Solid State Fermentation for Production of Food Enzymes. *Am. J. Biochem. Biotechnol.* **2008,** *4* (4), 354–366.

Akram, M.; Shahab-Uddin, Ahmed, Afzal; Usmanghani, Khan, Hannan, A.; Mohiuddin, E.; Asif, M. *Curcuma longa* and Curcumin: A Review Article. *Rom. J. Biol. Plant Biol.* **2010,** (55) 2, 65–70

Andrade-Damián, M. F.; Muñiz-Márquez, D. B.; Wong-Paz, J. E.; Veana-Hernández, F.; Reyes-Luna, C.; Aguilar-Zárate, P. Exploratory Study of Pigment Extraction from Curcuma longa L. by Solid-State Fermentation Using Five Fungal Strains. *Mexican J. Biotechnol.* **2019,** *4* (3), 1–11.

Bello-Pérez, L.; Jiménez-Aparicio, A. Alimentos Funcionales. *Inv. Hoy.* **2000,** *93,* 20–25.

Braga, M. E.; Leal, P. F.; Carvalho, J. E.; Meireles, M. A. Comparison of Yield, Composition, and Antioxidant Activity of Turmeric (*Curcuma longa* L.) Extracts Obtained Using Various Techniques. *J. Agric. Food Chem.* **2003,** *51* (22), 6604–6611. doi:10.1021/jf0345550

Chan, C. H.; Yusoff, R.; Ngoh, G. C.; Kung, F. W.-L. Microwave-Assisted Extractions of Active Ingredients from Plants. *J. Chromatogr. A* **2011,** *1218* (37), 6213–6225.

Chemat, F.; Huma, Z.-e.; Khan, M. K. Applications of Ultrasound in Food Technology: Processing, Preservation and Extraction. *Ultrasonics Sonochem.* **2011,** *18* (4), 813–835.

Elizalde, S. O. *Solubilidades de la capsaicina y pigmentos liposolubles (carotenoides) del chile poblano en CO supercrítico. Tesis de Doctorado en Ciencias en Ingeniería Química. Escuela Superior de Ingeniería Química e Industrias Extractivas*; Instituto Politécnico Nacional: México, D.F., 2008.

Fan, Z.; Yao, J.; Li, Y.; Hu, X.; Shao, H.; Tian, X. Anti-Inflammatory and Antioxidant Effects of Curcumin on Acute Lung Injury in a Rodent Model of Intestinal Ischemia Reperfusion by Inhibiting the Pathway of NF-Kb. *Int. J Clin. Exp. Pathol.* **2015,** *8* (4), 3451–3459.

García Ariza, L. L.; Montes Quim, J. H.; Sierra Acevedo, J. I.; Padilla Sanabria, L. Actividad biológica de tres Curcuminoides de *Curcuma longa* L.(Cúrcuma) cultivada en el Quindío-Colombia. *Revista Cubana de Plantas Medicinales* **2017,** *22* (1), 1–14.

Gil-Chavez, G.J.; Contreras-Angulo, L.; Valdez-Torres, B.; Gonzalez-Aguilar G. A.; Heredia J. B. Optimization Process for Recovering Phenolic Antioxidant Compounds from Low Quality Eggplant (*Solanum melongena* L.) Pulp by Modified Supercritical Carbon Dioxide Extraction. *Sep. Sci. Technol.* **2015,** *50* (6), 841–850.

Gupta, A. P.; Gupta, M. M.; Kumar, S. Simultaneous Determination of Curcuminoids in Curcuma Samples Using High Performance Thin Layer Chromatography. *J. Liquid Chromatogr. Relat. Technol.* **1999,** *22* (10), 1561–1569. doi:10.1081/JLC-100101751

Hirun, S.; Utama-ang, N.; Roach, P. D. Turmeric (*Curcuma longa* L.) Drying: An Optimization Approach Using Microwave-Vacuum Drying. *J. Food Sci. Technol.* **2014,** *51* (9), 2127–2133.

Janaki, N.; Bose, J. L. An Improved Method for the Isolation of Curcumin from Turmeric, *Curcuma longa* L. *J. Indian Chem. Soc.* **1967,** *44* (11), 985–986.

Jayaprakasha, G. K.; Mohan Rao, L. J.; Sakariah, K. K. Improved HPLC Method for the Determination of Curcumin, Demethoxycurcumin, and Bisdemethoxycurcumin. *J. Agric. Food Chem.* **2002,** *50* (13), 3668–3672. doi:10.1021/jf025506a

Jiang, H.; Timmermann, B. N.; Gang, D. R. Use of Liquid Chromatography–Electrospray Ionization Tandem Mass Spectrometry to Identify Diarylheptanoids in Turmeric (*Curcuma longa L.*) Rhizome. *J. Chromatogr. A* **2006**, *1111* (1), 21–31. doi:10.1016/j.chroma.2006.01.103

Kant, V.; Gopal, A.; Pathak, N. N.; Kumar, P.; Tandan, S. K.; Kumar, D. Antioxidant and Anti-Inflammatory Potential of Curcumin Accelerated the Cutaneous Wound Healing in Streptozotocin-Induced Diabetic Rats. *Int. Immunopharmacol.* **2014**, *20* (2), 322–330.

Krishnamurthy, N.; Mathew, A. G.; Nambudiri, E. S.; Shivashankar, S.; Lewis, Y. S.; Natarajan, C. P. Oil and Oleoresin of Turmeric. *Trop. Sci.* **1976**, *18* (1), 37.

Kuo, C. F.; Chyau, C. C.; Wang, T. S.; Li, C. R.; Hu, T. J. (2009). Enhanced Antioxidant and Anti-inflammatory Activities of Monascus pilosus Fermented Products by Addition of Turmeric to the Medium. *Journal of Agricultural and Food Chemistry.* **2009**, 57(23), 11397–11405. doi:10.1021/jf9027798

Lashkari, Z. A Story of Resurgence of Natural Products. *Finechem Nat. Prod.* **1999**, *1* (4), 131–157.

Li, M.; Ngadi, M. O.; Ma, Y. Optimisation of Pulsed Ultrasonic and Microwave-Assisted Extraction for Curcuminoids by Response Surface Methodology and Kinetic Study. *Food Chem.* **2014**, *165* (15), 29–34.

Liang, H.; Wang, W.; Xu, J.; Zhang, Q.; Shen, Z.; Zeng, Z.; Li, Q. Optimization of Ionic Liquid-Based Microwave-Assisted Extraction Technique for Curcuminoids from *Curcuma longa* L. *Food Bioprod. Process.* **2017**, *104*, 57–65.

Luque de Castro, M. D.; García-Ayuso, L. E. Soxhlet Extraction of Solid Materials: An Outdated Technique with a Promising Innovative Future. *Analyt. Chim. Acta.* **1998**, *369* (1–2), 1–10.

Maheshwari, R. K.; Singh, A. K.; Gaddipati, J.; Srimal, R. C. Multiple Biological Activities of Curcumin: A Short Review. *Life Sci.* **2006**, *78* (18), 2081–2087.

Mandal, V.; Mohan, Y.; Hemalatha, S. Microwave Assisted Extraction of Curcumin by Sample–Solvent Dual Heating Mechanism Using Taguchi L9 Orthogonal Design. *J. Pharm. Biomed. Analy.* **2008**, *46* (2), 322–327.

Martínez, M. *Catálogo de nombres vulgares y científicos de plantas mexicanas*; México. Fondo de Cultura Económica, 1970.

Martínez, L. J. *Supercritical Fluid Extraction of Nutraceuticals and Bioactive Compounds*; Taylor & Francis Group, LLC. CRC Press: Boca Ratón, FL, 2008; 424 p.

Martins, S.; Mussatto, S. I.; Martínez-Avila, G.; Montañez-Saenz, J.; Aguilar, C. N.; Teixeira, J. A. Bioactive Phenolic Compounds: Production and Extraction by Solid-State Fermentation. A review. *Biotechnol. Adv.* **2011**, *29* (3), 365–373.

Mohamed, S. A.; Saleh, R. M.; Kabli, S. A.; Al-Garni, S. M. Influence of Solid-State Fermentation by Trichoderma spp. on Solubility, Phenolic Content, Antioxidant, and Antimicrobial Activities of Commercial Turmeric. *Biosci., Biotechnol. Biochem.* **2016**, *80* (5), 920–928.

Mussatto, S. I.; Machado, E. M.; Martins, S.; Teixeira, J. A. Production, Composition, and Application of Coffee and Its Industrial Residues. *Food Bioprocess Technol.* **2011**, *4* (5), 661.

Pant, N.; Misra, H.; Jain, D.C. Phytochemical investigation of Ethyl Acetate Extract from Curcuma aromatica Salisb. Rhizomes. *Arab. J. Chem.* **2013**, *6*, 279–283.

Paramapojn, S.; Gritsanapan, W. Free Radical Scavenging Activity Determination and Quantitative Analysis of Curcuminoids in Curcuma Zedoaria Rhizome Extracts by HPLC Method. *Curr. Sci.* **2009**, *97* (7), 1069–1073.

Priyadarsini, K. I. The Chemistry of Curcumin: From Extraction to Therapeutic Agent. *Molecules.* **2014,** *19* (12), 20091–20112. doi:10.3390/molecules191220091

Revathy, S.; Elumalai, S.; Benny, M.; Antony, B. Isolation, Purification and Identification of Curcuminoids from Turmeric (*Curcuma longa L.*) by Column Chromatography. *J. Exp. Sci.* **2011,** *2* (7), 21–25.

Ríos V, E.; Duque C. A. L.; León R. D. F. Spectroscopy and Chromatography Characterization of Curcumin Extracted from the Rhizome of Turmeric Crops in the Department of Quindío (Colombia). *Rev. Invest. Univ. Quindio.* **2009,** *19,* 18–22.

Robles-Ozuna, L. E.; Ochoa-Martínez, L. A. Ultrasonido y sus aplicaciones en el procesamiento de alimentos. *Revista Iberoamericana de Tecnología Postcosecha.* **2012,** 13(2), 109–122.

Sahne, F.; Mohammadi, M.; Najafpour, G. D.; Moghadamnia, A. A. Extraction of Bioactive Compound Curcumin from Turmeric (*Curcuma longa L.*) via Different Routes: A Comparative Study. *Pak. J. Biotechnol* **2016,** *13* (3), 173–180.

Saiz de Cos, P.; Pérez-Urria, E. Cúrcuma I (*Curcuma longa L.*). *Reduca (Biología). Serie Botánica.* **2014,** *7* (2), 84–99.

Sastry, B. S. Curcumin Content of Turmeric. *Res. Ind.* **1970,** *15* (4), 258–260.

Schiborr, C.; Eckert, G. P.; Rimbach, G.; Frank, J. A Validated Method for the Quantification of Curcumin in Plasma and Brain Tissue by Fast Narrow-Bore High-Performance Liquid Chromatography with Fluorescence Detection. *Analyt. Bioanalyt. Chem.* **2010,** *397* (5), 1917–1925. doi:10.1007/s00216-010-3719-3

Shirsath, S. R.; Sable, S. S.; Gaikwad, S. G.; Sonawane, S. H.; Saini, D. R.; Gogate, P. R. Intensification of Extraction of Curcumin from *Curcuma amada* Using Ultrasound Assisted Approach: Effect of Different Operating Parameters. *Ultrasonics Sonochem.* **2017,** *38,* 437–445.

Soleno Wilches, R. Tecnologías no térmicas en el procesado y conservación de alimentos vegetales. Una revisión. Non Thermal Technologies in the Processing and Conservation of Vegetable Foods. A Review. *Revista Colombiana de Investigaciones Agroindustriales.* **2015,** *2,* 73–83.

Stransky, C. E. *U.S. Patent No. US4138212A.* 1979.

Tahani, A.; Ahmed, M.; Maulidiani, K. S.; Siti, M. M.; Faudzi., M.; Sukari, A. H.; Lajis., N. H.; Abas, F. Phytochemical Profiles and Biological Activities of Curcuma Species Subjected to Different Drying Methods and Solvent Systems: NMR-Based Metabolomics Approach. *Indust. Crops Prod.* **2016,** *94,* 342–352.

Tønnesen, H. H.; Karlsen, J.; Adhikary, S. R.; Pandey, R. Studies on Curcumin and Curcuminoids XVII. Variation in the Content of Curcuminoids in *Curcuma longa L.* from Nenal during One Season. *Zeitschrift für Lebensmittel-Untersuchung und Forschung.* **1989,** *189* (2), 116–118.

Torres Rodríguez, E.; Guillén González, Z.; Hermosilla Espinosa, R.; Arias Cedeño, Q.; Vogel, C.; Almeida Saavedra, M. Empleo de ultrasonido en la extracción de curcumina a partir de su fuente natural. *Revista Cubana de Plantas Medicinales* **2014,** *19* (1), 14–20.

Wakte, P. S.; Sachin, B. S.; Patil, A. A.; Mohato, D. M.; Band, T. H.; Shinde, D. B. Optimization of Microwave, Ultra-Sonic and Supercritical Carbon Dioxide Assisted Extraction Techniques for Curcumin from *Curcuma longa. Sep. Purif. Technol.* **2011,** *79* (1), 50–55.

Wang, H.; Ding, J.; Ren, N. Recent advances in microwave-assisted extraction of Trace Organic Pollutants from Food and Environmental Samples. *TrAC Trends Analyt. Chem.* **2016,** *75,* 197–208.

Xu, G.; Hao, C.; Tian, S.; Gao, F.; Sun, W.; Sun, R. A Method for the Preparation of Curcumin by Ultrasonic-Assisted Ammonium Sulfate/Ethanol Aqueous Two-Phase Extraction. *J. Chromatogr. B* **2017,** *1041,* 167–174.

Zhang, H. F.; Yang, X. H.; Wang, Y. Microwave Assisted Extraction of Secondary Metabolites from Plants: Current Status and Future Directions. *Trends Food Sci. Technol.* **2011,** *22* (12), 672–688.

Zorofchian Moghadamtousi, S.; Abdul Kadir, H.; Hassandarvish, P.; Tajik, H.; Abubakar, S.; Zandi, K. A Review on Antibacterial, Antiviral, and Antifungal Activity of Curcumin. *BioMed Res. Int.* **2014.**

Zou, L.; Liu, W.; Liu, C.; Xiao, H.; McClements, D. J. Utilizing Food Matrix Effects To Enhance Nutraceutical Bioavailability: Increase of Curcumin Bioaccessibility Using Excipient Emulsions. *J. Agric. Food Chem.* **2015,** *63* (7), 2052–2062. doi:10.1021/jf506149f

CHAPTER 10

Effect of LED Light in Biomass Production and Presence of Metabolites in Medicinal Plants

LUIS ENRIQUE ORDÓÑEZ LÓPEZ, HUMBERTO RODRÍGUEZ-FUENTES, ALEJANDRO ISABEL LUNA-MALDONADO, JULIA MARIANA MÁRQUEZ-REYES, and ROMEO ROJAS[*]

Universidad Autonoma de Nuevo Leon, School of Agronomy, 66054, General Escobedo, Nuevo León, México

[*]Corresponding author. E-mail: romeo.rojasmln@uanl.edu.mx

ABSTRACT

Over the years, a large number of plants have been used in traditional medicine. However, the production of these depends on the region, height, humidity, light, and all environmental factors. All this without forgetting that many are endemic and do not grow all year. In addition, all environmental factors influence the quantity and quality of the secondary metabolites they produce, since they are their defense mechanism against droughts, winters, microorganisms, and pests. The plant factory system consists of a closed system with all controlled conditions (nutrients, humidity, light, pest-free, etc.) to produce crops, which is why it is a viable alternative to produce species with medicinal properties since they can be produced all year round, anywhere in the world, promote the overproduction of their secondary metabolites, reduce water consumption, increase production levels, etc. However, one of the most important factors is to control the hours of light (intensity and type) to promote the production of secondary metabolites with medicinal properties. That is why the plant factory system is a viable alternative to produce biomass from medicinal plants.

10.1   INTRODUCTION

For so many years, even at the beginning of the civilization, the medicinal plants have been used as remedies to relieve some of the most commons diseases and disorders present in humans. Nevertheless, nowadays with the power and influence of pharmacological companies and pharmaceutical industries, the use of all these remedies has become less and less. Today, there is lack of knowledge and research on most of the medicinal plants. But, in recent years, it has been more interesting from the scientific community and investigators to study the principal benefits and main components of medicinal plants. It is known that more than the 80% of the world's population use some kind of traditional herbal medicine as first option, and sometimes they are the only source of medicine (Stutte, 2016). Medicinal and aromatic plants represent a primary resource for therapeutic and culinary purposes. There are too many ways to take advantage of all the components that medicinal plants have. Medicinal plants are available as either fresh or dry material (as the first and most common option), processed to yield essential oils (increasing significantly their value to the market) and sometimes processed to obtain extracts. The total number of products obtained from the nature is actually unknown, but there is a large list of products that can be obtained in the form of medical prescription and over the counter drugs.

The constant problem associated with the increase of population has caused the agricultural production to be compromised. So alternative systems to produce more and better food have been found in the last two or three decades. In addition, the climatic change, the limited space for agriculture, areas with limited daylight, and adverse environmental conditions have allowed new production systems to appear to counteract these adversities. New technologies, including culture management of crops, and techniques as greenhouses, hydroponic systems, aeroponic and aquaculture systems, vertical farms and plant factory systems with artificial lighting provide a total control of the environment conditions for the production of crops. However, the benefits of artificial lighting which include elevating the number of crop yield per year (decreasing or shortening the crop cycle), producing greater plant density per/m^2, efficient nutrient and water use, fewer crop losses and no pesticide or insecticide applications (due to the low or no appearance of pests and other pathogens such as undergrowth) make the technologies more efficient in the production of crops. All these new technologies can provide and produce standard high quality horticultural products. Nevertheless, in contrast to outdoor agriculture, these closed and indoor plant cultivation systems need artificial light sources, such as LEDs or fluorescent lamps capable of simulating energy from the sun and stimulating plant growth while drastically reducing the energy consumption, which is difficult in so many countries because of the high cost of the energy.

10.1.1   PLANT FACTORY WITH ARTIFICIAL LIGHTINGSYSTEMS

The term "Plant Factory" can be defined as an isolated and almost hermetic structure that allows a total control of the environmental conditions (Kozai, 2013). Thus, a Plant Factory System with Artificial Lighting (PFAL) has total control over some environmental factors that determine the accumulation of antioxidant compounds during the crop production cycle, such as temperature, quantity and quality of light, % of relative humidity, plant nutrition, and concentration of CO_2 (Johkan et al., 2010). The PFAL provides a spectrum and amount of light necessary for plants to perform all their photosynthetic processes efficiently. These photosynthetic processes sometimes are modified when the plants are grown under this artificial lighting, and one of the reasons is that the use of these lamps does not usually match with the normal spectrum and energy coming from the sun. The new lighting technologies, such as light emitting diodes (LED) provide the necessary quantity and wavelength to cover all the requirements from plants. From the biological point of view, sometimes specific questions arise about the use of these modern artificial lighting systems, most of these are related to their ability to mimic the results shown from the use of natural sunlight.

The light emitting diodes are considered as an innovate source to provide light for plants. These are used not only for the intensity that is essential to produce some of the most important growth characters as biomass and leaf area, but also for enhancing metabolic activities and improving antioxidant properties, etc. These characters are important, and the quality of the spectrum of light determines the total quantity of many metabolic compounds and their antioxidant activity. LEDs are commercially applicable and profitable mainly for vegetables as lettuce, coriander, arugula, spinach and chards, baby greens vegetables as broccoli, cauliflower, soybean, and more. In addition to this, LEDs have an enormous potential for generating oxygen and purify water, producing different plants and crops for the entire year and contribute to the efficient light consumption.

10.1.2 PHOTOPERIOD, INTENSITY, AND QUALITY OF LIGHT

When considering light as an environmental factor the light intensity and quality of light have an important influence in the growth of many plants and production of secondary metabolites and antioxidant capacity. Recent research indicates that medicinal plants cultivated in controlled environment such as plant factory increase the production of biomass and certain metabolites. Lin et al. (2013), Goto (2012), and Sabzalian et al. (2014) have shown that the combination of red, blue, and white LED lights can be an effective light source for plant growth and development, and that the spectra, intensities, and durations of photoperiod can be easily manipulated in production systems in controlled environments. The use of light-emitting diodes in the production of plants in controlled environments allows to provide higher photosynthetic photon flow levels (more than 500 $\mu mol/m^2/s$) and a higher intensity ratio from light to heat radiation compared with conventional lighting systems.

In other traditional systems outdoor, the sun provides the necessary light and energy for photosynthetic organisms. They use these light quality and intensity to respond to their environment. The growing systems with total controlled environment use artificial lighting to increase the production capacity and make the efficient use of electrical energy. Recent investigations have shown that LEDs have an enormous potential and efficiency for improving plant growth and making the indoor systems more sustainable (Darko et al., 2014).

Light is considered an indispensable environmental factor that influences the development and growth of plants, as well as being the main source of energy for photosynthesis and photomorphogenesis (Qian et al., 2016; Liu et al., 2011). The quality (emitted wavelength) and light intensity (photon flow per unit area over specific time period), as well as the photoperiod (number of hours of light/dark for 24 h) are the key elements of the condition of the light. This intensity and quality of light can manipulate the plant metabolism with the aim to produce some specific components as dry mater, secondary metabolites, and antioxidant compounds.

Intensity and quality of light can determine the biomass production, according to Folta (2004) and Kim et al. (2004), who claim that green light acts as a growth regulator, and Ahmad et al. (2016) reported that the white LED light proved to be the most effective factor for the accumulation of dry biomass in the *Stevia rebaudiana* Bert. crop (established in a system similar to a plant factory) compared with the rest of the treatments complemented only with blue, green, or red LED lights.

10.1.2.1   PHOTOPERIOD

It is the time (expressed in hours) to which an organism is exposed to a natural or artificial light source for a day. The photoperiod influences numerous metabolic processes in plants (Nitschke et al., 2017) such as photosynthesis, flowering, and growth. In production systems in controlled environments, there is total control of the photoperiod, obtaining greater production in less time with better quality (greater accumulation of biomass and antioxidant compounds; free of agrochemicals, more uniform fruits and larger caliber, etc.). This photoperiod does not determine the production of any specific component in plants, but is related to the quality and intensity of light. This can be explained because if we provide to the plant the necessary intensity and composition of light in a period of 4 h (taking into account both factors as the most important components of the production), there is no necessity to add more intensity of light and to change the quality of the light source. So, it is a term that depends directly on other factors and cannot be taken as an essential production component.

In the field of plants grown in controlled environments and assisted with artificial lighting sources such as LEDs or other traditional lighting systems as the fluorescent lamps, a term known as daily light integral or DLI is used. This refers to the amount of photosynthetically active photons (individual particles of light between 400 and 700 nm range) that are present within a specific surface area for a period of 24 h. This variable is used to describe or classify under which light environment a plant species can be grown.

10.1.2.2   LIGHT QUALITY

It can be defined as the spectrum of wavelengths that affect plants in order to carry out photosynthesis and other physiological processes, such as biosynthesis, accumulation and retention of plant phytochemicals (Ilić and Fallik, 2017). The quality of light is associated with the composition of the electromagnetic spectrum, the proportion of the different wavelengths that are visible for the human eye. The spectrum of visible light is found between the violet light (380 nm) till distant red light (to 800 nm approximately). The various spectra of visible light are depicted in Table 10.1 that lists all the different spectra, their composition, and the respective wavelength (approximately):

TABLE 10.1   Electromagnetic Radiation Composition (Light Quality).

Color of the visible light spectrum	Wavelength range approximately (nm)
Violet	380–450
Blue	450–495
Green	495–570
Yellow	570–590
Orange	570–620
Red	620–700
Distant Red	700–800

10.1.2.3   LIGHT INTENSITY

The light intensity can be described as the amount or flow of photons present on the surface unit in a specific period of time (in this case $\mu mol/m^2/s$) (Hunter and Burritt, 2004; Li and Kubota, 2009; Stutte et al., 2009; Johkan et al., 2010). Light intensity is a factor that influences plant growth, in addition to generating an increase in the biomass accumulation. It has been proved that the intensity of light works as a determining factor for the production of biomass and other agronomic variables in agricultural production. Within the production systems with total environmental control, we can modify the light intensity according to the plant species grown within the system, providing the necessary amount of light for each specific crop. It depends directly on the species, production objective (if we want to induce the growth of leaves, stems, or the appearance of flowers), and the crop cycle (shortening or decreasing the period).

10.2   IMPORTANCE OF LED LIGHT IN AGRICULTURAL PRODUCTION SYSTEMS

In recent years, more attention has been given to carry out research with red and blue light-emitting diodes, due to their role as essential energy sources for photosynthetic carbon assimilation (Lin et al., 2013). Bula et al. reported one of the first experiments on indoor culture plants, where they settled lettuce plants under red LED lamps and analyzed the vegetative growth (Bula et al., 1991). In other experiment, Martineau et al. (2012) estimated the amount of dry matter accumulated by lettuce plants per mole emitted from the artificial lighting system using red LEDs (650 nm) against traditional high-pressure sodium lamps, and concluding that difference between treatments were not significant. Chang et al. (2011) estimated the optimal consumption of photons for growth and culture of the green alga *Chlamydomonas reinhardtii*, which most efficiently utilized the light quality under red LEDs at 674 nm.

Recently, the LEDs have been successfully tested in different experiments and have proven to be agronomically sustainable for the cultivation of different plant species and vegetables (mainly from leaf crops, such as lettuce, coriander, arugula, broccoli, microgreens, among others), medicinal and floral plants, and even some fruit trees.

10.2.1 PHOTORECEPTORS

Plants have the ability to respond to the intensity and quality of light (Zhang and Folta, 2012) through different photoreceptors (phytochromes, cryptochromes, and phototropins) that are activated under specific wavelengths (Kozai, 2007). Therefore, lighting systems for crop production in controlled environments are of paramount importance (Sabzalian et al., in press).

The high efficiency of the red light spectrum (650–665 nm) supplemented with blue light for LED lamps on plant growth is pretty simple to understand because all these wavelengths are perfectly absorbed by the chlorophyll and phytochrome, which introduces the idea that the growth and culture of plants under blue and red LEDs can mimic the efficiency of the natural light. In addition to providing a better excitation of some different types of photoreceptors, blue and red combination allowed a higher photosynthetic activity than that under either monochromatic light (Goins et al., 1998). This effect can be due to increased nitrogen content of the blue light supplemented plants. Other authors suggested that blue light facilitates a better stomatal opening and thus provides higher concentration of CO_2 for photosynthesis. It is known that stomatal opening is controlled by the effect of photoreceptors present in the blue light spectrum (450–495 nm). This is possibly reflected in the increase of shoot dry matter with increasing levels of blue light (Goins et al., 1998). It is recommended that the supplementation of red and blue light could be combined with green light (495–570 nm) in some proportions (less than 30%). Illumination with more than 50% of green LED light causes a reduction in plant growth, although this depends on the cultivated plant species (Kim et al., 2007).

In a few experiments, it has been shown that lettuce grown under red LEDs has consequences on growth, stimulating the elongation of hypocotyls and cotyledons. This is a phenomenon characterized by being dependent on the phytochromes present in plants. The spectrum of red light produces an over excitation of phytochromes, so the elongation of organs such as hypocotyl can be reduced within a minimum of 15 µmol/m^2/s of blue light addition (Hoenecke et al., 1992). Although the above mentioned experiment was not demonstrated, based on bibliography cited by the author, it was concluded that supplemented blue light activated cryptochrome, which is another photoreceptor that responds exclusively under this light spectrum, and is capable of reducing or mediating the reduction in hypocotyl length (Ahmad et al., 2002).

According to some authors, the phytochromes present in the plant respond to light of specific wavelengths (between 660 and 730 nm) and can increase the accumulation of antioxidant compounds, such as vitamin C, carotenoids, anthocyanins, flavonoids, and total polyphenols, due to an overstimulation of phytochromes or over excitation of them. The plant in turn responds to that light stress, and increases the production and activity of antioxidant compounds as secondary metabolites (Lefsrud et al., 2008; Li and Kubota, 2009; Stutte et al., 2009).

All the photosynthetic organisms are capable of responding to some kind of stress, in this case, when they are exposed to high light, they develop or activate some short- and long-term response mechanisms to this stress. These mechanisms of defense are the specific and central topic of other papers and works because of their pharmaceutical importance and medicinal properties. They are commonly used in repairing mechanisms (Long et al., 1994), shielding (Lee and Gould, 2002), reactive oxygen species (ROS) quenching (Lee and Gould, 2002), or the production of specific storage compounds (Lemoine and Schoefs, 2010). The most typical examples about all these mechanisms can be found in studies with medicinal plants and herbs of pharmaceutical importance such as mint (*Mentha* sp.) (Sabzalian et al., 2014), rosemary (*Rosmarinus officinalis* L.), oregano (*Origanum* sp.), jewel orchid (*Anoectohilus* sp.) (Ma et al., 2010), cheat (*Artemisia ludoviciana* Nutt.), and even some ornamental plants as periwinkle (*Catharanthus roseus* L.). The increase in light intensity can also generate a decrease in the amount of secondary metabolites like flavonoids and other phenolic compounds, such is the case of the whiskers of medicinal plant cats (*Orthosiphon stamineus*) (Ibrahim and Jaafar, 2012), which may indicate that the amount of irradiance can have adverse consequences on the production and capacity of antioxidant compounds. The truth is that all these aspects about the accumulation of some antioxidant compounds, such as secondary metabolites and specific pigments or flavonoids content in studies with plants, depend directly on the species and growing conditions, genetic aspects and the interaction between all these components. Although the mechanistic aspects about LED light effects are not well understood.

The most important and interesting aspects of study for the scientific community on the use of lighting emitting diodes in controlled environments (in addition to the cultivation of plant species) are related to photosynthetic microorganisms (such as algae) due to their enormous biotechnological and economic potential in different fields from biofuels, pharmaceuticals, medicines, and additives to culinary and food purposes (Mimouni et al., 2012). Similar case has been reported by Wang et al. (2007), who evaluated the efficiency in the production and profitability of the cultivation of microalgae (*Spirulina platensis*) under different spectra of monochromatic LED lights. The objective of the experiment was to evaluate the conversion of the energy emitted by the LEDs in the biomass production and the yield in grams per liter of fuel produced, which translates into a profit in dollars. The results showed that at light intensities of 1500–3000 $\mu mol/m^2/s$, the red LED lamps had the lowest energy consumption and produced the highest economic profitability (greater than 110 g/L/dollar) compared with blue lamps with the same intensity (with a yield less than 10 g/L/dollar, respectively), which means that these controlled environment systems can have a greater exploitation and can be used on a larger scale with profitability that can be competitive to other production systems.

Fukuyama et al. (2017) established that the content of total alkaloids in *Catharanthus roseus* is influenced by environmental conditions, such as light intensity (Liu et al., 2011; Fukuyama et al., 2015) and nitrogen content in fertilization (Guo et al., 2014; Gholamhoss et al., 2011). The blue light (maximum length of 450 nm) and the irradiation of UV light (maximum length of 370 nm) in the culture of *Catharanthus roseus* cultivated in soil induced an increase of the vinblastine content and decreased the total content of vindoline and catarantin (Hirata et al., 1991, 1992, 1993). In addition to this, Fukuyama et al. (2017) reported that increases in the production of vinblastine present in leaves are due to irradiation by UV (ultraviolet) light in combination with red light at 150 $\mu mol/m^2/s$. However, a previous study showed that the production of biomass for *Catharanthus roseus* under two different light treatments showed that the production of biomass under the UV light treatment for 4 weeks was less than the red light treatment without the presence of UV light during the same period of time (Hirata et al., 1992), so that biomass production may be a function of other factors. Fukuyama et al. (2017) also noted that the vinblastine content increased in leaf discs (obtained from the leaves) that were incubated at different levels of irradiation with UV light. The treatment provided the highest light intensity ($10 W/m^2$) which was 54 times higher than the control treatment, with the leaf discs incubated at any radiation of UV light (control treatment). It has been previously shown that light is directly related to the quality of rosemary essential oil, increasing, or decreasing the concentration of various constituents, such as camphene and α-pinene. Mulas et al. (2006) evaluated the effect of distant red and red light on the production of rosemary essential oil, suggesting that the participation of phytochrome, which acts as a photoreceptor of red (600–700 nm) and distant red (700–800 nm) light, had an effect on the synthesis process of rosemary essential oil as opposed to the control treatment, which had no application of supplementary light.

In other experiments with green microalga (*Dunaliella salina*) using LEDs, it has been shown that the light stress (low intensity: 170–255 μmol/m^2/s) increases the amount of β-carotene, but while using conventional lights, such as fluorescent lamps and high-pressure sodium lamps (with a intensity higher than 1000 μmol/m^2/s), they obtained more biomass of microalga. In addition, this experiment showed that the supplement of red or blue (600–750 and 450–490 nm respectively) LED light caused some level of stress on plants, activating the xanthophyll cycle. The addition of red light caused more stress than the blue light.

10.2.2 EFFECT OF MONOCHROMATIC LIGHTING IN MODIFICATION OF THE METABOLISM OF PLANTS

It is proved that the spectra of blue and red light are not just related to certain agronomic variables of plant growth and development, but they are also involved in different phytochemical qualities, such as the concentration, activity and capacity of certain antioxidant compounds, and other chemicals substances that occur under specific stress conditions.

Supplementary LED light can be useful to increase the antioxidant contents in plants, especially with the use of blue or red light spectra. For example, red LED light (658–660 nm) can increase the concentration of the total phenolic compounds and the anthocyanin contents of green vegetables like lettuce or broccoli and for red arugula or cabbage leaves. Under this specific level, they can also increase the antioxidant content of many other vegetables. Colquhoun et al. (2013) used different treatments of LED light to modify the synthesis of some volatile compounds in flowers and fruits of petunia hybrid plants cultivated in a glass greenhouse. Meanwhile, in tomato plants, the red LED treatment (with a spectra of 668 nm and an intensity of 50 μmol/m^2/s) showed a significant increase of 2-methyl-butanol and 3-methyl-1-butanol levels, whereas the amount of cis-3-hexanol was reduced when compared with the levels reached with white LED light. The action of monochromatic light and its mechanism has not been studied yet, but some authors assume that the red light affects the terpenoid content in the chloroplast through phytochromes. Hence, more research is required to verify this hypothesis.

10.3 ADVANTAGES AND DISADVANTAGES OF PLANT FACTORY WITH ARTIFICIAL LIGHTING SYSTEMS

Before we decide the establishment of a PFAL system, it is necessary to take into account some important points, such as the source of energy to be used, the water supply, equipment, and technical material, among others. From many points, these technological systems look like a simple and effective way to obtain high yields and better production costs, which means higher profits. Nevertheless, in so many countries, the access to energy sources at low cost sometimes is difficult and not profitable.

10.3.1 ADVANTAGES OF PLANT FACTORY WITH ARTIFICIAL LIGHTING SYSTEMS

The closed environment systems guarantee optimum conditions to maximize the biomass and metabolite production, manipulating, inducing or avoiding the levels of stresses in plants, and also facilitating the reallocation of carbon to the production of medicinal compounds. All these adaptation systems allow plants to grow and facilitate their development to be safe, healthy, consistent, and produce high quality pharmaceutical product for the medicinal plants market. Some of the most important advantages of these technological systems are described below:

- **Sterilized food:** The plants can be grown in a clean area which is ideal for vegetables that are ready to eat (as lettuces, arugulas, coriander, tomatoes, broccoli, spinach, or cauliflower), this can mean an increase in value, quality, and of course and most importantly price of the product.
- **Standardized conditions:** The total control of the environment conditions including biotic and abiotic factors of the entire area, the system is free from many contaminations as undergrowth's and pests.

- **More uniform plants:** This is achieved because all plants established in the system receive the same conditions, both environmental and nutritional, since in this type of systems, there is no competition for space or elements present in the nutrient solution, always obtaining the same vegetables with physicochemical characteristics similar in each crop cycle. This is similar to what happens in a conventional factory. That is the reason why they are called plant factories to these production systems.
- **Always producing the same product:** Similar to the last point, the biochemical profiles are consistent, uniform, and similar by one period crop to another, always obtaining the same production of secondary metabolites. This gives to the farmer even the ease of placing a label on their products with the security that it will always get the same quality parameters in each one of the plants produced under these systems.

10.3.2   *DISADVANTAGES OF THE PLANT FACTORY WITH ARTIFICIAL LIGHTING SYSTEMS*

The biggest challenge now for these controlled environment systems is the need to find a sustainable light source to provide energy for the 365 days of the year without the problematic of the high cost for this service. The lighting industry must offer to the consumers a viable, ecologically and sustainable lamps to be adapted to the different request and needs from consumers, providing the best levels of consumption, quality and intensity of light with some additional skills as ultraviolet radiation (UV), and a long life cycle. This is clearly necessary in a PFAL system to replace the light coming from the sun, which is the main source of energy for plants.

- **Exorbitant costs (in many cases):** One of the most important requirements that can stop the establishment of these types of systems in some developing countries is undoubtedly the cost of the establishment, because in most of these countries, there is limited access to these technologies and in most cases, the costs are usually exorbitant and almost impossible to pay, thus decreasing the profitability of these systems. This clearly contrast with the most developed countries as Netherlands, Japan or USA, that in recent years has established plant factory systems with solar panels as source of energy, which is more sustainable.

- **High level of technology:** Other disadvantage of these systems is the level of knowledge that is required for the whole operation of the system. There are critical points and some aspects that must be taken in account before establishing a plant factory. Is necessary to have knowledge about agronomical aspects, management stuff, and the use of technology, but is pretty important to have an improvement in the use of special equipment to take specific data of all environmental conditions inside the building. Monitoring of conditions must be constant and precise with the objective of having a total control of the environmental conditions, and thus be able to establish the optimal levels of each one of these conditions before establishing a plant species.

- **Low quality and insufficient quantity of light:** It is not always enough to get any LED lamps and establish inside the plant factory system, it has to be effective and efficient to provide to the plants all the necessary resources needed for the growth and development. In some cases, this essential part of the system is not taken into account before establishing the system. This brings a consequence as deficient production.

10.4 RESULTS OBTAINED IN EXPERIMENTS ON PLANT FACTORY WITH ARTIFICIAL LIGHTING OR SIMILAR (Table 2)

TABLE 10.2 Main Results Obtained Using LED light for Cultivation in Plant Factory.

Plant (Specie)	Treatments and conditions	Principal results and observations	Reference
Petunia x hybrid Juss	White, blue, red, and far-red light LED. All treatments had the same intensity of light: 50 mmol/m²/s	Levels of 2-phenylethanol (floral volatile compound) increased under red and far-red light radiation	Colquhoun et al. (2013)
Amaranthus cruentus, Ocimum basilicum, Brassica oleracea, y juncea, Atriplex hortensis L., *Borago officinalis, Beta vulgaris, Petroselinum crispum and Pisum sativum*	Plants were illuminated with the basal red 640 nm light-emitting diodes, supplemented with 450, 660, and 735 nm components for 3–5 days in different developmental phases	Light intensity and quality affected the concentration of phenolic compounds, anthocyanin, and ascorbic acid	Urbonavičiūtė et al. (2008)

TABLE 10.2   *(Continued)*

Plant (Specie)	Treatments and conditions	Principal results and observations	Reference
Brassica juncea, Lactuca sativa, Ocimum gratissimum, Coleus blumei, and *Tagetes patula*	Plants were grown in controlled environment. PPF was provided by LED panels which were composed of blue and red light LEDs with the peak wavelengths at 460, 635, and 660 nm, respectively	Effects on chlorophyll a/b, carotenoids and anthocyanin content and concentration	Tarakanov et al. (2012)
Panax ginseng C.A. Mey.	Using different wavelengths of light (spectra: 380, 450, 470, or 660 nm).	Blue light spectra (450 and 470 nm) showed a significant increase in the production of ginsenosides in the root tissue	Park et al. (2013)
Hypericum perforatum L.	Plants were cultivated under three different treatments of light: red, blue, and white fluorescent lights in a controlled environment	Plants grown under red (600–700 nm) light increased the biomass, but, under blue (400–500 nm) or white (400–700 nm) light, showed a significant increment in the concentration of bioactive constituents	Nishimura et al. (2007)
Coriandrum sativum L.	Five different light qualities, red (R), blue (B), green (G), red:blue (RB), and red:blue:far-red (RBFr)	Chlorophyll index, total biomass, and ascorbic acid content were significantly higher in RB or RBFr light treatments. The antioxidant capacity and total phenolic content were the highest under B light. The total phenolic content per plant was the highest under RBFr light due to the high growth rate and biomass production	Nguyen, et al. (2019)
Crepidiastrum denticulatum (Houtt.) Pak & Kawano.	*C. denticulatum* seedlings were transplanted to a hydroponic system in a plant factory equipped	Shoot fresh and dry weight, leaf area, length, and width under R/FR were 2.1 times higher than control treatment. R/FR irradiation	Bae et al. (2017)

TABLE 10.2 *(Continued)*

Plant (Specie)	Treatments and conditions	Principal results and observations	Reference
	with Red (R), Blue (B), and Far-Red (FR) LEDs with different ratios of light	reduced in general the total phenolic content per dry weight compared with the control, but was not significant. Under R/FR some phenolic compounds levels per shoot increased 1.6 times compared with control treatment	
Nopalea cochenillifera (L.) Mill.	The effects of different colored light-emitting diodes (LEDs) on the growth and quality of daughter cladodes of the edible cactus *Nopalea cochenillifera* were investigated	The total fresh weight (FW) of the daughter cladodes was also lowest under blue light, but the average FW was the highest under blue light, or under simultaneous irradiation with red and blue light. The antioxidant activity was the highest under blue light or under simultaneous irradiation with red and blue light	Horibe et al. (2018)
Perilla frutescens	Three different levels of intensity of light were investigated (100, 200 and 300 mmol/m²/s) using LED lamps and other three different nutrient solution concentrations	The treatment with 200 mmol/m²/s and the highest level of nutrient solution showed the maximum level of apigenin concentration for green perilla, while for red perilla, it was with 300 mmol/m²/s. This result suggests that not only depends on the intensity of light and nutrition, but also for the species. For luteolin, other organic compound, the highest level was found under the lower nutrient solution concentration, and also affected by the increase of light intensity	Lu et al. (2018)

TABLE 10.2   *(Continued)*

Plant (Specie)	Treatments and conditions	Principal results and observations	Reference
Catharanthus roseus L.	*Catharanthus roseus* plants were established under red light treatment (R, 660 nm) for a period of 28 days and cultivated under three light quality treatments: UV supplemented with Red, Blue, or just Red light for 7 days	The total vinblastine content in leaves increased sharply under UV supplemented with R compared with R alone. The vinblastine content under B was 1/6 of that under UV supplemented with R. Vinblastine content increased as the UV intensity was increased from 0 to 10 W/m^2	Fukuyama et al. (2017)

10.5   CONCLUSIONS AND DISCUSSION

In the last few years, there has been a significant progress on the production and investigation for medicinal plants under controlled environments such as plant factories, which represents an important step for the pharmaceutical production of natural medicines. It is recommended to keep searching new ways and alternatives to produce medicinal plants, not only because of its economic importance at a global level or due to the growing demand of this type of products by the population to treat multiple diseases but also to generate new knowledge of multiple plant species of which there is still a lot to study and discover (Kozai et al., 2016).

Most of the results obtained for these kind of works and investigations with medicinal plants cultivated in controlled environments suggest that supplementary irradiation with LEDs of different colors (but especially: blue and red light) could be useful to improve the growth, biomass accumulation, total content of phenolic compounds, and secondary metabolites in plants. Some of these results show that the quality and quantity of light has a strong effect on development and phytochemical composition of medicinal plants or plants in general, so controlling the light and other factors of the environment represents an effective method for improving the growth and quality of farm foods and vegetables.

The use of LEDs can be an exceptional option to produce pesticide-free vegetables in the middle of the cities, providing additional benefits as more efficiency, reliability, environment control, and intelligence for plant factories with artificial lighting, and other systems as greenhouses. However, it will depend of the improvements and efficiency of the lighting system, taking into account the total consumption and final cost per lumens. In spite of few systems of this type established around the world, it is predicted that the new production systems or farms will be under the control of environmental conditions. It is expected that this horticultural management will expand in the next few years. Thanks to these new technologies, it will be possible to provide new possibilities and better options for economically efficient consumption of the light energy used for horticultural cultivation of crops, and not just on Earth, but also in the space, in a not too far future. It may contribute to feeding a big part of the human population and maintaining or preserving the ecosystems outdoor, thus protecting the species in danger, making a better planet for the present and future generations.

ACKNOWLEDGMENTS

Authors thank to Sectorial fund for research, development and forest technological innovation CONAFOR-CONACYT projects 2018-B-S-65769 and 2018-2-B-S-131466.

KEYWORDS

- **LED light**
- **biomass**
- **secondary metabolites**
- **plant factory**
- **medicinal**

REFERENCES

Ahmad, M.; Grancher, N.; Heil, M.; Blac, R.; Giovani, B.; Galland, P.; Lardemer D. Action Spectrum for Cryptochrome Dependent Hypocotyl Growth Inhibition in Arabidopsis. *Plant Physiol.* **2002**, *129*, 774–785.

Ahmad, N.; Rab, A.; Ahmad, N. Light-Induced Biochemical Variations in Secondary Metabolite Production and Antioxidant Activity in Callus Cultures of *Stevia rebaudiana* (Bert). *J. Photochem. Photobiol. B Biol.* **2016**, *154*, 51–56.

Bae, J.; Park, S.; Oh, M. Supplemental Irradiation with Far-Red Light-Emitting Diodes Improves Growth and Phenolic Contents in *Crepidiastrum denticulatum* in a Plant Factory with Artificial Lighting. *Horticult. Environ. Biotechnol.* **2017**, *58* (4), 357–366.

Bula, R.; Morrow, R.; Tibbitts, T.; Barta, D. J.; Ignatius, R.; Martin, T. Light Emitting Diodes as a Radiation Source for Plants. *HortScience* **1991**, *26*, 203–205.

Chang, L. et al. Metabolic Network Reconstruction of *Chlamydomonas* Offers Insight into Light-Driven Algal Metabolism. *Mol. Syst. Biol.* **2011**, *7*, 518.

Colquhoun, T.; Schwieterman, M.; Gilbert, J.; Jaworski, E.; Langer, K.; Jones, C.; Folta, K. Light Modulation of Volatile Organic Compounds from Petunia Flowers and Select Fruits. *Postharvest Biol. Technol.* **2013**, *86*, 37–44.

Darko, E.; Heydarizadeh, P.; Schoefs, B.; Sabzalian, M. Photosynthesis under Artificial Light: The Shift in Primary and Secondary Metabolism. *Phil. Trans. R. Soc. B Biol. Sci.* **2014**, *369* (1640).

Folta, K. Green Light Stimulates Early Stem Elongation, Antagonizing Light-Mediated Growth Inhibition. *Plant Physiol.* **2004**, *135* (3), 1407–1416.

Fukuyama, T.; Ohashi-Kaneko, K.; Watanabe, H. Estimation of Optimal Red Light Intensity for Production of the Pharmaceutical Drug Components, Vindoline and Catharan- Thine, Contained in *Catharanthus roseus* (L.) G. *Don. Environ. Cont. Biol.* **2015**, *53*, 217–220.

Fukuyama, T.; Ohashi-Kaneko, K.; Hirata, K.; Muraoka, M.; Watanabe, H. Effects of Ultraviolet a Supplemented with Red Light Irradiation on Vinblastine Production in *Catharanthus roseus*. *Environ. Control Biol.* **2017**, *55* (2), 65–69.

Gholamhoss, Z.; Hemati, K.; Dorodian, H.; Bashiri-Sadr, Z. Effect of Nitrogen Fertilizer on Yield and Amount of Alkaloids in Periwinkle and Determination of Vinblastine and Vincristine by HPLC and TLC. *Plant Sci. Res.* **2011**, *3*, 4–9.

Goins, G.; Yorio, N.; Sanwo, M.; Brown, S. Life Cycle Experiments with Arabidopsis under Red Light-Emitting Diodes (LEDs). *Life Support Biosph. Sci.* **1998**, *5*, 143–149.

Goto, E. Plant Production in a Closed Plant Factory with Artificial Lighting. *Acta Horticult.* **2012**, *956*, 37–49.

Guo, X.; Chang, B.; Zu, Y.; Tang, Z. The Impacts of Increased Nitrate Supply on *Catharanthus roseus* Growth and Alkaloid Accumulations under Ultraviolet B Stress. *J. Plant Interact.* **2014**, *9*, 640–646.

Hirata, K.; Asada, M.; Yatani, E.; Miyamoto, K.; Miura, Y. Effects of Near-Ultraviolet Light on Alkaloid Production in *Catharanthus roseus* Plants. *Plant Med.* **1993**, *59*, 46–50.

Hirata, K.; Horiuchi, M.; Ando, T.; Asada, M.; Miyamoto, K.; Miura, Y. Effect of Near Ultraviolet Light on Alkaloid Production in Multiple Shoot Cultures of *Catharanthus roseus*. *Plant Med.* **1991**, *57*, 499–500.

Hirata, K.; Horiuchi, M.; Asada, M.; Ando, T.; Miyamoto, K.; Miura, Y. Stimulation of Dimeric Alkaloid Production by Near-Ultraviolet Light in Multiple Shoot Cultures of *Catharanthus roseus*. *J. Fermentation Bioeng.* **1992**, *74*, 222–225.

Hoenecke, M.; Bula, R.; Tibbitts, T. Importance of 'blue' Photon Levels for Lettuce Seedlings Grown under Red-Light-Emitting Diodes. *HortScience* **1992**, *27*, 427–430.

Horibe, T.; Imai, S.; Matsuoka, T. Effects of Light Wavelength on Daughter Cladode Growth and Quality in Edible Cactus *Nopalea cochenillifera* Cultured in a Plant Factory with Artificial Light. *J. Horticult. Res.* **2018**, *26* (2), 71–80.

Hunter, D.; Burritt, D. Light Quality Influences Adventitious Shoot Production from Cotyledon Explants of Lettuce (*Lactuca sativa* L.). *Vitro Cell. Dev. Biol.-Plant* **2004**, *40* (2), 215–220.

Ibrahim, M.; Jaafar, H. Primary, Secondary Metabolites, H_2O_2, Malondialdehyde and Photosynthetic Responses of *Orthosiphon stamineus* Benth. to Different Irradiance Levels. *Molecules* **2012,** *17*, 1159–1176.

Ilić, Z.; Fallik, E. Light Quality Manipulation Improves Vegetable Quality at Harvest and Postharvest: A Review. *Environ. Exp. Bot.* **2017,** *139*, 79–90.

Johkan, M.; Shoji, K.; Goto, F.; Hashida, S.; Yoshihara, T. Blue Light-Emitting Diode Light Irradiation of Seedlings Improves Seedling Quality and Growth after Transplanting in Red Leaf Lettuce. *HortScience* **2010,** *45* (12), 1809–1814.

Kim, H.; Goins, G.; Wheeler, R.; Sager, J. Green-Light Supplementation for Enhanced Lettuce Growth under Red-And Blue-Light-Emitting Diodes. *HortScience* **2004,** *39* (7), 1617–1622.

Kim, H.; Wheeler, R.; Sager, C.; Goins, G.; Norikane, H. Evaluation of Lettuce Growth Using Supplemental Green Light with Red and Blue Light-Emitting Diodes in a Controlled Environment: A Review of Research at Kennedy Space Center. *Acta Hort.* **2006,** *711*, 111–119.

Kozai, T. Resource Use Efficiency of Closed Plant Production System with Artificial Light: Concept, Estimation and Application to Plant Factory. *Proc. Jpn. Acad. Ser. B Phys. Biol. Sci.* **2013,** *89*, 447.

Kozai, T. Propagation, Grafting and Transplant Production in Closed Systems with Artificial Lighting for Commercialisation in Japan. *Propagation of Ornamental Plants* **2007,** *7*, 145–149.

Kozai, T.; Niu, G.; Takagaki, M. *Plant Factory: An Indoor Vertical Farm System for Efficient Quality Food Production*; Nikky Levi, Elsevier Inc., Book Aid International: London, UK, 2016.

Lee, D.; Gould, K. Why Leaves Turn Red: Pigments Called Anthocyanins Probably Protect Leaves from Light Damage by Direct Shielding and by Scavenging Free Radicals. *Am. Sci.* **2002,** *90*, 524–531.

Lefsrud, M..; Kopsell, D.; Sams, C. Irradiance from Distinct Wavelength Light-Emitting Diodes Affect Secondary Metabolites in Kale. *HortScience* **2008,** *43* (7), 2243–2244.

Lemoine, Y.; Schoefs, B. Secondary Ketocarotenoid Astaxanthin Biosynthesis in Algae: A Multifunctional Response to Stress. *Photosynth. Res.* **2010,** *106*, 155–177.

Li, Q.; Kubota, C. Effects of Supplemental Light Quality on Growth and Phytochemicals of Baby Leaf Lettuce. *Environmental and Experimental Botany* **2009,** *67* (1), 59–64.

Lin, K.; Huang, M.; Huang, W.; Hsu, M.; Yang, Z.; Yang, C. The Effects of Red, Blue, and White Light-Emitting Diodes on the Growth, Development, and Edible Quality of Hydroponically Grown Lettuce (*Lactuca sativa* L. var. capitata). *Scientia Horticulturae* **2013,** *150*, 86–91.

Liu, X.; Chang, T.; Guo, S.; Xu, Z.; Li, J. Effect of Different Light Quality of LED on Growth and Photosynthetic Character in Cherry Tomato Seedling. *Acta Horticulturae* **2011,** *907*, 325–330.

Long, S.; Humphries, S.; Falkowski, G. Photoinhibition of Photosynthesis in Nature. *Annu. Rev. Plant Physiol. Plant Mol. Biol.* **1994,** *45*, 633–662.

Lu, N.; Takagaki, M.; Yamori, W.; Kagawa, N. Flavonoid Productivity Optimized for Green and Red Forms of *Perilla frutescens* via Environmental Control Technologies in Plant Factory. *Journal of Food Quality* **2018,** *2018*.

Ma, Z.; Sh, L.; Zhang, M.; Jiang, S.; Xiao, Y. Light Intensity Affects Growth, Photosynthetic Capacity, and Total Flavonoid Accumulation of Anoectohilus Plants. *HortScience* **2010,** *45*, 863–867.

Martineau, V.; Lefsrud, M.; Naznin, M. T. Comparison of Light-Emitting Diode and High-Pressure Sodium Light Treatments for Hydroponics Growth of Boston Lettuce. *HortScience* **2012,** *47,* 477–482.

Mimouni, V.; Ulmann, L.; Pasquet, V.; Mathieu, M.; Picot, L.; Bougaran, G.; Cadoret, J.; Morant-Manceau, A.; Schoefs, B. The Potential of Microalgae for the Production of Bioactive Molecules of Pharmaceutical Interest. *Curr. Pharm. Biotechnol.* **2012,** *13,* 2733–2750.

Mulas, G.; Gardner, Z.; Craker, L. Effect of Light Quality on Growth and Essential Oil Composition in Rosemary. *Int. Symp. Labiatae: Adv. Prod. Biotechnol. Utilisation* **2006,** *723,* 427–432.

Nguyen, D.; Kitayama, M.; Lu, N.; Takagaki, M. Improving Secondary Metabolite Accumulation, Mineral Content, and Growth of Coriander (*Coriandrum sativum* L.) by Regulating Light Quality in a Plant Factory. *J. Horticult. Sci. Biotechnol.* **2019,** 1–8.

Nishimura, R.; Zobayed, S.; Kozai, T.; Goto, E. Medicinally Important Secondary Metabolites and Growth of *Hypericum perforatum* L. Plants as Affected by Light Quality and Intensity. *Environ. Control Biol.* **2007,** *45,* 113–120.

Nitschke, S.; Cortleven, A.; Schmülling, T. Novel Stress in Plants by Altering the Photo Period. *Trends Plant Sci.* **2017,** *22* (11), 913–916.

Park, S.; Lee, J.; Cho, H.; Seong, E.; Kim, H.; Yu, C.; Kim, J. Metabolite Profiling Approach for Assessing the Effects of Colored Light-Emitting Diode Lighting on the Adventitious Roots of Ginseng (*Panax ginseng* C. A. Mayer). *Plant Omics J.* **2013,** *6,* 224–230.

Qian, H.; Liu, T.; Deng, M.; Miao, H.; Cai, C.; Shen, W.; Wang, Q. Effects of Light Quality on Main Health-Promoting Compounds and Antioxidant Capacity of Chinese Kale Sprouts. *Food Chem.* **2016,** *196,* 1232–1238.

Sabzalian, M.; Heydarizadeh, P.; Zahedi, M.; Boroomand, A.; Agharokh, M.; Sahba, M.; Schoefs, B. High Performance of Vegetables, Flowers, and Medicinal Plants in a Red-Blue LED Incubator for Indoor Plant Production. *Agronomy for Sustainable Development* **2014,** *34* (4), 879–886.

Sabzalian, M.; Heydarizadeh, P.; Zahedi, M.; Boroomand, A.; Agharokh, M.; Sahba, M.; Schoefs, B. High Performance of Vegetables, Flowers and Medicinal Plants in a Red–Blue LED Incubator for Indoor Plant Production. *Agron. Sustain. Dev.* In press.

Stutte, G. W. Controlled Environment Production of Medicinal and Aromatic Plants. In *Medicinal and Aromatic Crops: Production, Phytochemistry, and Utilization*; American Chemical Society, 2016; pp. 49–63.

Stutte, G.; Edney, S.; Skerritt, T. Photoregulation of Bioprotectant Content of Red Leaf Lettuce with Light-Emitting Diodes. *HortScience* **2009,** *44* (1), 79–82.

Tarakanov, I.; Yakovleva, O.; Konovalova, I.; Paliutina, G.; Anisimov, A. Light-Emitting Diodes: On the Way to Combinatorial Lighting Technologies for Basic Research and Crop Production. *Acta Hortic* **2002,** *956,* 171–178.

Urbonavičiuūtė, A.; Samuolienė, G.; Brazaitytė, A.; Ulinskaitė, R.; Jankauskienė, J.; Duchovskis, P.; Žukauskas, A. The Possibility to Control the Metabolism of Green Vegetables and Sprouts Using Light Emitting Diode Illumination. *Sodininkyste ir Darz ininkyste* **2008,** *27,* 83–92.

Wang, C.; Fu, C.; Liu, Y. Effects of Using Light-Emitting Diodes on the Cultivation of *Spirulina platensis*. *Biochem. Eng. J.* **2007,** *37,* 21–25.

Zhang, T.; Folta, K. Green Light Signaling and Adaptive Response. *J. Plant Signal. Behav.* **2012,** *7,* 1–4.

CHAPTER 11

Innovation in Agriculture: A Review of Plant Factory

VICTOR H. AVENDAÑO ABARCA,
DULCE CONCEPCIÓN GONZÁLEZ-SANDOVAL,
GUILLERMO CRISTIAN GUADALUPE MARTÍNEZ-ÁVILA, and
ROMEO ROJAS*

Universidad Autonoma de Nuevo Leon, School of Agronomy. 66054, General Escobedo, Nuevo León, México.

Corresponding author. E-mail: romeo.rojasmln@uanl.edu.mx

ABSTRACT

In recent years, food production has been affected due to a shortage caused by population growth worldwide. To this end, various alternatives have been generated, ranging from protected agriculture to a total production control system (plant factory). In this, all the variables for production are controlled, from temperature, relative humidity, light, nutrients, CO_2, pH, oxygen, among others. Additionally, the use of large extensions of land is not necessary since it is carried out on shelves and a homogeneous production is achieved throughout the year, mainly vegetables. In addition, it is possible to modify all the variables to promote an increase in the concentration of anti-oxidant components and/or nutrients that promote a healthier life, as well as being an environmentally friendly process. Therefore, continuous research is a viable, profitable, healthy, and environmentally friendly alternative to improve food production systems day by day.

11.1 INTRODUCTION

Protected agriculture is a production system that uses various structures for the protection of crops in order to provide adequate conditions for their growth and development; this form of production has been developed over time, generating multiple advantages, because of which it is possible to obtain products "off season" and in less time (precocity), in addition to having a better control of pests and diseases, greater production in less space, and products of greater commercial value. Greenhouses (Reséndiz, 2011) that allow partial climate control (Arellano-García et al., 2011) stand out among the best-known structures used in protected agriculture due to the low efficiency of the control systems used and their interaction with the environment outside (open system); from this disadvantage arises the concept of agricultural production under controlled environments called the plant factory production system (PFPS) (Kozai, 2013).

Plant factory (PF) is the most-advanced agricultural production system due to the total control of climatic variables that influence crop growth, presenting multiple advantages over open field and greenhouse production as the efficient use of all inputs for agricultural production (Kitaya et al., 1998; Kozai, 2013; Kozai et al., 2015).

The PFPS maximizes crop production using artificial lighting based on the light-emitting diode (LED), a technological device that has greater advantages than conventional devices, such as being able to emit specific and adjusted wavelengths to provide plants with a source of light capable of providing the energy necessary to optimally stimulate photosynthetic processes (wavelengths of red color: 610–750 nm and blue: 400–520 nm), as well as the possibility of promoting the synthesis of certain secondary metabolites of commercial interest such as antioxidant compounds from the induction of a photo-oxidative stress. However, the magnitude of this promoter effect will depend on the spectrum of light incident in the crop, its intensity, and photoperiod, as well as the plant species, variety, and stage of growth (Fan y Jao, 2000; Lefsrud et al., 2008; Li y Kubota, 2009; Mizuno et al., 2011; Žukauskas et al., 2011; Samouline et al., 2012; Tarakanov et al., 2012).

Among the most-cultivated horticultural species in this type of production system, leafy vegetables stand out, *Lactuca sativa* L. being the species of greatest interest, because of its ability to grow at low-light intensity, high-population density, low bearing, rapid growth, and profitability (Kozai, 2013).

11.2 PLANT FACTORY PRODUCTION SYSTEM

Since the 1990s, there is a tendency to obtain high productions out of season, anticipated and harmless. This trend has created the need to use various elements, tools, materials, and structures of protection for crops in order to modify environmental conditions in a way that favors the growth of plants (Juárez et al., 2011).

The PFPS is a model of protected horticulture, whose objective is to optimize the production of some plant species based on the knowledge and optimum control of their climatic, water, and nutritional requirements (Ohyama et al., 2000; Yokoi et al., 2005; Kozai et al., 2015).

The system consists of six main elements: (1) hermetically and thermally insulated structure, (2) multi-level system or floors with lighting devices, (3) electronic temperature and relative humidity control system, (4) environmental CO_2 injection unit, (5) hydroponic system, and (6) automatic system control unit (Kozai et al., 2006; Kozai, 2007). The advantages that this system present are: (1) reduction of the production cycle of up to 50% with uniform growth and high-quality products in lettuce cultivation, (2) planning of continuous production, and (3) less use of resources compared to conventional systems, such as the elimination of pesticide use, water use efficiency of up to 95–98% in lettuce (due to the high tightness of the PFPS that allows the water transpired by plants to be contained and through the system of cooling condense it, collect it, and return it to the irrigation system), and lower use of labor (Ohyama et al., 2000; Yoshinaga et al., 2000; Yokio et al., 2003; Kozai, 2013; Hu et al., 2014).

11.2.1 MAIN ELEMENTS TO CONSIDER IN A PLANT FACTORY PRODUCTION SYSTEM

11.2.1.1 LIGHTING SYSTEM

Currently, the electricity requirement for the production of 1 kg of fresh lettuce in a PFPS amounts to 10 kWh (36 MJ), consumption that represents 25–30% of the total production cost, of which 70–80% is due to artificial lighting. This energy consumption is currently high due to the low efficiency of the systems used for its production (high energy losses). Kozai et al. (2015) mention that energy consumption values can be reduced to less than 5 kWh/kg through the use of more efficient lighting devices and the ability to emit light with wavelengths related to the absorption range of photosynthetic pigments; and by means of the use of elements or systems that avoid the loss of light improving the distribution of this in the canopy of the leaf, to generate a greater absorption in the plant making more efficient the use of the energy.

Various forms and sources of electromagnetic stimulation have been used in recent years for the optimal development of the photosynthesis process of plants in a natural and artificial way (Gonzalías y Lasso, 2016). In the case of production systems in controlled environments, the main sources of artificial lighting used are (Table 11.1): fluorescent lamps, high pressure sodium lamps (HPS), and, as of 2010, the incorporation of LED because these are more efficient elements that can be located very close to the plants and configured to emit a high specific light flow with lower-energy consumption (Gonzalías y Lasso, 2016).

TABLE 11.1 Comparison of Parameters of Artificial Lighting Devices: Light Emitting Diode (LED), High-Pressure Sodium Lamp (HPS), Incandescent, and Fluorescent Lamp.

	LED	**HPS**	**Incandescent**	**Fluorescent**
Light efficiency (lumens/Watts)	80	100	20	80
Lighting source feature	Solid-state lighting	Gas-discharge lighting	Solid-state lighting	Gas-discharge lighting
Useful life (hours)	50,000	8000	3000	10,000
Environment friendly	It does not pollute	Contamination by mercury and lead	It does not pollute	Contamination by mercury and lead
Power consumption of the transformer (loss)	15 W	50 W	0 W	45 W

The application of LED lighting in plant growth was documented for the first time in the cultivation of *Lactuca sativa* L. cv. Grand Rapids (Bula et al., 1991), a study in which the growth and development of the crop under red monochromatic LED lighting (660 nm), supplemented with blue fluorescent lamps (400–500 nm) was comparable against fluorescent and incandescent lights of cold white color. From this, the interest of the application of LED lighting in horticulture increased (Morrow, 2008) and it is being evaluated in other plant species such as spinach (*Spinacia oleracea*), cabbage (*Brassica oleracea*), cucumber (*Cucumis sativus*), medicinal plants, among others (Yorio et al., 2001; Li et al., 2012; Hernández y Kubota, 2016; Wang et al., 2016).

The innovations in artificial lighting based on light-emitting diodes for horticultural production in this type of systems focus on the development of high-quality devices, capable of emitting wavelengths similar to those absorbed by photosynthetic pigments (455 and 660 nm), as well as a greater efficiency in the processes in the conversion of electrical energy to light energy, mainly in the red light-emitting diodes (660 nm), because these present a lower efficiency compared to the emitting diodes of blue light.

11.2.2   TYPE OF HYDROPONIC SYSTEM

11.2.2.1   FLOATING ROOT SYSTEM

The floating root system was one of the first hydroponic systems used both experimentally and commercially, optimizing the production area (Chang et al., 2000). It is a system in which the roots of the plants are partially or totally submerged in nutrient solution (NS), generally using perforated plates of expanded polystyrene, of high and medium density that act as mechanical support of a certain number of plants (Cruz, 2016). In this type of system, it is very important to keep track of the dissolved oxygen content in the NS. Also, it is important to mention that the floating root system is not a complex mechanism; therefore, it is easy to operate and it is very simple to produce through it. It does not even require a large investment and the results are permanently reflected over time. Like other hydroponic techniques, it is also possible to obtain large productions. This technique can be adapted to any space. It also allows the use of different materials; for this reason, it does not harm the environment.

The crops that best adapt to this system are those of low size, highlighting the cultivation of lettuce, spinach, and some aromatic plants (Tigrero, 2018).

11.2.2.2   NFT SYSTEM

The nutrient film technique (NFT) production system was developed in England by Allen Cooper in the year 1965 and consists of maintaining an NS strand in the root zone of plants to provide water and nutrients. It is considered as one of the most viable systems for growing crops, mainly leafy vegetables (Cometti, 2003), highlighting the cultivation of lettuce, because this system provides an effective method in the control of water, nutrition, and mainly by allowing an intermittent circulation regime if a zero slope is maintained that helps to obtain reductions in electricity consumption (Carrasco et al., 1999); an important aspect to consider under the PFSP.

The NFT system does not use any type of substrate and is classified as a closed system, because the SN that circulates through the pipes or growth channels is recirculated (Rodríguez et al., 2002). According to Urrestarazu (2004), the NFT system comprises the following elements: (1) culture channels, (2) SN distribution network, (3) hydraulic pump, (4) collection pipe system, and (5) a tank for the NS.

Among the most important aspects to consider in the design of an NFT hydroponic system, the channel length, flow velocity, oxygenation, temperature, NS strip, and the population density of the crop stand out.

11.2.3   CONTROL VARIABLES IN A PLANT FACTORY PRODUCTION SYSTEM

11.2.3.1   ENVIRONMENTAL VARIABLES

11.2.3.1.1   Temperature

The closed agricultural production system offers the possibility of controlling the ambient temperature and the NS. The first through an air-conditioning system and the second through temperature sensors, tubular resistors (heater), and heat sinks (cooler) regulated by an electronic control system. Among the climatic conditions, the environmental temperature has an important effect on the growth, development, and quality of the crops (Choi et al., 2000; Savvas, 2002; Lee et al., 2013; Moon et al., 2014), this for having direct implications in the processes of photosynthesis and respiration, permeability of the cell membrane, absorption of water and nutrients, perspiration, and enzymatic activities (Baudoin et al., 2002). Choi et al. (2000) report that until 25 days after transplantation, the optimum temperature for lettuce growth under PFPS is 22–26°C and subsequently 20–24°C until harvest and an optimal night temperature of 15–20°C throughout the crop cycle. The respiratory rate of plants that grow under a high concentration of environmental CO_2 (700 ppm) have a respiratory rate (expressed as dry weight) between 15% and 20% higher compared to others that have grown to 350 ppm of CO_2 (Drake et al., 1999), so lowering the temperature during the night may decrease the effects of the high respiration rate.

11.2.3.1.2   Relative Humidity

The relative humidity (RH) between 55% and 90% (1.0–0.2 kPa of vapor pressure deficit at a temperature of 20°C) has little effect on the physiology and growth of horticultural crops. Low RH (<75%) produces a reduction in growth due to a high perspiration rate, while high RH (> 80%) promotes the presence of diseases causing growth and development disorders.

In the case of lettuce, the requirements of HR in the open field are between 30% and 70% (Alvarado et al., 2001), while in the SPFP the values established in the research work are in the range of 60–80% (1.19–0.59 kPa of vapor pressure deficit at 24°C) (Choi et al., 2000; Park et al., 2012; Kang et al., 2013) having positive effects on water potential (no stress due to evapotranspiration demand of the environment) and in the control of the physiological disorder of the burned tips in leaves.

11.2.3.1.3   *CO_2 Concentration*

The concentration of CO_2 in the air is one of the most studied environmental factors due to the increase in its concentration and possible impact on agricultural ecosystems (Ainsworth et al., 2007). A high concentration of CO_2 in the air can cause two physiological effects in plants: (1) increase in the photosynthetic rate due to the increase in the concentration of CO_2 in the mesophyll (Lee et al., 1999; Kang et al., 2013) and (2) stomatic closure, which reduces water loss due to perspiration. Under conditions of high concentration of environmental CO_2 (1000–2000 ppm), long-term effects on plant phytochemical composition (Poorter et al., 2003) may occur as increases in the content of antioxidant compounds (Park et al., 2012; Samuoliené et al., 2012) and carbohydrates (Wang, 2006; Sirtautas et al., 2012), thereby improving the nutritional properties of the crop for human consumption.

Investigations on the effects of the concentration of CO_2 in the air in different crops report increases in photosynthetic rate and growth in combination with artificial lighting under concentrations of up to 2000 ppm CO_2, allowing a greater production of dry biomass in less time, which allows to shorten the production cycle (Yoshinaga et al., 2000; Park et al., 2012; Kozai et al., 2015).

11.2.3.1.4   *Light*

The wavelengths of the visible electromagnetic spectrum are not of total utility for plants; in general, photoreceptors absorb violet (390–450 nm), blue (450–490 nm), orange (590–620 nm), and red (620–750 nm). Therefore, light quality (Zhang and Folta, 2012), light intensity (photon flow per unit area over time), and photoperiod (light/dark hours) are key elements of the light condition.

In the aspect of artificial lighting, LED-based lamps are the most recently used technology in agriculture in a controlled environment, due to their high light efficiency; allowing to reduce the costs generated by the consumption of electrical energy (Bourget, 2008). This technology has the advantage of emitting monochromatic wavelengths (specific wavelengths), offering plants a light source tailored to their needs, and increasing the efficiency of the entire production process (Fan and Jao, 2000). It should be noted that each plant species has specific light requirements.

Quality of light. The quality of light is understood as the wavelengths that affect the crop in order to carry out photosynthesis. Red and blue light show greater efficiency in the photosynthetic processes in all plant species (Lefsrud et al., 2008; Mizuno et al., 2011; Žukauskas et al., 2011; Samuolienė et al., 2012; Tarakanov et al., 2012). In addition to this, certain wavelengths can generate increases in the concentrations of antioxidant compounds such as vitamin C, carotenoids, anthocyanins, and flavonoids (Lefsrud et al., 2008; Li and Kubota, 2009; Stutte et al., 2009; Žukauskas et al., 2011; Li et al., 2012).

Intensity of the light. The light intensity represents the amount or flow of photons that fall in an area unit in one second and its unit of measurement is $\mu mol/m^2/s$. Light intensity has important effects on crop growth such as the increase and speed in biomass accumulation (Hunter and Burritt, 2004; Li and Kubota, 2009; Stutte et al., 2009; Johkan et al., 2010), greater precocity, and less accumulation of nitrates in leaves in lettuce cultivation (Samuolienė et al., 2009; Li and Kubota, 2009).

Photoperiod. The photoperiod is the time (expressed in hours) of exposure to daily natural or artificial lighting at which a crop grows. The PFPS allows to manipulate this variable of light condition mainly to provide the optimal requirements of daily light integral (DLI) to the cultivated plant species (Table 11.2) and, in this way, gives an agronomic management to the crop to obtain a production in less time and of better quality (high content of antioxidants, low content of nitrates in leaves, disease free, greater fresh and dry weight, and uniformity), integrating aspects of quality and intensity. Lettuce cultivation requires an accumulation of daily photons between 12 and 17 $mol/m^2/day$, which represents a photoperiod of 12 light h and 12 dark h at a light intensity of 300 $\mu mol/m^2/s$ (Albright et al., 2000).

11.2.4 NUTRITIVE SOLUTION VARIABLES

Rodríguez et al. (2016) defines the SN as the mixture of chemical fertilizers, which contain all the essential nutrients for plants (N, P, K, Ca, Mg, S, Fe, Mn, B, Cu, Zn, Cl, and Mo) and that when dissolved in water (ionic form) provide complete nutrition for plants. The aspects of the nutritive solution with the greatest influence on production are: the mutual relationship between cations and anions, the concentration of nutrients, pH, electrical conductivity (EC), and temperature (Sonneveld and Straver, 1999).

TABLE 11.2 Requirements of Light Intensity, Photoperiod, and Daily Integral Light, for the Production of Crops in an PFPS.

Crop	Finish	Light intensity ($\mu mol/m^2/s$)	Photoperiod (h/day)	DLI ($mol/m^2/day$)
Rice	Commercial production	1000–1200	12	43.2–51.84
Tomato	Commercial production	520.83–607	16	30–35
Strawberry	Commercial production	225–337.5	24–16	19.4
Seedlings	Commercial production	200–300	18–12	13
Spinach	Commercial production	200–300	16.6–11.1	12–13
Lettuce	Commercial production	200–300	16.6–15.74	12–17
Japanese mint	Production of secondary metabolites	200	12	8.64

11.2.4.1 NUTRIENT CONCENTRATION: MASS BALANCE PRINCIPLES

One way to determine the nutritional requirements for each plant species is based on the concept of mass balance, which is based on the fact that the composition of the dry matter of a plant is formed by the 16 essential chemical elements (C, H, O, N, P, K, Ca, Mg, S, Fe, Mn, B, Cu, Zn, Mo, and Cl) of which three are contributed by air and water (C, H, and O) and the rest 13 must be incorporated into the NS to be absorbed through the roots of the plants (Etchevers-Barra, 2000).

By knowing the amount of dry matter produced and concentration of each of the elements in each of its phenological stages of the crop, you can estimate the total amount of each nutrient that the plant absorbed and that is necessary to apply in the SN during this period of growth, that is, the application of fertilizers must be throughout the production cycle and proportional to the growth of the crop (Alpízar-Vargas et al., 2006). Considering the efficiency of nutrient absorption as the source of these (fertilizer), a fertilization plan can be established for each stage of the crop.

11.2.4.2   TEMPERATURE OF THE NUTRIENT SOLUTION

Temperature mainly affects the content of dissolved oxygen in the NS (Steiner, 1968; Vestergaard, 1984). As the temperature rises, there are negative effects such as inhibition of cell division and reduction of root elongation causing decrease in the absorption rate of Ca^{2+} ions, an essential element that causes low physiological disorder called burnt leaf tips on lettuce (Collier et al., 1984). The nitrogen absorption rate in the form of NO_3^- is less affected under high-temperature conditions of the NS compared to other elements, because the plant obtains oxygen from the reduction of NO_3^- to NO_2^-, generating higher concentrations of this element in the lettuce leaves, a relevant factor in terms of the quality of the final product (Morard et al., 2000) and undesirable for human consumption. For all this, it is recommended that the temperature of the NS does not exceed 20°C (Magalhães, 2006).

11.2.4.3   ELECTRIC CONDUCTIVITY

The EC is the most used way to express the total salt content in the water and its measurement unit is the decisiemens per meter (dS/m) referenced at a temperature of 25°C, because Ion activity depends on it. The plant's response to high EC is dependent on the age, environmental conditions, crop management, and characteristics of the plant species, directly damaging the plants from the early stages of development until the culmination of their biological cycle, affecting the production and even causing death in the most susceptible species. Some of the most common and harmful effects caused by salinity are: (1) specific toxicity (normally associated with excessive absorption of Na^+ and Cl^-), (2) nutritional imbalance due to the interference of saline ions with essential nutrients, and (3) osmotic effect (Prieto, 2008).

According to Rhoades et al. (1992), an EC equal 2.1 mS/cm generates a 10% decrease in the production of lettuce and 50% when it has a value of 5.1 mS/cm.

11.2.4.4 PH

The pH of the NS is determined by the concentration of H^+ and is important as it controls the availability of nutrients. Rodríguez et al. (2016) recommends that the pH of the NS be adjusted to a value of 5 to 5.5 to avoid that the solubility of Fe^{2+} and Mn^{2+} does not decrease and is available to plants, because nutrients and others such as Ca^{2+} and P having pH values greater than 6 can precipitate. The pH of the NS should be monitored frequently since it is not static, because it depends on the concentration of environmental CO_2, the rate of nutrient absorption, the source of nitrogen used, among other variables (Favela et al., 2006).

Alvarado et al. (2001) points out that lettuce is a vegetable classified as slightly tolerant of acidity, reporting an optimal growth pH equal to 5.8. Under the PFPS, the recommended pH value for baby lettuce production is in the range of 5.5–6 (Shimizu et al., 2011; Kang et al., 2013; Kobayashi, et al., 2013; Kozai et al., 2015).

11.2.4.5 DISSOLVED OXYGEN CONCENTRATION

Water, in addition to dissolving mineral salts, also does it with oxygen. The temperature of the NS is directly related to the amount of oxygen consumed by the plants and inverse to the oxygen dissolved in it. An NS with a temperature of 10°C maintains an Oxygen concentration of 10.93 and 6 ppm at 45°C (Steiner, 1968; Vestergaard, 1984). Favela et al. (2006) report that at a temperature below 22°C the dissolved oxygen content is sufficient to meet the demand for this nutrient; but factors such as population density, growth status, photosynthetic, and respiratory rates that directly influence the total oxygen requirements of the crop should be considered (Papadopoulous et al., 1999). Alvarado et al. (2001) report that the minimum requirement of dissolved oxygen in the NS for lettuce production is 4 ppm.

11.3 CONCLUSION

Plant factories have now taken a lot of importance because of the efficient use of inputs being an innovation for the production of various crops; research is being initiated to continue improving this agricultural innovation and leave traditional production methods in a moment. Alternative uses of LEDs are making it possible to obtain better-quality products from plant factories, and advances in LED efficiency will make plant factories with artificial lighting even more desirable.

ACKNOWLEDGMENTS

All authors thank to Sectorial fund for research, development and forest technological innovation CONAFOR-CONACYT projects 2018-B-S-65769 and 2018-2-B-S-131466.

CONFLICTS OF INTEREST

The authors declare no conflict of interest.

KEYWORDS

- **LED Light**
- **vegetables**
- **controlled production**
- **nutrients**
- **production system**

REFERENCES

Ainsworth, E. A.; Rogers, A. The Response of Photosynthesis and Stomatal Conductance to Rising (CO_2): Mechanisms and Environmental Interactions. *Plant Cell Environ.* **2007,** *30,* 258–270.

Albright, L. D.; Both, A. J.;Chiu, A. J. Controlling Greenhouse Light to a Consistent Daily Integral. *Trans. ASAE* **2000,** *43* (2), 421.

Alpízar-Vargas, M. E.; González-Abaunza, D. F.; Spaans, E. J.; Tabora, P. Plan dinámico de fertilización para escalopine verde (*Cucurbita pepo*). Dynamic Fertilization Plan for the Scallop Gourd (*Cucurbita pepo*). *Tierra Trop. Sostenibilidad, Ambiente y Sociedad* **2006,** *2* (1), 39–47.

Alvarado, D.; Chavez, F.; y Anna, K. Seminario de Agronegocios: Lechugas Hidropónicas. Universidad del Pacífico. 2001. http://files.aiagronegocios.webnode.es/200000072-ae45bb 039e/Lechuga%20Hidroponica.pdf.

Arellano-García, M.; Valera-Martínez, D.; Urrestarazu-Gavilán, M.; Quezada-Martín, M.; Murguía-López, J.; Zermeño González, A. Ventilación natural y forzada de invernaderos tipo almería y su relación con el rendimiento de tomate. *Terra Latinoamericana* **2011**, *29*, 379–386.

Baudoin, W.; Nisen, A.; Grafiadellis, M.; Verlodt, H.; Jiménez, R.; De Villele, O.; La Malfa, G.; Martínez-García P.; Verlodt H.; Monteiro, A. El cultivo protegido en clima mediterráneo. In *Medios y Técnicas de Producción. Suelo y Sustratos*; FAO: Roma, 2002; pp 143–182.

Bourget, C. M. An Introduction to Light-Emitting Diodes. *HortScience* **2008**, *43*, 1944–1946.

Bula, R. J.; Morrow, R. C.; Tibbitts, T. W.; Barta, D. J.; Ignatius, R. W.; Martín, T. S. Light-Emitting Diodes as a Radiation Source for Plants. *HortScience* **1991**, *26* (2), 203–205.

Carrasco, G.; Rodríguez, E.; Escobar, P.; Izquierdo, J. Development of Nutrient Film Technique "NFT" in Chile: The Use of Intermittent Recirculation Regimes. *Acta Horticulturae* **1999**, 481, 305–310.

Chang, M.; Hoyos, M.; Rodríguez, A. *Manual práctico de hidroponía: sistema de raíz flotante y sistema de sustrato sólido*; SE: Perú, 2000.

Choi, K. Y.; Paek, K. Y.; Lee, Y. B. Effect of Air Temperature on Tipburn Incidence of Butterhead and Leaf Lettuce in a Plant Factory. In *Transplant Production in the 21st Century*; Springer Netherlands, 2000; pp 166–171.

Collier, G. F.; Tibbitts, T. W. Effects of Relative Humidity and Root Temperature on Calcium Concentration and Tipburn. *J. Am. Soc. Horticult. Sci.* **1984**, *109*, 128–131.

Cometti, N. N. *Nutrição Mineral da Alface (Lactuca sativa L.) em Cultura Hidropônica-Sistema NFT.* (Tesis doctoral) Universidad Federal Rural de Rio de Janeiro, Brasil, 2003.

Drake, B. G.; Azcon-Bieto, J.; Berry, J.; Bunce, J.; Dijkstra, P.; Farrar, J.; Gifford, R. M.; Gonzalez-Meler, M.; Koch, A. G.; Lambers, H.; Wullschleger, S.; Siedow, J. Does Elevated Atmospheric CO_2 Concentration Inhibit Mitochondrial Respiration in Green Plants? *Plant Cell Environ.* **1999**, *22* (6), 649–657.

Etchevers, B. J. D. Técnicas de diagnóstico útiles en la medición de la fertilidad del suelo y el estado nutrimental de los cultivos. *Terra Latinoamericana* **2000**, *17* (3), 209–219.

Fang, W.; Jao, R. C. A Review on Artificial Lighting of Tissue Cultures and Transplants. In *Transplant Production in the 21st Century*; Springer Netherlands, 2000; pp 108–113.

Favela, C. E.; Preciado, R. P.; Adalberto, B. *Manual para preparar soluciones nutritivas*; México: UAAAN, 2006.

Gonzalías, Y. R.; Lasso, E. R. Desarrollo de un sistema de iluminación artificial LED para cultivos en interiores-Vertical Farming (VF). *Informador técnico* **2016**, *80* (2), 111–120.

Hernández, R.; Kubota, C. Physiological Responses of Cucumber Seedling Sunder Different Blue and Red Photon Flux Ratios Using LEDs. *Environ. Exp. Bot.* **2016**, *121*, 66–74.

Hu, M. C.; Chen, Y. H.; Huang, L. C. A Sustainable Vegetable Supply Chain Using Plant Factories in Taiwanese Markets: A Nash–Cournot Model. *Int. J. Prod. Econ.* **2014**, *152*, 49–56.

Hunter, D. C.; Burritt, D. J. Light Quality Influences Adventitious Shoot Production from Cotyledon Explants of Lettuce (*Lactuca sativa* L.). *In Vitro Cell. Dev. Biol.-Plant* **2004**,*40* (2), 215–220.

Johkan, M.; Shoji, K.; Goto, F.; Hashida, S. N.; Yoshihara, T. Blue Light-Emitting Diode Light Irradiation of Seedlings Improves Seedling Quality and Growth after Transplanting in Red Leaf Lettuce.*HortScience* **2010**, *45* (12), 1809–1814.

Juárez-López, P.; Bugarin-Montoya, R.; Castro-Brindis, R.; Sánchez-Monteon, A. L.; Cruz-Crespo, E.; Juárez-Rosete, C. R.; Alejo-Santiago, G.; Balois Morales, R. Estructuras utilizadas en la agricultura protegida. *Revista Fuente* **2011**, *3* (8), 21–27.

Kang, J. H.; KrishnaKumar, S.; Atulba, S. L. S.; Jeong, B. R.; Hwang, S. J. Light Intensity and Photoperiod Influence the Growth and Development of Hydroponically Grown Leaf Lettuce in a Closed-Type Plant Factory System. *Horticult. Environ. Biotechnol.* **2013**, *54*, 501–509.

Kitaya, Y.; Niu, G.; Kozai, T.; Ohashi, M. Photosynthetic Photon Flux, Photoperiod, and CO_2 Concentration Affect Growth and Morphology of Lettuce Plug Transplants. *HortScience* **1998**, *33*, 988–991.

Kobayashi, K.; Amore, T.; Lazaro, M. Light-Emitting Diodes (LEDs) for Miniature Hydroponic Lettuce. *Opt. Photon. J.* **2013**, *3* (01), 74.

Kozai T.; Ohyama K.; Chun C. Commercialized Closed Systems with Artificial Lighting for Plant Production. *Acta Horticult.* **2006**, *711*, 61–70.

Kozai, T. Propagation, Grafting and Transplant Production in Closed Systems with Artificial Lighting for Commercialization in Japan. *Propagat. Ornament. Plants* **2007**, *7*, 145–149.

Kozai, T. Resource Use Efficiency of Closed Plant Production System with Artificial Light: Concept, Estimation and Application to Plant Factory. *Proc. Jpn Acad. Ser. B Phys. Biol. Sci.* **2013**, *89*, 447.

Kozai, T.; Niu, G.; Takagaki, M., Eds. *Plant Factory: An Indoor Vertical Farming System for Efficient Quality Food Production*; Academic Press, 2015.

Lee, J. G.; Choi, C. S.; Jang, Y. A.; Jang, S. W.; Lee, S. G.; Um, Y. C. Effects of Air Temperature and Air Flow Rate Control on the Tipburn Occurrence of Leaf Lettuce in a Closed-Type Plant Factory System. *Horticult. Environ. Biotechnol.* **2013**, *54* (4), 303–310.

Lee, Y. B.; Park, M. H. Effects of CO_2 Concentration, Light Intensity and Nutrient Level on Growth of Leaf Lettuce in a Plant Factory. *Int. Symp. Grow. Media Hydropon.* Aug **1999**, *548*, 377–384.

Lefsrud, M. G.; Kopsell, D. A.; Sams, C. E. Irradiance from Distinct Wavelength Light-Emitting Diodes Affect Secondary Metabolites in Kale. *HortScience* **2008**, *43* (7), 2243–2244.

Li, H.; Tang, C.; Xu, Z.; Liu, X.; Han, X. Effects of Different Light Sources on the Growth of Non-Heading Chinese Cabbage (*Brassica campestris* L.). *J. Agric. Sci.* **2012**, *4* (4), 262–273.

Li, Q.; Kubota, C. Effects of Supplemental Light Quality on Growth and Phytochemicals of Baby Leaf Lettuce. *Environ. Exp. Bot.* **2009**, *67* (1), 59–64.

Magalhães, A. G. *Caracterização de genótipos de alface (Lactuca sativa L.) em cultivo hidropônico sob diferentes valores de condutividade elétrica da solução nutritiva*(Tesis Doctoral). Universidad Federal Rural de Pernambuco, Brasil, 2006.

Mizuno, T.; Amaki, W.; Watanabe, H. Effects of Monochromatic Light Irradiation by LED on the Growth and Anthocyanin Contents in Leaves of Cabbage Seedlings. *Acta Horticult.* **2011**, *907*, 179–184.

Moon, S. M.; Kwon, S. Y.; Lim, J. H. Minimization of Temperature Ranges between the Top and Bottom of an Air Flow Controlling Device through Hybrid Control in a Plant Factory. *Sci. World J.* **2014**, *2014*, 1–7.

Morard, P.; Lacoste, L.; Silvestre, J. Effect of Oxygen Deficiency on Uptake of Water and Mineral Nutrients by Tomato Plants in Soilless Culture. *J. Plant Nutr.* **2000**, *23*, 1063–1078.

Morrow, R. C. LED Lighting in Horticulture. *HortScience* **2008**, *43* (7), 1947–1950.

Ohyama, K.; Yoshinaga, K.; Kozai, T. Energy and Mass Balance of a Closed-Type Transplant Production System (Part 2). Water Balance. *J. Soc. High Technol. Agric.* **2000**, *12*, 217–224.

Park, G. Y.; Park, E. J.; Hwang, J. S.; Jeong, R. V. Light Source and CO_2 Concentration Affect Growth and Anthocyanin Content of Lettuce under Controlled Environment. *J. Horticult. Environ. Biotechnol.* **2012**, *53*, 460–466.

Prieto, D. Riego con aguas salinas y aguas de drenaje, control e impacto de salinidad. *Jornadas sobre "Ambiente y riego: modernización y ambientalidad", La antigua, Guatemala. Red de riegos. CYTED y AECI*, 2008.

Reséndiz, M.; Aguilar, J.; Luévano, A. Características de la agricultura protegida y su entorno en México. *Revista Mexicana de Agronegocios* **2011,** *15*, 763–774.

Rodríguez, D.; la Rosa, A.; Rojas, M. H.; Gutiérrez, M. F. *Manual práctico de hidroponía*; Perú: Universidad Agraria la Molina, 2002.

Rodríguez, H.; Rodríguez, J. C.; Vidales, J. A.; Luna, A. I. *Cultivo hidropónico de Lilium (Azucena): para flor de corte y en maceta*; México: Trillas S.A. de C.V, 2016.

Samuolienė, G.; Sirtautas, R.; Brazaitytė, A.; Duchovskis, P. LED Lighting and Seasonality Effects Antioxidant Properties of Baby Leaf Lettuce. *Food Chem.* **2012,** *134* (3), 1494–1499.

Savvas, D.; Passam, H. C. *Hydroponic Production of Vegetable and Ornamentals*; Embryo Publications: Greece, 2002.

Shimizu, H.; Saito, Y.; Nakashima, H.; Miyasaka, J.; Ohdoi, K. Light Environment Optimization for Lettuce Growth in Plant Factory. *IFAC Proc. Volumes* **2011,** *44* (1), 605–609.

Sirtautas, R.; Samouline, G.; Brazaityte, A.; ShaKalauskaite, J.; Shakalauskiene, S.; Virsile, A.; Jankau-skiene, J.; Vastakaite, V.; Duchovskis, P. Impact of CO_2 on Quality of Baby Lettuce Grown under Optimized Light Spectrum. *J. Acta Scientiarum Polonorum Hortorum Cultus* **2012,** *13*, 109–118.

Sonneveld, C.; Straver, N. Nutrient Solutions for Vegetables and Flower Grow in Water or Substrates. Serie: Voedingsoplossingen Glastuinbouw (Netherlands), 1999.

Steiner, A. A. Soilles Culture. In *Proceedings of the 6th Colloquium of the International Potash Institute*; Italy, 1968; pp 324–341.

Stutte, G. W.; Edney, S.; Skerritt, T. Photoregulation of Bioprotectant Content of Red Leaf Lettuce with Light-Emitting Diodes. *HortScience* **2009,** *44* (1), 79–82.

Tarakanov, I.; Yakovleva, O.; Konovalova, I.; Paliutina, G.; Anisimov, A. Light-Emitting Diodes: On the Way to Combinatorial Lighting Technologies for Basic Research and Crop Production. *VII Int. Symp. Light Horticult. Syst.* Oct **2012,** *956*, 171–178.

Urrestarazu, G. M. *Tratado de cultivo sin suelo*; Mundi-Prensa: España, 2004.

Vestergaard, B. Oxygen Supply to the Roots in Different Hydroponic Systems. In *6. International Congress on Soilless Culture*; Lunteren (Netherlands), 29 Apr–5 May 1984.

Wang, S. Y. Effect of Pre-Harvest Conditions on Antioxidant Capacity in Fruits. *Journal Acta Horticulturae* **2006,** *712*, 299–306.

Yokoi, S.; Kozai, T.; Hasegawa, T.; Chun, C.; Kubota, C. CO_2 and Water Utilization Efficiencies of a Closed Transplant Production System as Affected by Leaf Area Index of Tomato [*Lycopersicon esculentum*] Seedling Populations and the Number of Air Exchanges. *J. Soc. High Technol. Agric. (Japan)*, 2005.

Yokoi, S.; Kozai, T.; Ohyama, K.; Hasegawa, T.; Chun, C.; Kubota, C. Effects of Leaf Area Index of Tomato Seedling Populations on Energy Utilization Efficiencies in a Closed Transplant Production System. *J. Soc. High Technol. Agri. (Japan)*, 2003.

Yorio, N. C.; Goins, G. D.; Kagie, H. R.; Wheeler, R. M.; Sager, J. C. Improving spinach, Radish, and Lettuce Growth under Red Light-Emitting Diodes (LEDs) with Blue Light Supplementation. *HortScience* **2001,** *36* (2), 380–383.

Yoshinaga, K.; Ohyama, K.; Kozai, T. Energy and Mass Balance of a Closed-Type Transplant Production System (Part 3). Carbon Dioxide Balance. *J. Soc. High Technol. Agric.* **2000,** *12*, 225–231.

Zhang, T.; Folta, K. M. Green Light Signaling and Adaptive Response. *J. Plant Signal. Behav.* **2012**, *7*, 1–4.

Žukauskas, A.; Bliznikas, Z.; Breivė, K.; Novičkovas, A.; Samuolienė, G.; Urbonavičiūtė, A.; Brazaitytė, A.; Jankauskienė, J.; Duchovskis, P. Effect of Supplementary Pre-Harvest LED Lighting on the Antioxidant Properties of Lettuce Cultivars. *Acta Horticulturae* **2011**, *907*, 87–90.

CHAPTER 12

Sorghum and Aphid (*Melanaphis sacchari*) Interaction: Plant Physiology, Breeding, and Molecular Overview

B. RINCÓN-LÓPEZ[1], ANTONIO FLORES NAVEDA[1,2], U. ARANDA LARA[3], MARTIN E. TIZNADO HERNÁNDEZ[4], and JULIO CÉSAR TAFOLLA-ARELLANO[1,5,*]

[1]*Programa de Maestría en Ciencias en Fitomejoramiento, Universidad Autónoma Agraria Antonio Narro 25315, Buenavista, Saltillo, Coahuila, México*

[2]*Centro de Capacitación y Desarrollo en Tecnología de Semillas, Universidad Autónoma Agraria Antonio Narro, 25315, Buenavista, Saltillo, Coahuila, México*

[3]*INIFAP-Campo Experimental Río Bravo 88900, Río Bravo, Tamaulipas, México*

[4]*Research Center in Food and Development A.C. 83304, Hermosillo, Sonora, México*

[5]*Departamento de Ciencias Básicas, Laboratorio de Biotecnología y Biología Molecular, Universidad Autónoma Agraria Antonio Narro, 25315, Buenavista, Saltillo, Coahuila, México*

Corresponding author. E-mail: jtafare@uaaan.edu.mx

ABSTRACT

Sorghum (*Sorghum bicolor* (L.)) is considered one of the most important cereals in the world. Among the various biotic factors that are reducing and limiting sorghum production is insect pest. Around 150 species of insects have been reported to negatively affect world sorghum production; Sugarcane Aphid (*Melanaphis sacchari* Zehntner) being the major pest, which reduces yield by 50–100%, grain quality, and marketability depending on tolerance of the sorghum genotype. To face this pest, different strategies has been tried such as modifying the planting date, best agronomic practices, utilization of parasitoid species, genetic improvement, insecticide application, among others. However, the physiological and molecular basis for sorghum resistance is not well understood; scarce information is available on the mechanism of tolerance, which limit the strategies designed from these perspectives. In this chapter, overviews of sorghum plant physiology, breeding, and molecular structure about the defense mechanism during sugar cane aphid interaction are presented.

12.1   INTRODUCTION

Sorghum, *Sorghum bicolor* L., is an important cereal domesticated in Africa and a multi-functioning staple food crop grown in the world, valued for its grain, stalks, and leaves and is drought-tolerant optimal for arid climates. The sorghum grains and other plant constituents, such as as sugar and cellulose-material, has been used in livestock feed, building material, fencing, or for brooms. Their other novel use is in foods and beverages because they are gluten-free and are suitable for coeliacs. The sorghum has the highest phenolic compounds content among cereals, which has been explored and successfully produced for human comsuption in some by-products such as cakes, cookies, pasta, tortillas, snack foods, and beers. Also, sorghum has been used for ethanol production and to maintain the shelf-life through biodegradable, edible, bioplastic film, and coatings (Dahlberg et al., 2011; Taylor et al., 2006).

The worlwide production of sorghum was estimated to be 59.53 million metric tons for 2019. The largest producer countries are USA, Nigeria, Ethiopia, India, and Mexico (USDA, 2019). However, several biotic and abiotic factor are limiting their production. Currently, insect pest is a major problem; it has been reported that around 150 insect species affect the sorghum (Sharma, 1993). The sugarcane aphid *Melanaphis sacchari* (Zehnt.) has become the major pest, which reduces yield by 30–100%, grain quality, and marketability depending on tolerance of the sorghum genotype. Another factor to consider is the season of year; during rainy season, the loss increased to 32% compared to the post-rainy season with 26% loss (Sharma et al., 2013).

The *Melanaphis sacchari* (Zehtner, 1897) (Homoptera, Aphididae, see Table 12.1)., commonly know as yellow sugarcane aphid, has been reported in various synomyn scientific as well common names (see Table 12.2). *M. sacchari*, considered to be mainly anholocyclic, is spread worlwide. It is also known as sugarcane aphid (Nibouche et al., 2014). The sugarcane aphid exhibits an exponential growth rate during the transition from vegetative to reproductive growth triggering unmanageable populations in a short period of time (Limaje et al., 2018). It has been reported that each female produces around 60–100 nymphs in 12–20 days and the adults live from 10 to 16 days. Aphid infestation in sorghum causes reduction in nutrients and plant growth; it also induces leaf chlorosis. Sugarcane aphid can induce mold growth due to the honeydew excreted and influence plant respiration. This pest is critical during the flowering and grain-filling stages (Fang, 1990). To face this pest damage, the use of sugarcane aphid-resistant sorghum germplasm, including integrated pest management strategies, has been proposed. It is crucial to identify the plant mechanism defense during sugarcane aphid interaction. Some strategies to understand these mechanism has been the differences between the susceptible and resistant plants responses to sugarcane aphid and how aphids respond to these plants (Tetreaul et al., 2019).

TABLE 12.1 Sugar Cane Aphid Taxonomy.

Kingdom	Animalia
Phylum	Arthropoda
Subphylum	Hexapoda
Class	Insecta
Order	Hemiptera
Suborder	Sternorrhyncha
Family	Aphididae
Genus	Melanaphis
Species	*Melanaphis sacchari*

Source: EPPO (2020).

12.2 MECHANISM OF PLANT DEFENSE AGAINST APHID: SORGHUM AND *M. SACCHARI*

Plants have evolved a range of mechanisms to defend against, tolerate, or avoid insect herbivory attack. Plant resistance is induced by the expression of traits that affect the herbivore's interaction with the host plant and with other plant-associated organisms. In other words, it is all the genetically inherited qualities'that influence the damage degree, as well yield loss by the herbivore (Stout, 2014).

TABLE 12.2 Yellow Sugar Cane Aphid Identity.

Preferred scientific name	Melanaphis sacchari (Zehntner)	
Preferred common name	Yellow sugarcane aphid	
Other scientific names	Name	**Authority**
	Aphis pheidolei	Theobald, 1916
	Aphis sacchari	Zehntner
	Aphis sorghella	Schouteden, 1906
	Aphis sorghi	Theobald
	Doralis sorghi	Theobald
	Longiunguis sacchari	Zehntner
	Melanaphis sorghi	Theobald
	Rhopalosiphum sacchari	Zehntner
	Sipha sacchari	Zehntner
	Uraphis sorghi	Theobald
English common names	cane aphid, dura asyl fly, green sugarcane aphid, grey aphid, sorghum aphid, sugar cane aphid, yellow sugarcane aphid	
French common names	puceron du sorgho, puceron jaune du mil	
German common names	gelbe Hirseblattlaus, grüne Zuckerrohrblattlaus, weiße Hirseblattlaus	
Japanese common names	kansyo-aburamusi, カンショアブラムシ	
Persian common name	schatte neyschekar	
Portuguese common name	pulgão da cana-de-açúcar	
Spanish common names	pulgón amarillo del sorgo, pulgón gris, áfido de la caña de azúcar	
EPPO code	RHOPSA	

Source: EPPO (2020).

The resistant categorizacion defined by Painter, 1951, are (a) antibiosis, (b) antixenosis or non-preference, and (c) tolerance. Plant resistance may include a combination of these three mechanisms. However, some authors points out that this categorization is misleading and it represents their potential role in a management programme and suggests dividing into "constitutive" or "inducible" and "direct" or "indirect" subcategories. Thus, the constitutive plant resistance is that which is expressed regardless of the prior history of the plant and the inducible resistance is resistance only expressed, or expressed to a greater extent, after prior injury. (Stout, 2013). Plants developed direct defense through specialiced morphological structures such as waxes, hairs, tichomes, etc. Also, secondary metabolites and proteins have toxic chemicals, repellent, and/or antinutitional effects on the herbivores. In the case of indirect defenses, the release of a blend of volatiles that specifically attract natural enemies, for example, has been reported (War et al., 2012).

12.2.1 ANTIBIOSIS

Antibiosis in plant resistance is expressed as prolonged or incomplete life cycle, reducing the progeny production, fecundity, viability, and mortality of immature stages. These may arise from physical factors and may result in the inability of the insect to utilize the host for feeding or shelter leading to starvation. Both chemical and morphological plant defense mediate antibiosis. The death of early instars, reduced size or low weight, prolonged periods of development of the immature stages, reduced adult longevity and fecundity, and death in the prepupal or pupal stage are the effects of antibiosis (Padmaja, 2016; Kalaisekar et al., 2017). Antibiosis has an impact on pests when they first attack and/or, subsequently, consume the plant; here, tissue hardness, phenology, toxins and deterrents, and nutritional resistance are important factors (Vänninen, 2005). *M. sacchari* feed primarily on phloem sap; however, some chemical factors of the leaf can impede access to it. In wheat, *Triticum aestivum* L., and sugar cane, *Saccharum spp.*, it has been reported that the secondary metabolites such as phenolic compounds can be chemical barriers in the leaves (Akbar et al., 2014).

12.2.2 ANTIXENOSIS OR NON-PREFERENCE

Antixenosis denotes plant traits affecting herbivore behavior in ways that reduce the preference for, or acceptance of, a plant as a host by a herbivore. Antixenosis factors include the pubescence, surface waxes, and leaf size and width that adversely affect the behavior of the insect (Eickhoff et al., 2008). Resistant plants may be less preferred by the pest (antixenosis that has an impact on pests upon their arrival/first attack on the plant by color, palatability, hairiness, waxiness, morphology, gummosis, and necrosis) (Vänninen, 2005).

In wheat, *Triticum aestivum* L., and sugar cane, *Saccharum spp.*, it has been reported that cuticular components (cuticle) and foliar pubescence can be physical barriers that reduce the attack of aphids (Akbar et al., 2014). The cuticle is the protective layer that is found on the outermost surface of plants and that interacts with the environment. It is constituted mainly of two types of polymers lipophilic: cutin and cuticular waxes. The cuticle plays an important role by acting as a barrier and participates in plant–insect interactions (Tafolla-Arellano et al., 2018) (see Fig. 12.1). The anti-adhesive properties of the cuticle influence plant–insect interactions, for example, cuticle waxes can physically hinder movement and reduce the adherence of insects to the plant surface. It has been reported that epicuticular waxes can be barriers to reduce adhesion by insect adsorption (Gorb et al., 2017). There is evidence that more three-dimensional deposition of epicuticular waxes, in general, reduces adhesion by insect adsorption, which are based on wet adhesion (Eigenbrode and Jetter, 2002). Plant "shiny" phenotype have been reported to be often less susceptible to insect attack due to a variation in the chemical composition and wax morphology (Tafolla-Arellano et al., 2018). For example, in *Eucalyptus globulus*, psyllid adhesion was reported to decrease when the wax content of juvenile leaves changes to bright adult leaves (Brennan and Weinbaum, 2001).

Epicuticular waxes (EW) are important factors in the antixenotic resistance of cabbage (*Brassica oleracea* L.) to the beetle (*Phyllotreta spp.*) and the insect (*Eurydema spp*). Although epicuticular wax levels differ between genotypes, a high negative correlation was observed between wax abundance and the size of lesions caused by both groups of harmful pests (Bohinc et al., 2014). In addition, studies with Brassica mutants with low levels of cuticle waxes demonstrated that epicuticular leaf waxes influence the incidence and feeding pattern of the beetle *Phyllotreta cruciferae* (Bodnaryk, 1992). For the specific case of sorghum, Nwanze et al. (1992) reported that sorghum fly tolerant genotypes to *Atherigona soccata* are characterized by a soft amorphous wax layer and dispersed wax crystals, while susceptible genotypes exhibit a dense layer of crystalline epicuticular wax. On the other hand, sorghum mutants of the *Bloomless2* (Bm2) locus, such as bm2–3 and bloom-cuticle (blmc), lack visible EW and have inhibited synthesis of the cuticle. The results suggest that the Bm2 locus is on sorghum chromosome 10 and could have pleiotropic effects, such as highly susceptibility to greenbug biotype C (*Schizaphis graminum* Rondani) (Punnuri et al., 2017). This clearly demonstrates the strongly physiological importance of the cuticle in plant–insect interaction, even, the inclusion of cuticle as a selection criterion in plant breeding programs has been suggested (Tafolla-Arellano et al., 2018). In this sense, our research group are working in a sorghum breeding program based on cuticle analysis.

12.2.3 TOLERANCE

Conversely to antoibiosis and antixenosis, tolerance does not interfere with the insect pests' physiology or behavior. Basically, the plant may tolerate the damage without an economic loss in yield or quality maintaining insect populations similar to those seen on susceptible plants (Painter, 1951;Vänninen, 2005).

12.3 PLANT BREEDING

Nowadays, best agronomic practices, biological control such as natural enemies, host plant resistance, and synthetic insecticides have been used as strategies for controlling sugarcane aphid in different regions of Mexico. The use of chemical insecticides for aphid control is not ecology and economically viable. Therefore, is it important identify and develop sugarcane aphid-resistant sorghum germplasm. Several efforts have been carried out to screen sorghum germplasm for their resistance to the sugarcane aphid, sorghum shoot fly, shoot bug, sorghum midge, and head bugs. There has been effort in different institutions worldwide to identify sorghum genotypes for their resistance to the sugarcane aphid (See Table 12.3). Therefore, it still remains necessary to increase the sorghum germoplasm development with resistance to *Melanaphis sacchari*.

12.3.1 BREEDING FOR INSECT RESISTANCE

Breeding for resistance requires a capability to screen large numbers of lines in a Sorghum Breeding Program. The rate of change achieved depends on several factors: mode of inheritance and breeding procedures, characteristic or trait identification for intensity of selection pressure that can be applied and environmental factors. In this sense, it crucial to study the inheritance of the desired trait in promising lines.

TABLE 12.3 Sorghum Resistant Genotypes.

Genotype	Country	Method breeding	Main results	References
BTx3408	USA	Pedigree	Exhibited high levels of tolerance to sugarcane aphid infestation and fewer phenotypic effects typically associated with SCA infestation.	Mbulwe, L.; Peterson, G. C.; Scott-Armstrong, J.; Rooney, W. L. Registration of Sorghum Germplasm Tx3408 and Tx3409 with Tolerance to Sugarcane Aphid [Melanaphis sacchari (Zehntner)].-*J. Plant Reg.* **2016,** *10* (1), 51–56.
BTx3409	USA	Pedigree	Consistent tolerance in different environments in phenotypic expression	Mbulwe, L.; Peterson, G. C.; Scott-Armstrong, J.; Rooney, W. L. Registration of Sorghum Germplasm Tx3408 and Tx3409 with Tolerance to Sugarcane Aphid [Melanaphis sacchari (Zehntner)].-*J. Plant Reg.* **2016,** *10* (1), 51–56.
H13073	Chromatin Inc, USA	Pedigree	Expression mechanisms to antibiosis in defense to *M. sacchari*	Paudyal, S.; Armstrong, J. S.; Giles, K. L.; Payton, M. E.; Opit, G. P.; Limaje, A. Categories of Resistance to Sugarcane Aphid (Hemiptera: Aphididae) among Sorghum Genotypes. *J. Econ. Entomol.* **2019.** doi:10.1093/jee/toz077
GW1489	Advanta, USA	Pedigree	Expressed mechanisms of tolerance and antibiosis in the interaction aphid-sorghum, decreasing the damage in the plants	Paudyal, S.; Armstrong, J. S.; Giles, K. L.; Payton, M. E.; Opit, G. P.; Limaje, A. Categories of Resistance to Sugarcane Aphid (Hemiptera: Aphididae) among Sorghum Genotypes. *J. Econ. Entomol.* **2019,** *112* (4), 1932–1940.
HG35W	Heartland Genetic, USA	Pedigree	Presents a tolerance Med-Early in response to *M. sacchari*	Serba, D. D.; Michaud, J. P. Sugarcane Aphid Resistance in Pearl Millet. *Kansas Field Res.* **2019,** 146.
SC112-14	USA	Single-seed descent	The line present resistance for the o chromosome 6, presenting good phenotypic appearance to aphids	Cuevas, H. E.; Prom, L. K.; Erpelding, J. E. Inheritance and Molecular Mapping of Anthracnose Resistance Genes Present in Sorghum Line SC112-14. *Mol. Breeding* **2014,** *34* (4), 1943–1953.

TABLE 12.3 *(Continued)*

Genotype	Country	Method breeding	Main results	References
OL2042	Chromatin, USA	Pedigree	High-resistance low-damage range, in phenotypic expression	Armstrong, J. S.; Paudyal, S.; Limaje, A.; Elliott, N.; Hoback, W. Plant Resistance in Sorghums to the Sugarcane Aphid (Hemiptera: Aphididae). *J. Entomol. Sci.* **2018**, *53* (4), 478–485.
SP7715	S&W, USA	Pedigree	Hybrid tolerance to aphids, added resistance to leaf disease resistance to biotypes (C and E) to greenbug	Armstrong, J. S.; Paudyal, S.; Limaje, A.; Elliott, N.; Hoback, W. Plant Resistance in Sorghums to the Sugarcane Aphid (Hemiptera: Aphididae). *J. Entomol. Sci.* **2018**, *53* (4), 478–485.
GRS1	INDIA	Recurrent selection	Line white resistance selection into local materials in India for showing good reaction to aphids	Deshpande, S. K.; Biradar, B. D.; Gangashetty, P. I.; Salimath, P. M. Studies on Inheritance of Charcoal Rot Resistance and Aphid Resistance in Rabi Sorghum [Sorghum bicolor (L.) Moench]. *Plant Arch.* **2011**, *11* (2), 635–643.
RTx2783	USA	Pedigree	Generate mediums to antibiosis and antixenosis such as method to resistance to *Melanaphis sachari*	Tetreault, H. M.; Grover, S.; Scully, E. D.; Gries, T.; Palmer, N. A.; Sarath, G., Sattler, S. E. Global Responses of Resistant and Susceptible Sorghum (*Sorghum bicolor*) to Sugarcane Aphid (*Melanaphis sacchari*). *Front. Plant Sci.* **2019**, *10*, 145.
PI 550610	USA	Pedigree	Is used as a recurrent method parental in the insertion of resistance to *Melanaphis sacchari*	Bayoumy, M. H.; Perumal, R.; Michaud, J. P. Comparative Life Histories of Greenbugs and Sugarcane Aphids (Hemiptera: Aphididae) Coinfesting Susceptible and Resistant Sorghums. *J. Econ. Entomol.* **2016**, *109* (1), 385–391.
R.LBK1	Texas, USA	Pedigree	High degree of resistance to the sugarcane aphid *Melanaphis sacchari* (Zehntner)	Hayes, C. M.; Armstrong, J. S.; Limaje, A.; Emendack, Y. E.; Bean, S.; Wilson, J.; Xin, Z. Registration of R. LBK1 and R. LBK2 Sorghum Germplasm with Resistance to the Sugarcane Aphid [*Melanaphis sacchari* (Zehntner)]. *J. Plant Reg.* **2019**, *13* (1), 91–95.

TABLE 12.3 (*Continued*)

Genotype	Country	Method breeding	Main results	References
R.LBK2	Texas, USA	Pedigree	Expressed strong resistance to *Melanaphis sacchari* (Zehntner) pressures at both seedling and mature plant stages	Hayes, C. M.; Armstrong, J. S.; Limaje, A.; Emendack, Y. E.; Bean, S.; Wilson, J.; Xin, Z. Registration of R. LBK1 and R. LBK2 Sorghum Germplasm with Resistance to the Sugarcane Aphid [*Melanaphis sacchari* (Zehntner)]. *J. Plant Reg.* **2019**, *13* (1), 91–95.
RTx3410	Texas, USA	Pedigree	They expressed high to higher levels of resistance to SCA in seedling stages and adult plants	Peterson, G. C.; Armstrong, J. S.; Pendleton, B. B.; Stelter, M.; Brewer, M. J. Registration of RTx3410 through RTx3428 Sorghum Germplasm Resistant to Sugarcane Aphid [*Melanaphis sacchari* (Zehntner)]. *J. Plant Reg.* **2018**, *12* (3), 391–398.
RTx3428	Texas, USA	Pedigree	The lines present resistant to damage caused by the sugarcane aphid *Melanaphis sacchari* (Zehntner)	Peterson, G. C.; Armstrong, J. S.; Pendleton, B. B.; Stelter, M.; Brewer, M. J. Registration of RTx3410 through RTx3428 Sorghum Germplasm Resistant to Sugarcane Aphid [*Melanaphis sacchari* (Zehntner)]. *J. Plant Reg.* **2018**, *12* (3), 391–398.
ATx2752	USA	Pedigree	Expresses mechanisms of antibiosis. The plants were significantly less damaged by aphid feeding *Melanaphis sacchari* (Zehntner)	Gonzales, J. D.; Kerns, D. L.; Brown, S. A.; Beuzelin, J. M. Evaluation of Commercial Sorghum Hybrids for Resistance to Sugarcane Aphid, *Melanaphis sacchari* (Zehntner). *Southwestern Entomol.* **2019**, *44* (4), 839–851.
HN16	USA	Pedigree	Dominant gene (RMES1) conferring resistance to *M. sacchari*	Wang, F.; Zhao, S.; Han, Y.; Shao, Y.; Dong, Z.; Gao, Y.; Wang, D. Efficient and Fine Mapping of RMES1 Conferring Resistance to Sorghum Aphid *Melanaphis sacchari*. *Mol. Breeding* **2013**, *31* (4), 777–784.
DKS 37-07	USA	Recurrent selection	Significant effects of sorghum resistance to Melanaphis	Neupane, S. B.; Kerns, D. L.; Szczepaniec, A. The Impact of Sorghum Growth Stage and Resistance on Life History of Sugarcane Aphids (Hemiptera: Aphididae). *J. Econ. Entomol.* **2020**, *113* (2), 787–792.

TABLE 12.3 *(Continued)*

Genotype	Country	Method breeding	Main results	References
ICSR 165	India	Recurrent selection	Exhibited moderate levels of resistance to aphid damage	Sharma, H. C.; Bhagwat, V. R.; Daware, D. G.; Pawar, D. B.; Munghate, R. S.; Sharma, S. P.; Gadakh, S. R. Identification of Sorghum Genotypes with Resistance to the Sugarcane Aphid Melanaphis Sacchari under Natural and Artificial Infestation. *Plant Breed.* **2014,** *133* (1), 36–44.
R13219	Texas, USA	Pedigree	Expressed significant degrees of tolerance, antibiosis, and antixenosis	Armstrong, J. S.; Mbulwe, L.; Sekula-Ortiz, D.; Villanueva, R. T.; Rooney, W. L. Resistance to Melanaphis Sacchari (Hemiptera: Aphididae) in Forage and Grain Sorghums. *J. Econ. Entomol.* **2017,** *110* (1), 259–265.

12.3.2 SELECTION

For the plant breeder, selection is considered the most important force available for changing gene frequency due to production of a population that has a mean value greater (or less) than the mean value of the parent population. The sorghum germoplasm is useful to increase the opportunity of better selection.

Genetic variation can be sustained if crosses are made between selected lines. Segregation will occur in the F2 and advanced generations, and pedigree selection for the development of superior lines is effective in the segregating generations. Selection of lines entering into composites is based on criteria such as yield, disease and pest resistance, drought and heat tolerance, etc. After entries have been selected, they are crossed and backcrossed to a source of genetic sterility. Therefore, we evaluated a diverse array of sorghum genotypes to identify cultivars with resistance to Sugarcane Aphid under natural infestation. The experiments were conducted at the Experimental Station Universidad Autonoma Agraria Antonio Narro (UAAAN) in Xalostoc, Morelos, and Buenavista, Saltillo, Coahuila, Mexico during the rainy seasons at the summer. The seed planted in the experiments had chemical treatment. In both experiments, one week after seedling emergence, thinning was carried out to maintain a spacing of 12 cm between the plants. No insecticide was applied in the experimental plots.

In the 2018 year, a set of 34 lines was evaluated under a randomized complete block design with three replicates, with the seed harvested from sorghum genotypes that had shown resistance to *M. sacchari*, high grain yield potential, and bird resistant. We evaluated in the 2019 year in Saltillo, Coahuila, under natural infestation and based on the expression of resistance to sugarcane aphid, a set of 46 lines including the resistant genotypes and the susceptible checks. Recently, we continued to evaluate the agronomic performance of sorghum genotypes and the resistance to *M. sacchari* in Tamaulipas, Mexico, to find new sorghum genotypes. In the experiments conducted on field conditions, we observed that sorghum plants selected with a greater height, longer distance between the leaves, smaller leaf angle, stay-green traits, and presence of waxy bloom has been reported to be less susceptible to aphid damage.

12.4 PLANT MOLECULAR OVERVIEW

The modes of the sorghum resistant to the aphid attack can be grouped in antibiosis or antixenosis (Perales-Rosas et al., 2019). Further, the control of the level of sugarcane aphid infestation over sorghum had been tried by changing the planting date, utilization of parasitoid species, genetic improvement, insecticide application to the leaf, and the seeds with some success (Singh et al., 2004; Haar et al., 2019). In this regard, the utilization of tools from the molecular and DNA recombinant technology can help in the stepwise design of a better control protocol.

12.4.1   MAPPING OF THE MELANAPHIS SACCHARI RESISTANT GENE IN SORGHUM GENOME

The first attempts to study the resistance of sorghum against the attack of M. sacchari were carried out by artificially inoculating the plant and recording the resistance phenotype by observation. In this study, 5000 sorghum accesions were evaluated, which showed that 0.3% of them showed the resistance phenotype (Lu and Dahlberg, 2001).

The studies using the molecular biology approach to study the *M. sacchari* interaction are rather few. The first attempt to find the presence of genes conferring to sorghum the phenotype of *M. sacchari* attack resistance was carried out in 2003 by using random-amplified polymorphic DNA markers. This study reported for the first time the presence of a gene conferring M. sacchari attack resistance in sorghum. After that, in 2006, the mapping of the sorghum resistance gene to the aphid attack was carried out using simple sequence repeats (SSR) or microsatellites. This work was able to locate the resistance gene in chromosome 6. In this work, the resistance gene was named RMSE1, which stands for resistance to *Melanaphis sacchari*.

The information was obtained from the introduction of a more recent work reporting a study to map the RMES1 gene (Wang et al., 2013). The first attempt to map the gene conferring resistance to the M. sacchari attack was carried out by studying the segregation progeny of the cross between the M. sacchari attack susceptible sorghum variety named Qiansan and the resistant sorghum variety named Henong 16 (HN16) by using amplified-fragment-length polymorphism. The progeny was artificially infested to find individuals resistant and susceptible to the aphid attack. In this work, it was found two markers with a genetic distance from the resistance gene of 6 and 10 centimorgan. These two markers were isolated and sequenced to develop a sequence characterized amplified region-marker linked to the resistance gene. The authors reported efficiency of 95.9 % to predict the susceptible or resistant phenotype with the utilization of the two markers simultaneously (Chang et al., 2012). In this last study, the closest distance between the resistance gene locus and the markers was of 6 cM, which is rather a large recombination distance. Because of that, another study was carried out to get a better linkage map of the resistance gene. In this work, it was carried out a cross between the sorghum variety HN16 y the variety BTx623, which is the variety chosen by the international program of sorghum genome sequencing. Thereafter, it was obtained progeny by selfing F1 and F2 progeny. Individual seedlings from the resultant progeny were inoculated with adult aphids to test the phenotype. The markers were developed by analyzing the presence of SSR in the chromosomal region from 1.9 to 4.9 Mb obtained from the sorghum genomic database. Besides the SSR, several repeat junction markers were also developed from the same genomic region. The work was able to generate 11 molecular markers close to the RMES1 gene. Out of these, the markers Sb6rj2776, which is a repeat junction marker and the Sb6m2650, which is a SSR marker, are at 1 and 2 cM away from the resistance gene. Furthermore, it was found that the Sb6rj2776 marker and Sb6m2650 showed an efficiency of 99.4% and 99.1%, respectively (Wang et al., 2013). Despite all the efforts mentioned, the RMES1 gene had not been cloned and studied, to our knowledge. Furthermore, most likely it is a gene with a domain for nucleotide binding site and leucine rich repeats (NBS-LRR) because most of the resistant genes to aphid resistance located in different plants belong to this gene family (Tetreault et al., 2019). The utility of the scar marker developed was tested by analyzing 561 sorghum accessions. Out of these, the scar maker was found in the genome of 91 accessions. It has chosen 26 accessions to test the resistance in the field and 12 were found to show high resistance and 13 show high and medium-high resistance (Guden et al., 2019).

12.4.2 MELANAPHIS SACCHARI GENETIC DIVERSITY

Knowing the genetic diversity of an insect makes feasible the design of strategies intelligently driven to control the infestation. A study was undertaken to know the genetic diversity of the aphid population in the United States. Using the Ion-Torrent platform, it was possible to generate 1.14 Gb de nucleotide sequences of the M. sacchari genome, after the elimination of short reads from the symbiont. With this information, the authors generated 52 SSR markers. The analysis of genetic diversity using these markers along with 14 already published was carried out with samples from 17 locations of the United States. It was concluded that all the individuals belong to the same multilocus lineage with except one sample obtained from Sinton, Texas, USA. Furthermore, all aphids showed the presence of the symbiont

B. aphidicola, except for the aphids sampled from Texas. These results are in agreement with the fact that the reproduction of this specie is asexual (Harris-Shultz et al., 2017). The presence of a genetically similar individuals in the aphid population can make easier to design a strategy to control the sorghum infestion.

Reactive oxygen species (ROS) had been related to the plant resistant to insect attack, although the studies in this field are scarce (Liu et al., 2010). A study was carried out to know the ROS role in the sorghum resistant to the aphid attack. To know the role of hydrogen peroxide in sorghum resistant, plant seedlings of the resistant line HN16 (which carries the RMES1 gene) and susceptible line asm1 were infected with 20 adult aphids. Samples of leaves were taken at 0, 24, 48, 72, and 96 h after infestation. Quantification of hydrogen peroxide as well as the enzymatic activity of superoxide dismutase, catalase, peroxidase, ascorbate peroxidase, and glutathione peroxidase was carried out. Differentially expressed genes were analyzed by RNAseq at 0, 24, and 48 h after treatment. It was recorded a higher concentration of hydrogen peroxide in the resistant line throughout all the experiments. Further, 22 differentially expressed genes were found. Out of these, a gene encoding a glutathione peroxidase isoenzyme (SbGPx1) showed upregulation and downregulation in the susceptible line (asm1) and resistant (HN16) line, respectively. It was concluded that the higher amount of hydrogen peroxide plays a very important role in the resistance phenotype of the HN16 line. Further, it appears that the changes in gene expression in the resistant line are related with the induction of higher concentration of hydrogen peroxide at the right time but also in the ROS detoxification to avoid membranes cell damage whereas the susceptible line fails to control the ROS levels and the negative effects (Shao et al., 2019).

12.4.3 STUDIES OF SORGHUM RESISTANT MECHANISM IN THE OMIC ERA

The utilization of the next-generation sequencing technology is very important to elucidate the resistant phenotype molecular mechanism. The transcriptome in response to aphid attack was studied in 2 weeks and 4 weeks plants of a susceptible sorghum genotype (DKS 44-20) and resistant phenotype (DKS 37-07) by using RNA-seq approach. The challenge was carried out by exposing two plant leaves to 15 aphids during 24 h. It utilized the platform Illumina HiSeq4000 to generate between 43.9 and 68.68 millions of short reads with a size of 75 bp in a paired end mode. The generation of total reads was much higher but these are the reads that mapped to the sorghum genome. The differential expression calculation was carried out by mapping the short reads to the annotated genes of the sorghum reference genome. The in silico calculations were validated in biology by quantifying with qRT-PCR the expression of five genes playing a role in hormone signaling using as a reference the cyclophilin/peptidylprolyl isomerase gene.

In 2-weeks old plants, it was found that there were 3525 and 986 differentially expressed genes in the resistant and susceptible sorghum lines, respectively. Furthermore, in 4 weeks old plants, it was recorded that there were 3820 and 116 differentially expressed genes in the resistant and susceptible sorghum lines, respectively. From this data, it is clear that the transcriptional responses is much higher in the resistant lines. Considering the gene-specific functions, it was found that the resistant phenotype showed the activation of much higher number and expression level of genes playing a role in hormone signaling, ROS detoxification, plant pathogen interaction, and genes playing a role in phenylpropanoid and flavonoid biosynthetic pathways. In the case of the transcription factors, it was found that there were 52 differentially expressed genes in the younger resistant lines, which is a much higher number as compared with the susceptible line. Out of these, 18 belong to the WRKY family of transcription factors, which had been found to be related with the response to biotic stress (Gao et al., 2020). In the case of the 6 weeks old plants, the number of transcription factors differentially expressed were lower although the amount was higher in the resistant as compared with the susceptible line. Furthermore, the differential expressions of WRKY1, WRKY19, WRKY28, and WRKY72 transcription factors only in the 2 weeks old shorghum plant was recorded (Kiani and Szczepaniec, 2018). It will be interesting to consider the utilization of these four transcription factors as markers in the selection of sorghum lines resistant to the aphid attack.

The next-generation sequencing technology was utilized in another experiment in which the genetic response comparison in response to aphid infestation of the resistant sorghum line RTx2783 and the susceptible sorghum line A/BCK60 was carried out. Each plant was challenged with five aphids and the samples for RNA extraction were taken after 5, 10, and 15 days. The transcriptomes were sequenced using the platform Illumina HiSeq 2500 generating short reads of 100 bp in mode single end. The calculation of changes in gene expression was carried out by mapping the high-quality short reads to the sorghum genome ver 3.1 (phytozome.jgi.doe.gov/pz/portal. html). The authors does not mention the validation of the gene expression calculated in silico with the determination of gene expression using quantitative real-time PCR. Besides, a principal component analysis was carried out.

The antixenosis and antibiosis mode of sorghum plant resistant were also evaluated in this work, but this part of the experiment of the experiment will not be discussed taken into account that the focus of this chapter section is the utilization of the molecular biology approach.

The principal component analysis using the transcriptomic data was able to separate the control and infested plants in the different days of sampling, which suggest that many genes showed differential expression due to changes in development the plant rather than as a result of the response to aphid attack.

Considering the expression of genes, it was found at day five that in the resistant sorghum line, nine genes encoding WRKY transcription factors as well as four genes encoding proteins participating in the jasmonate signal transduction pathway showed an increased expression as compared with the susceptible line. Further, at day 10, it was recorded an increased expression of 72 genes, which encodes NB-LRR proteins in the resistant line. It is clear that the resistant line showed the expression of a much higher number of genes related with pathogen response. Besides, the changes in expression strongly suggest that the resistant line show the ability to utilize at the right time and control the levels of ROS in response to the aphid attack. Likewise, the resistance line is able to maintain the same level of photosynthetic activity to generate the energy needed to respond and resist the Melanaphis sacchari attack (Tetreault et al., 2019).

12.5 CONCLUDING REMARKS

Sugarcane Aphid (*Melanaphis sacchari* Zehntner), the major pest, reduces yield by 50–100%, grain quality, and marketability depending on tolerance of the sorghum genotype. To face this pest, different strategies has been tried such as modifying the planting date, best agronomic practices, utilization of parasitoid species, genetic improvement, insecticide application, among other. It is clear that the number of experiments using the tools derived from the DNA recombinant technology with the goal to study the interaction sorghum and *Melanaphis sacchari* are rather scarce. The studies carried out range from the utilization of SSR markers to map a resistant gene, and the study of the aphid genetic diversity to the sequencing of the aphid attack transcriptome using the next-generation sequencing technologies. From these studies, it appears that the aphid resistant phenotype is linked to one gene, which most likely belong to the family of NBS-LRR genes, although more scientific data is needed to support this statement. Also, the genetic diversity of Melanaphis sacchari is low and the individuals are clones. Furthermore, it seems that the expression of WRKY transcription factors is important for the resistant phenotype. However, the ability to maintain the photosynthetic apparatus active and the optimal utilization of the reactive oxygen species are also important to survive this biotic stress, which suggests that most likely the sorghum resistant phenotype is multigenic. In this regard, we suggest that a program for the developing of sorghum resistant lines should focus in the search for a quantitative trait loci rather than in one gene. Also, the literature demonstrates the strongly physiological importance of the cuticle in plant–insect interaction, and even the inclusion of cuticle as a selection criterion in plant breeding programs has been suggested.

KEYWORDS

- **sorghum**
- **aphid**
- **defense mechanism**
- **physiology**
- **breeding**
- **molecular**

REFERENCES

Akbar, W.; Showler, A. T.; Reagan, T. E.; Davis, J. A.; Beuzelin, J. M. Feeding by Sugarcane Aphid, *Melanaphis sacchari*, on Sugarcane Cultivars with Differential Susceptibility and Potential Mechanism of Resistance. *Entomol. Exp. Appl.* **2014,** *150* (1), 32–44.

Bodnaryk, R. P. Leaf Epicuticular Wax, an Antixenotic Factor in Brassicaceae That Affects the Rate and Pattern of Feeding of Flea Beetles, *Phyllotreta cruciferae* (Goeze). *Can. J. Plant Sci.* **1992,** *72* (4), 1295–1303.

Bohinc, T.; Markovič, D.; Trdan, S. Leaf Epicuticular Wax as a Factor of Antixenotic Resistance Of Cabbage to Cabbage Flea Beetles and Cabbage Stink Bugs Attack. *Acta Agric. Scand. Sect. B—Soil Plant Sci.* **2014,** *64* (6), 493–500.

Brennan, E. B.; Weinbaum, S. A. Effect of Epicuticular Wax on Adhesion of Psyllids to Glaucous Juvenile and Glossy Adult Leaves of *Eucalyptus globulus* Labillardière. *Aust. J. Entomol.* **2001,** *40* (3), 270–277.

Chang, J. H.; Cui, J. H.; Xue, W.; Zhang, Q. W. Identification of Molecular Markers for a Aphid Resistance Gene in Sorghum and Selective Efficiency Using These Markers. *J. Integr. Agric.* **2012,** *11* (7), 1086–1092.

Dahlberg, J.; Berenji, J.; Sikora, V.; Latković, D. Assessing Sorghum [*Sorghum bicolor* (L) Moench] Germplasm for New Traits: Food, Fuels & Unique Uses. *Maydica* **2012,** *56* (2).

Eigenbrode, S. D.; Jetter, R. Attachment to Plant Surface Waxes by an Insect Predator. *Integr. Comp. Biol.* **2002,** *42* (6), 1091–1099.

Fang, M. N. Population Fluctuation and Timing for Control of Sorghum Aphid on Variety Taichung 5. *Bull. Taichung Dist. Agric. Improv. Stn.* **1990,** *28,* 59–71.

Gao, Y. F.; Liu, J. K.; Yang, F. M.; Zhang, G. Y.; Wang, D.; Zhang, L.; Ou, Y. B.; Yao, Y. A. The WRKY Transcription Factor WRKY8 Promotes Resistance to Pathogen Infection and Mediates Drought and Salt Stress Tolerance in *Solanum lycopersicum. Physiol. Plant.* **2020,** *168* (1), 98–117.

Gorb, E. V; Hofmann, P.; Filippov, A. E.; Gorb, S. N. Oil Adsorption Ability of Three-Dimensional Epicuticular Wax Coverages in Plants. *Sci Rep.* **2017,** *7,* 45483.

Guden, B.; Yol, E.; Ikten, C.; Erdurmus, C.; Uzun, B. Molecular and Morphological Evidence for Resistance to Sugarcane Aphid (*Melanaphis sacchari*) in Sweet Sorghum Sorghum Bicolor (L.) Moench. *Biotech* **2019,** *9* (6), 7.

Haar, P. J.; Buntin, G. D.; Jacobson, A.; Pekarcik, A.; Way, M. O.; Zarrabi, A. Evaluation of Tactics for Management of Sugarcane Aphid (Hemiptera: Aphididae) in Grain Sorghum. *J. Econ. Entomol.* **2019,** *112* (6), 2719–2730.

Harris-Shultz, K.; Ni, X. Z.; Wadl, P. A.; Wang, X. W.; Wang, H. L.; Huang, F. N.; Flanders, K.; Seiter, N.; Kerns, D.; Meagher, R.; Xue, Q. W.; Reisig, D.; Buntin, D.; Cuevas, H. E.; Brewer, M. J.; Yang, X. B. Microsatellite Markers Reveal a Predominant Sugarcane Aphid (Homoptera: Aphididae) Clone is Found on Sorghum in Seven States and One Territory of the USA. *Crop Sci.* **2017,** *57* (4), 2064–2072.

House, L. R. A Guide to Sorghum Breeding, 2nd edn. *Patencheru. Int. Crop. Res. Inst. Semi-Arid Trop.* **1985,** 2–10.

Kiani, M.; Szczepaniec, A. Effects of Sugarcane Aphid Herbivory on Transcriptional Responses of Resistant and Susceptible Sorghum. *BMC Genomics* **2018,** *19,* 18.

Koch, K. G.; Chapman, K.; Louis, J.; Heng-Moss, T.; Sarath, G. Plant Tolerance: A Unique Approach to Control Hemipteran Pests. *Front. Plant Sci.* **2016,** *7,* 1363.

Limaje, A.; Hayes, C.; Armstrong, J. S.; Hoback, W.; Zarrabi, A.; Paudyal, S.; Burke, J. Antibiosis and Tolerance Discovered in USDA-ARS Sorghums Resistant to the Sugarcane Aphid (*Hemiptera: Aphididae*) *J. Entomol. Sci.* **2018,** *53* (2), 230–241.

Liu, X. M.; Williams, C. E.; Nemacheck, J. A.; Wang, H.; Subramanyam, S.; Zheng, C.; Chen, M. S. Reactive Oxygen Species are Involved in Plant Defense against a Gall Midge. *Plant Physiol.* **2010,** *152* (2), 985–999.

Lu, Q. S.; Dahlberg, J. A. Chinese Sorghum Genetic Resources. *Econ. Bot.* **2001,** *55* (3), 401–425.

Mote, U. N.; Shahane, A. K. Biophysical and Biochemical Characters of Sorghum Varieties Contributing Resistance to Delphacid, Aphid and Leaf Sugary Exudations. *Indian J. Entomol.* **1994,** *56,* 113–122.

Nwanze, K. F.; Pring, R. J.; Sree, P. S.; Butler, D. R.; Reddy, Y. V. R.; Soman, P. Resistance in Sorghum to the Shoot Fly, *Atherigona soccata*: Epicuticular Wax and Wetness of the Central Whorl Leaf of Young Seedlings. *Ann. Appl. Biol.* **1992,** *120* (3), 373–382.

Perales-Rosas, D.; Hernández-Pérez, R.; López-Martínez, V.; Andrade-Rodríguez, M.; Alia-Tejacal, I.; Juárez-López, P.; Perdomo-Roldán, F.; Guillén-Sánchez, D. Evaluación de la Antibiosis, Antixenosis, y Tolerancia de *Melanaphis sacchari* en Híbridos de Sorgo. *Southwest. Entomol.* **2019,** *44* (3), 8, 763–770.

Punnuri, S.; Harris-Shultz, K.; Knoll, J.; Ni, X.; Wang, H. The Genes Bm2 and Blmc That Affect Epicuticular Wax Deposition in Sorghum Are Allelic. *Crop Sci.* **2017**, *57*, 1552–1556.

Rosenow, D. T. Breeding for Resistance to Root and Stalk Rots in Texas. *Sorghum Root Stalk Rots, a Crit. Rev. Patancheru, AP, India ICRISTAT* **1983**, 209–217.

Shao, Y. T.; Guo, M. X.; He, X. F.; Fan, Q. X.; Wang, Z. J.; Jia, J.; Guo, J. B. Constitutive H$_2$O$_2$ Is Involved in Sorghum Defense against Aphids. *Braz. J. Bot.* **2019**, *42* (2), 271–281.

Sharma, H. C. Host-Plant Resistance to Insects in Sorghum and Its Role in Integrated Pest Management. *Crop Prot.* **1993**, *12* (1), 11–34.

Sharma, H. C.; Sharma, S. P.; Munghate, R. S. Phenotyping for Resistance to the Sugarcane Aphid *Melanaphis sacchari* (Hemiptera: Aphididae) in *Sorghum bicolor* (Poaceae). *Int. J. Trop. Insect Sci.* **2013**, *33* (4), 227–238.

Singh, B. U.; Padmaja, P. G.; Seetharama, N. Biology and Management of the Sugarcane Aphid, Melanaphis sacchari (Zehntner) (Homoptera: Aphididae), in Sorghum: A Review. *Crop Prot.* **2004**, *23* (9), 739–755.

Tafolla-Arellano, J. C.; Báez-Sañudo, R.; Tiznado-Hernández, M. E. The Cuticle as a Key Factor in the Quality of Horticultural Crops. *Sci. Hortic. (Amsterdam).* **2018**, *232*.

Taylor, J. R. N.; Schober, T. J.; Bean, S. R. Novel food and Non-Food Uses for Sorghum and Millets. *J. Cereal Sci.* **2006**, *44* (3), 252–271.

Tetreault, H. M.; Grover, S.; Scully, E. D.; Gries, T.; Palmer, N. A.; Sarath, G.; Louis, J.; Sattler, S. E. Global Responses of Resistant and Susceptible Sorghum (Sorghum bicolor) to Sugarcane Aphid (Melanaphis sacchari). *Front. Plant Sci.* **2019**, *10*, 19.

Tsumuki, H.; Kanehisa, K.; Moharramipour, S. Sorghum Resistance to the Sugarcane Aphid, Melanaphis sacchari (Zehntner) Amounts of Leaf Surface Wax and Nutritional Components. *Bull. Res. Inst. Bioresour. Univ.* **1995**.

Wang, F. M.; Zhao, S. M.; Han, Y. H.; Shao, Y. T.; Dong, Z. Y.; Gao, Y.; Zhang, K. P.; Liu, X.; Li, D. W.; Chang, J. H.; Wang, D. W. Efficient and Fine Mapping of RMES1 Conferring Resistance to Sorghum Aphid *Melanaphis sacchari*. *Mol. Breed.* **2013**, *31* (4), 777–784.

CHAPTER 13

Effect of LED Light on Plants and Microgreens Production in a Plant Factory System

NATIELY GALLO DE LA PAZ, C. MARTÍNEZ-ÁVILA,
HUMBERTO RODRÍGUEZ-FUENTES, NATIELY GALLO DE LA PAZ,
ALEJANDRO ISABEL LUNA-MALDONADO, and ROMEO ROJAS*

Universidad Autonoma de Nuevo Leon, School of Agronomy. 66054, General Escobedo, Nuevo León, México.

*Corresponding author. E-mail: ROMEO.ROJASMLN@uanl.edu.mx

ABSTRACT

With the growing demand for natural, healthy, minimally processed products, reducing environmental damage and increasing production worldwide, a range of opportunities opens up for the field of closed systems (factory plan) for food production where production can be controlled throughout the year. All these have a goal of ending malnutrition, reducing chronic diseases, and having products with a high content of bioactive compounds. Plant factory is a closed food production system that allows to produce food products all year-round in small spaces and with high yields, this is the case of microgreens (they can be defined as seedlings with developed cotyledons as well as their first true leaves and are harvested between 7 and 20 days after sowing in function of the species). To achieve this, it is necessary to find the best growing conditions in a closed system, such as humidity, temperature, nutrients, planting density, and LED light, choosing the specific spectral characteristics for each crop.

13.1 INTRODUCTION

In the world, food has always been of great importance but has become more relevant because of the risks to human health caused by malnutrition, including undernutrition and overnutrition. Malnutrition causes early death in mothers, infants, and children as well as deficiencies in the physical and mental development of young people, contributing to one-third of child mortality. On the other hand, obesity and overweight caused by excessive eating are linked to the increase in chronic diseases, with an estimated 1.5 billion people being overweight, including 500 million who are obese (OMS. Nutricion). Because of the importance of healthcare and its relationship to food consumption, our society has preferred those that provide benefits at the physiological level. Among all healthy food choices, vegetables are consumed throughout human history and are forming a fundamental part of the human diet. They exist in great variety and nutritional quality providing macronutrients and micronutrients that are fundamental for the body. Nowadays, greater importance has been given to bioactive compounds (phytochemicals) that are secondary metabolites of the plant.

Within the great number of forms and varieties in which vegetables are consumed are the microgreens, which are considered functional foods with an excellent nutritional contribution and greater number of bioactive compounds due to the physiological stage in which they are harvested. They can be defined as seedlings with developed cotyledons as well as their first true leaves (Xiao et al., 2012; Vastakaite Lithuanian Research Centre for Agriculture and Forestry, 2015) and are harvested between 7 and 20 days after sowing of the species (Lee et al., 2004; Bulgari et al., 2017). Leaf vegetables and microgreens are most often produced in greenhouses located in or near cities due to significant savings in terms of cost and time of transportation from production sites to the consumer (Ohyama et al., 2000). The quality and quantity loss of fresh vegetables with 90% humidity is reduced due to the long distances required for transportation (Atanda, 2011).

Protected agriculture is carried out by means of structures built to optimize environmental conditions that restrict plants. The different types of these structures have been developed in order to provide alternatives that generate optimal conditions for the growth and development of plants. An alternative solution to problems such as the complication of supplying food to expanding cities (Kennedy et al., 2007; Lambin and Meyfroidt, 2011), the loss of arable land due to urbanization and contaminated soils, the use of systems like plant factory (vertical agriculture) is a promising proposal. Plan factory is a system that controls all the factors necessary to produce crops, such as lighting, relative humidity, temperature, CO_2, ventilation, hermetic, and thermally isolated structure. Recently, the use of LED light has been employed with which the specific spectral characteristics of light can be chosen for each species to be produced (Kozai et al., 2006; Kozai, 2013; Abarca et al., 2019).

Three types of plant factories are known: greenhouses using sunlight, greenhouses that use both sunligh and artificial lighting, and closed rooms with artificial lighting (Pessarakli, 2016). There are many benefits of growing crops in plant factories, for instance, the factory can be built anywhere, including near urban areas to reduce transportation costs and preserve freshness; the crops are not affected by the outside climate and soil fertility, so the production can be year-round, unaffected by season or weather hazards; the productivity is higher than the field through manipulation of the growing environment; and the produced quality is good because it is pesticide-free and it has less bacterial load (Kozai et al., 2016). For this reason, the plant factory system is an alternative for large-scale cultivation in a reduced space where it is not necessary to wait for the seasonality of the crops.

13.1.1 PLANT FACTORY SYSTEM AND ITS MAIN ELEMENTS

A plant factory production system starts from a trend that seeks optimal crop production in which the aim is to totally control environmental, water, and also nutritional factors in the plants (Ohyama et al., 2000) It consists of several basic elements that must be controlled: a hermetically and thermally insulated structure, a multilevel or multistory system with lighting devices, a hydroponic system and for environmental variables, a temperature and relative humidity control system, a CO_2 injection unit and an automatic system control unit (Huang et al., 2020). With these factors controlled, large-scale production of any product is possible. In addition, it does not limit seasonality, optimizes space, and does not require the use of large amounts of land to achieve cultivation including microgreens.

13.1.1.1 TEMPERATURE AND RELATIVE HUMIDITY

Most of the physiological processes of the plants are influenced by the temperature, which is given by the temperature of the environment in which they develop, and because of this, it is very important to control the temperature of the environment. This is precisely what a plant factory seeks to maintain a constant temperature (Niu et al., 2016). Plant growth is related to temperature has been documented in the last decades with a general increase of biomass. Also, plants growing in elevated temperatures have a different phenology and physiology that may, for example, flower earlier. Moreover, according to the metabolic scaling theory, the metabolism that is increasing with temperature drives the ecological processes at different levels of biological organization (Chauvat and Forey, 2021). By these ways, temperature measurement is considered a very important element. Within the plant factory, it is possible to control the temperature with an air conditioning system as well as temperature sensors with tubular resistors, such as heaters and heat sinks that function as coolers.

Some physiological and morphological processes of plants are modified by relative humidity (RH) (Rodrigues et al., 2016). Relative humidity is used to measure the humidity of the air, that is, RH is used to express the amount of water vapor contained in the air. Based on the maximum amount of water, the air can hold at a specific pressure and temperature. Variations in the RH have influence on plant growth, CO_2 assimilation rate, stomatal aperture, transpiration, and nutrient uptake (Cha-um et al., 2010). However, vapor pressure deficits lower than 0.2 kPa and higher than 1.0 kPa are frequently reached in controlled conditions such as in greenhouses. RH between 55% and 90% has little effect on the physiology and growth of horticultural crops. Low RH (<75%) produces a reduction in growth due to a high transpiration rate, while high RH (>80%) promotes the diseases causing growth and development disorders (Abarca et al., 2019).

13.1.1.2 CO_2 CONCENTRATION

The air in the atmosphere is composed of 78% nitrogen, 21% oxygen, 0.93% argon, and 0.04% carbon dioxide (CO_2), as well as water vapor and other gases in amounts that are not constant. CO_2 is used for photosynthesis, 40% of the dry biomass of plants is carbon-fixed by this process. An atmosphere with more CO_2 can have a favorable response in the physiological aspects of plants (Yepes et al., 2011; EA and Rogers). Due to the above facts, the concentration of CO_2 is very relevant since high concentrations can cause an increase in the photosynthetic rate in the mesophyll tissue as well as a closing of stomata producing water loss by transpiration (Suslov, 2020; Hee and Beom, 2001). If the concentration of CO_2 in the environment is elevated, increases in the content of antioxidants can be observed (Park et al., 2013; Samuoliene et al., 2013), which leads to better properties for human consumption. For this reason, the control of CO_2 in a PF system is of utmost importance to promote photosynthesis in crops.

13.1.2 PLANT FACTORY WITH ARTIFICIAL LIGHTING

Plant factory is an agricultural system where the environment in which crops are grown is planned and controlled for the production of plant species, with the aim of optimizing the environmental conditions that generate the highest production of biomass or metabolites of commercial interest (Ohyama et al., 2000; Chauvat and Forey, 2021; Yokoi Matsudo et al.; Kozai, 2008). The plant factory is classified as open or closed system. In the latter system, in addition to precisely controlling conditions, such as temperature, humidity and nutrients, it is possible to control the condition of light (quality, intensity, and photoperiod) (Niu et al., 2016; Shiina et al., 2011; Malayeri et al., 2010; Moon et al., 2011). In a plant factory with artificial light, a heat pump is generally used to control the environmental parameters. The light is the unique source or energy required for photosynthesis (Ahmed et al., 2020).

13.1.2.1 LED LIGHT AND ITS RELATIONSHIP TO PLANTS

Light regulates many biological processes and some metabolic pathways in plants (Carvalho et al., 2008),because they react to the intensity and quality of light (Zhang and Folta, 2012) through their photoreceptors, which are activated under specific wavelengths of light (Li et al., 2012), These photoreceptors absorb mostly PAR light corresponding to blue-violet (450–490 and 390–450 nm) and red-orange (620–750 and 590–620 nm) and do not use all wavelengths of the electromagnetic spectrum. The rest is reflected and is responsible for the visible color of the leaves (Martín-Ramos et al., 2010). Therefore, quality, intensity, and photoperiod are all elements that must be considered in the light factor to obtain good quality plant production (Zhang and Folta, 2012). The lighting with light-emitting diodes (LED) commonly used in SPF (Johkan et al., 2010) makes it possible to generate a specific spectral composition (Tamulaitis et al., 2005), for example, red and blue monochromatic light individually or in combination increases the accumulation of primary and secondary metabolites, such as sugars and soluble proteins, polyphenols, vitamins C, tocopherols, and carotenoids compared with the full spectrum of visible light (380–700 nm) (Li et al., 2012; Samuoliene et al., 2013; Kook, 2013). LED lighting is the latest technology used for crop production in intensively controlled environments, has a high light efficiency, which helps to reduce energy costs (Bourget, 2008). An advantage of the use of LED light is that it can provide a light source according to the needs of the crop and that this type of lighting can emit monochromatic wavelengths. It is important to mention that the need for light of each plant species is different, so it makes more convenient to use these lights because other light sources do not provide this benefit.

13.1.2.2   QUALITY

In agronomic terms, the quality of light is defined as the wavelengths of the electromagnetic spectrum that affect the plants with which photosynthesis takes place. Improving the quality, that is, specifically choosing the light requirements for the crop being worked with accelerates the photosynthesis of plants, especially when they are illuminated by irradiation with the red and blue wavelengths of the spectrum (Casierra-Posada and Peña Olmos, 2015).

13.1.2.3   LIGHT INTENSITY

Light intensity is defined as the number of particles or photons that impinge on a unit of surface area (m^2) per second, expressed in $\mu mol/m^2/s$. This element of light has a very important influence on crops such as the increase and speed of biomass accumulation (Johkan et al., 2010; Stutte et al., 2009).

13.1.2.4   PHOTOPERIOD

Photoperiod is defined as the time in hours when the crop is exposed to a source of light either natural or artificial. This can be controlled according to the need of each species in order to give the plant its requirement of daily integral light (LID). The concept of LID integrates the aspects of intensity and quality of light along with the photoperiod. This seeks to give proper management of crops to obtain the best production in the aspect of biomass and the best nutritional quality and mineral content.

13.1.2.5   LIGHTING SYSTEMS

Within agriculture, different forms of lighting have been used in order to provide the plants with the optimum light requirement and thus to effect the photosynthesis in the best possible way. In controlled environment systems, fluorescent and incandescent lamps, high pressure sodium (HPS) lamps, and LED light are used.

According to Avendaño et al. (Abarca et al., 2019), the following comparison of artificial lighting parameters is given in Table 13.1.

TABLE 13.1 Comparison of Artificial Lighting Parameters.

	LED	**HPS**	**Incandescent**	**Fluorescent**
Light efficiency (lumenes-Watts-1)	80	100	20	80
Light source feature	Solid-state lighting	Gas discharge lighting	Solid-state lighting	Gas discharge lighting
Lifetime (hours)	50,000	8000	3000	10,000
Environmentally friendly	Does not pollute	Mercury and lead contamination	Does not pollute	Mercury contamination
Transformer power consumption (loss)	15 W	50 W	0 W	45 W

13.1.2.6 EFFECT OF LED LIGHT ON THE PRODUCTION OF SECONDARY METABOLITES IN PLANTS

Similarly, Ohashi et al. (2007) mention that in the *Eruca sativa* crop, blue light increases the content of vitamin C, chlorophyll, and carotenoids compared with white light. On the other hand, Lee et al. (2016) mention that the combination of red:blue light favors the production of biomolecules that have anticarcinogenic properties such as glucosinolates (GLS) in *Brassica oleracea* var. acephala and *Brassica rapa* subsp. chinensis (van Dam et al., 2009; Holst and Williamson, 2004; Das et al., 2000). In addition to metabolite production, studies by Matsuda et al. (2008) show that blue light or blue:red combination increases the biomass production.

Optimal light intensity will favor photosynthesis and dry matter accumulation, whereas excess light can limit them (Hu et al., 2007). It is worth mentioning that the intensity provided to leaf vegetables that have been produced in controlled environments is in the range of 200–300 µmol/m^2/s (Samuoliene et al., 2013; Johkan et al., 2010; Bian et al., 2015), and providing 100–200 µmol/m^2/s of blue light in *Lactuca sativa* increases the biosynthesis of phenolic compounds, vitamin C, tocopherols, and carotenoids. Providing 90 µmol/m^2/s of the blue:red mixture in *Lactuca sativa* increases the synthesis of anthocyanins, polyphenols, flavonoids and GLS.

The morphology and physiology of the plant are also influenced by the photoperiod (Jackson, 2009), accumulation of phytochemicals has also been found (Bian et al., 2015). Ali et al. (2009) mention that with a 12 h photoperiod, high concentrations of chlorophyll, polyphenols, and total antioxidants were obtained, while in a 24 h photoperiod, they were lower in *Amaranthus cruentus, Beta vulgaris,* and *Spinacia oleracea.* On the contrary, Soffe et al. (Stutte et al., 2009) mention that extending the photoperiod increases the dry weight in biomass of *Lactuca sativa* and *Spinacia oleracea* L.

13.1.2.7   *EFFECT OF LED LIGHT ON THE MINERAL CONTENT OF PLANTS*

Like the light and its factors described, the control of the nutrition of the plant is primordial in SPF. Due to the fact that by means of this, the essential elements for its life cycle are provided, but next to them beneficial elements can also be provided that are essential for certain vegetal species, and in the case of the humans, they contribute in favorable reactions, Therefore, they are recommended in the daily diet, due to this, the content of minerals in plants for human consumption takes greater importance, and it is worth noting that a controlled environment system, as SPF, allows effectively to manage the amount and time of supply of these elements (Tsukagoshi and Shinohara, 2016). The main factor that determines the mineral content in the plant is genetic, as well as the availability of nutrients in the solution. Other factors are also involved. Shin et al. (2012) evaluated the effect of light on the inorganic elements of *Lactuca sativa* L. pink roll, finding that for N, Ca, Mg, and Fe the content was increased in plants grown under red:blue LED light when compared with those grown under fluorescent light conditions, concluding that the red:blue combination increased biomass production as well as the content of these nutrients.

Crop	Conditions	Main results	Reference
638 nm LED (170 μmol/ m²/s) supplement for HPS (130 μmol/m²/s) in green house	*Lactuca sativa*	Increased DPPH-free radical scavenging. Increased phenolic compound	Žukauskas et al. (2011)
Blue LED (30 μmol/m²/s) in combination with red (270 μmol/m²/s).	Red leaf lettuce	Increased concentration of anthocyanins	Stutte et al. (2009)
Higher irradiance level 545–440 μmol/m²/s	Red pak choi and mustard	Leaf chlorophyll index increased.	Samoliné et al. (2013)
Blue LED (20%, 470 nm) and Red (80%, 630 nm. Intensity 250 μmol/m²/s.	Microgreens broccoli	Higher concentration of tissue Ca, K, Mg, P, S, B, Cu, Fe, Mn, Mo and Zn.	Kopsell et al. (2015)

13.2 CONCLUDING REMARKS

Knowing, understanding, and choosing the best growing conditions for each crop such as microgreens are essential for the proper development of the plant. However, the proper use of LED light allows to promote the stress of the crop so that it produces in greater quantity and in less time the bioactive compounds beneficial to health. Hence, the importance of this study.

ACKNOWLEDGMENTS

Authors thank to Sectorial fund for research, development and forest technological innovation CONAFOR-CONACYT projects 2018-B-S-65769 and 2018-2-B-S-131466.

KEYWORDS

- **microgreens**
- **plant factory**
- **environmental damage reduction**
- **controlled conditions**
- **LED light**

REFERENCES

Abarca, V.; Niño-Medina, G.; Rodriguez, H.; Munguia-López, J. P.; Luna-Maldonado, A.; Vidales, J.; Márquez-Reyes, J. Influence of Light-Emittng Diodes on Phenolic Content and Antioxidant Capacity Level in Romaine Lettuce (Lactuca Sativa L. Var. Longifolia Lam). *Fresenius Environ. Bull.* **2019,** *28,* 7945.

Ahmed, H. A.; Yu-xin, T.; Qi-chang, Y. Lettuce Plant Growth and Tipburn Occurrence as Affected by Airflow Using a Multi-Fan System in a Plant Factory with Artificial Light. *J. Therm. Biol.* **2020,** *88,* 102496. https://doi.org/https://doi.org/10.1016/j.jtherbio.2019.102496.

Ali, Md. B.; Khandaker, L. –O.; Shinya, M. B. A.-A. Comparative Study on Functional Components, Antioxidant Activity and Color Parameters of Selected Colored Leafy Vegetables as Affected by Photoperiods. *J. Food Agric. Environ.* **2009,** *7* (3–4), 392–398–2009.

Atanda, S. The Concepts and Problems of Post-Harvest Food Losses in Perishable Crops. 2011.

Bian, Z. H.; Yang, Q. C.; Liu, W. K. Effects of Light Quality on the Accumulation of Phytochemicals in Vegetables Produced in Controlled Environments: A Review. *J. Sci. Food Agric.* **2015**, *95* (5), 869–877. https://doi.org/10.1002/jsfa.6789.

Bourget, C. M. An Introduction to Light-Emitting Diodes. *HortScience Horts* **2008**, *43* (7), 1944–1946. https://doi.org/10.21273/HORTSCI.43.7.1944.

Bulgari, R.; Baldi, A.; Ferrante, A.; Lenzi, A. Yield and Quality of Basil, Swiss Chard, and Rocket Microgreens Grown in a Hydroponic System. *New Zeal. J. Crop Hortic. Sci.* **2017**, *45* (2), 119–129. https://doi.org/10.1080/01140671.2016.1259642.

Carvalho, L. C.; Santos, S.; Jorge Vilela, B.; Amâncio, S. Solanum Lycopersicon Mill. and Nicotiana Benthamiana L. under High Light Show Distinct Responses to Anti-Oxidative Stress. *J. Plant Physiol.* **2008**, *165* (12), 1300–1312. https://doi.org/https://doi.org/10.1016/j.jplph.2007.04.009.

Casierra-Posada, F.; Peña Olmos, J. E. Modificaciones Fotomorfogénicas Inducidas Por La Calidad de La Luz En Plantas Cultivadas. *Rev. la Acad. Colomb. Ciencias Exactas, Físicas y Nat.* **2015**, *39*, 84–92. https://doi.org/10.18257/raccefyn.276.

Cha-um, S.; Ulziibat, B.; Kirdmanee, C. Effects of Temperature and Relative Humidity During in "vitro" Acclimatization, on Physiological Changes and Growth Characters of "Phalaenopsis" Adapted to in "Vivo." *Aust. J. Crop Sci.* **2010**, *4* (9), 750–756.

Chauvat, M.; Forey, E. Temperature Modifies the Magnitude of a Plant Response to Collembola Presence. *Appl. Soil Ecol.* **2021**, *158*, 103814. https://doi.org/https://doi.org/10.1016/j.apsoil.2020.103814.

Das, S.; Tyagi, A. K.; Kaur, H. Cancer Modulation by Glucosinolates: A Review. *Curr. Sci.* **2000**, *79* (12), 1665–1671.

EA, A.; Rogers, A. The Response of Photosynthesis and Stomatal Conductance to Rising [CO_2]: Mechanisms and Environmental Interactions. 2007. PG-258-270 LID-10.1111/j.1365-3040.2007.01641.x [Doi]. USDA/ARS Photosynthesis Research Unit and Department of Plant Biology, University of Illinois Urbana-Champaign, 147 ERML, 1201 W. Gregory Drive, Urbana, IL 61801, Department of Environmental Sciences, Brookhaven National Laboratory, Upton, NY 11973-5000 a.

Hee, P.; Beom, L. Effects of CO_2 Concentration, Light Intensity and Nutrient Level on Growth of Leaf Lettuce in a Plant Factory. *Acta Hortic.* **2001**, *548*, 377–383.

Holst, B.; Williamson, G. A Critical Review of the Bioavailability of Glucosinolates and Related Compounds. *Nat. Prod. Rep.* **2004**, *21* (3), 425–447. https://doi.org/10.1039/B204039P.

Hu, Y.; Sun, G.; Wang, X.-C. Induction Characteristics and Response of Photosynthetic Quantum Conversion to Changes in Irradiance in Mulberry Plants. *J. Plant Physiol.* **2007**, *164*, 959–968. https://doi.org/10.1016/j.jplph.2006.07.005.

Huang, K.-L.; Yang, C.-L.; Kuo, C.-M. Plant Factory Crop Scheduling Considering Volume, Yield Changes and Multi-Period Harvests Using Lagrangian Relaxation. *Biosyst. Eng.* **2020**, *200*, 328–337. https://doi.org/https://doi.org/10.1016/j.biosystemseng.2020.10.012.

Lee, J.S.; Pill, W.; Cobb, B. B.; Olszewski, M. Seed Treatments to Advance Greenhouse Establishment of Beet and Chard Microgreens. *J. Hortic. Sci. Biotechnol.* **2004**, *79*, 565–570. https://doi.org/10.1080/14620316.2004.11511806.

Jackson, S. Plant Responses to Photoperiod. *New Phytol.* **2009**, *181*, 517–531. https://doi.org/10.1111/j.1469-8137.2008.02681.x.

Johkan, M.; Shoji, K.; Goto, F.; Hashida, S.; Yoshihara, T. Blue Light-Emitting Diode Light Irradiation of Seedlings Improves Seedling Quality and Growth after Transplanting in Red Leaf Lettuce. *HortScience* **2010**, *45* (12), 1809–1814. https://doi.org/https://doi.org/10.21273/HORTSCI.45.12.1809.

Kennedy, C.; Cuddihy, J.; Engel-Yan, J. The Changing Metabolism of Cities. *J. Ind. Ecol.* **2007,** *11* (2), 43–59. https://doi.org/https://doi.org/10.1162/jie.2007.1107.

Kook, K. The Effect of Blue-Light-Emitting Diodes on Antioxidant Properties and Resistance to Botrytis Cinerea in Tomato. *J. Plant Pathol. Microbiol.,* **2013,** *04.* https://doi.org/10.4172/2157-7471.1000203.

Kopsell, D. A.; Sams, C. E.; Morrow, R. C. Blue Wavelengths from LED Lighting Increase Nutritionally Important Metabolites in Specialty Crops. *HortScience Horts* **2015,** *50* (9), 1285–1288. https://doi.org/10.21273/HORTSCI.50.9.1285.

Kozai, T. Closed Systems For High Quality Transplants Using Minimum Resources. *Plant TissueCult. Eng.* **2008,** *6,* 275–312. https://doi.org/10.1007/978-1-4020-3694-1_15.

Kozai, T. Resource Use Efficiency of Closed Plant Production System with Artificial Light: Concept, Estimation and Application to Plant Factory. *Proc. Japan Acad. Ser. B* **2013,** *89* (10), 447–461. https://doi.org/10.2183/pjab.89.447.

Kozai, T.; Niu, G.; Takagaki, M. *Plant Factory: An Indoor Vertical Farming System for Efficient Quality Food Production*; Elsevier Academic Press, 2016. https://doi.org/https://doi.org/10.1016/C2014-0-01039-8.

Kozai, T.; Ohyama, K.; Chun, C. Commercialized Closed Systems with Artificial Lighting for Plant Production. In *Acta Horticulturae*; International Society for Horticultural Science (ISHS), Leuven, Belgium, 2006; pp 61–70. https://doi.org/10.17660/ActaHortic.2006.711.5.

Lambin, E. F.; Meyfroidt, P. Global Land Use Change, Economic Globalization, and the Looming Land Scarcity. *Proc. Natl. Acad. Sci.* **2011,** *108* (9), 3465–3472. https://doi.org/10.1073/pnas.1100480108.

Li, H.; Xu, Z.; Liu, X.; Han, X. Effects of Different Light Sources on the Growth of Non-Heading Chinese Cabbage (Brassica Campestris L.). *Nat. Sci. Fund.* **2012,** *4.* https://doi.org/10.5539/jas.v4n4p262.

Malayeri, S.; Hikosaka, S.; Goto, E. Effects of Light Period and Light Intensity on Essential Oil Composition of Japanese Mint Grown in a Closed Production System. *Environ. Control Biol.* **2010,** *48,* 141–149. https://doi.org/10.2525/ecb.48.141.

Matsuda, R.; Ohashi, K.; Fujiwara, K.; Kurata, K. Effects of Blue Light Deficiency on Acclimation of Light Energy Partitioning in PSII and CO_2 Assimilation Capacity to High Irradiance in Spinach Leaves. *Plant Cell Physiol.* **2008,** *49,* 664–670. https://doi.org/10.1093/pcp/pcn041.

Moon, A.; Li, S.; Kim, K. *Components Based Integrated Management Platform for Flexible Service Deployment in Plant Factory BT–HCI International 2011—Posters' Extended Abstracts*; Stephanidis, C., Ed.; Springer Berlin Heidelberg: Berlin, Heidelberg, 2011; pp 524–528.

Niu, G.; Kozai, T.; Sabeh, N. *Chapter 8—Physical Environmental Factors and Their Properties*; Kozai, T., Niu, G., Takagaki, M. B. T.-P. F., Eds.; Academic Press: San Diego, 2016; pp 129–140. https://doi.org/https://doi.org/10.1016/B978-0-12-801775-3.00008-1.

Ohyama, K.; Yoshinaga, K.; Kozai, T. Energy and Mass Balance of a Closed-Type Transplant Production System (Part 2). *Shokubutsu Kojo Gakkaishi* **2000,** *12* (4), 217–224. https://doi.org/10.2525/jshita.12.217.

OMS, 2020. Obesidad y sobrepeso https://www.who.int/es/news-room/fact-sheets/detail/obesity-and-overweight (accessed Jan 16, 2022).

Park, Y.; Park, J.; Hwang, S.; Jeong, B. R. Light Source and CO_2 Concentration Affect Growth and Anthocyanin Content of Lettuce under Controlled Environment. *Hortic. Environ. Biotechnol.* **2013,** *53.* https://doi.org/10.1007/s13580-012-0821-9.

Pessarakli, M. *Handbook of Photosynthesis*; CRC Press: Boca Raton, 2016. https://doi.org/ https://doi.org/10.1201/9781315372136.

Rodrigues, C. R. F.; Silveira, J. A. G.; Viégas, R. A.; Moura, R. M.; Aragão, R. M.; Silva, E. N. Combined Effects of High Relative Humidity and K+ Supply Mitigates Damage Caused by Salt Stress on Growth, Photosynthesis and Ion Homeostasis in J. Curcas Plants. *Agric. Water Manag.* **2016**, *163*, 255–262. https://doi.org/https://doi.org/10.1016/j.agwat.2015.09.027.

Samuolienė, G.; Brazaitytė, A.; Jankauskienė, J.; Viršilė, A.; Sirtautas, R.; Novičkovas, A.; Sakalauskienė, S.; Sakalauskaitė, J.; Duchovskis, P. LED Irradiance Level Affects Growth and Nutritional Quality of Brassica Microgreens. *Cent. Eur. J. Biol.* **2013**, *8* (12), 1241–1249. https://doi.org/10.2478/s11535-013-0246-1.

Samuoliene, G.; Brazaitytė, A.; Sirtautas, R.; Viršilė, A.; Sakalauskaitė, J.; Sakalauskienė, S.; Duchovskis, P. LED Illumination Affects Bioactive Compounds in Romaine Baby Leaf Lettuce. *J. Sci. Food Agric.*, **2013**, *93*. https://doi.org/10.1002/jsfa.6173.

Shiina, T.; Hosokawa, D.; Roy, P.; Nakamura, N.; Thammawong, M.; Orikasa, T. Life Cycle Inventory Analysis of Leafy Vegetables Grown in Two Types of Plant Factories. *Acta Hortic.* **2011**, *919*, 115–122. https://doi.org/10.17660/ActaHortic.2011.919.14.

Shin, Y. S.; Lee, M. J.; Lee, E.; Ahn, J.; Lim, J. H.; Kim, H. J.; Park, H. W.; Um, Y. G.; Park, S. D.; Chai, J. H. Effect of LEDs (Light Emitting Diodes) Irradiation on Growth and Mineral Absorption of Lettuce (*Lactuca Sativa* L. 'Lollo Rosa'), 2012.

Stutte, G.; Edney, S.; Skerritt, T. Photoregulation of Bioprotectant Content of Red Leaf Lettuce with Light-Emitting Diodes. *HortScience* **2009**, *44*, 79–82. https://doi.org/10.21273/HORTSCI.44.1.79.

Suslov, M. A. Dynamics of Intercellular Water Transfer in the Roots of Intact Zea Mays L. Plants under Elevated Concentrations of Atmospheric CO2. *Plant Physiol. Biochem.*, **2020**, *151*, 516–525. https://doi.org/https://doi.org/10.1016/j.plaphy.2020.04.007.

Tamulaitis, G.; Duchovskis, P.; Bliznikas, Z.; Breivė, K.; Ulinskaitė, R.; Brazaitytė, A.; Novickovas, A.; Arturas, Z. High-Power Light-Emitting Diode Based Facility for Plant Cultivation. *J. Phys. D Appl. Phys.*, **2005**, *38*, 3182–3187. https://doi.org/10.1088/0022-3727/ 38/17/S20.

Tsukagoshi, S.; Shinohara, Y. *Nutrition and Nutrient Uptake in Soilless Culture Systems*; 2016; pp 165–172. https://doi.org/10.1016/B978-0-12-801775-3.00011-1.

van Dam, N. M.; Tytgat, T. O. G.; Kirkegaard, J. A. Root and Shoot Glucosinolates: A Comparison of Their Diversity, Function and Interactions in Natural and Managed Ecosystems. *Phytochem. Rev.* **2009**, *8* (1), 171–186. https://doi.org/10.1007/s11101-008-9101-9.

Vastakaite Lithuanian Research Centre for Agriculture and Forestry, Akademija, Kedainiai distr. (Lithuania), V.; Virsile Latvia Univ. of Agriculture, Jelgava (Latvia), A. Light - Emitting Diodes (LEDs) for Higher Nutritional Quality of Brassicaceae Microgreens. *Latvia Univ. Agric.* **2015**, 111–117.

Xiao, Z.; Lester, G. E.; Luo, Y.; Wang, Q. Assessment of Vitamin and Carotenoid Concentrations of Emerging Food Products: Edible Microgreens. *J. Agric. Food Chem.*, **2012**, *60* (31), 7644–7651. https://doi.org/10.1021/jf300459b.

Yepes, A.; Buckeridge, M. S. Respuestas De Las Plantas Ante Los Factores Ambientales Del Cambio Climático Global: Revisión. *Colombia Forestal*. scieloco **2011**, 213–232.

Yokoi S.; Kozai T.; Hasegawa T.; Chun C.; Kubota C. CO_2 and Water Utilization Efficiencies of A Closed Transplant Production System as Affected by Leaf Area Index of Tomato Seedling Populations and the Number of Air Exchanges. *J. SHITA* **2005**, 18, 182–186.

Zhang, T.; Folta, K. M. Green Light Signaling and Adaptive Response. *Plant Signal. Behav.* **2012,** *7* (1), 75–78. https://doi.org/10.4161/psb.7.1.18635.

Zukauskas, A.; Bliznikas, Z.; Breivė, K.; Novičkovas, A.; Samuolienė, G.; Urbonavičiūtė, A.; Brazaitytė, A.; Jankauskienė, J; Duchovskis, P. Effect of Supplementary Pre-Harvest Led Lighting on the Antioxidant Properties of Lettuce Cultivars. In *Acta Horticulturae*; International Society for Horticultural Science (ISHS): Leuven, Belgium, 2011; pp 87–90. https://doi.org/10.17660/ActaHortic.2011.907.8.

CHAPTER 14

New Trends in the Analysis of Abiotic Stress Resistance in Corn: Selected Secondary Metabolites

CESAR DE JESUS AYALA-MEZA, FRANCISCO ZAVALA-GARCÍA*, MARISOL GALICIA-JUÁREZ, and GUILLERMO NIÑO-MEDINA

Facultad de Agronomía, Universidad Autónoma de Nuevo León. Francisco Villa S/N Col. Ex Hacienda El Canadá, General Escobedo, Nuevo León

Corresponding author. E-mail: francisco.zavalagr@uanl.edu.mx

ABSTRACT

The secondary metabolism in plants contributes to proper functioning against insects, pests, as well as nonbiological conditions such as high radiation or temperature. This part of the metabolism is controlled by the expression of genes associated with external signals that affect the development of the plant, generating secondary metabolites, which have been of interest due to their beneficial properties for human health caused by their antioxidant activity. Phenolic compounds are the secondary metabolites with the greatest abundance in plants. These have gained attention due to their antioxidant potential, as well as their protective functions to UV rays, diseases and pests, as well as a characteristic and attractive coloration in plants. Flavonoids are considered part of the main antioxidant-capable compounds in plants; they are synthesized primarily in chloroplasts or cytoplasm through the route of phenylpropanoids and have a wide range of functions in plants. One of the coloring pigments found in colored corns is anthocyanin, which are glycosides responsible for the red, violet, blue, and purple pigmentation seen in flowers, fruits, leaves, and other plant tissues, they can confer added value on products for human consumption. In yellow or orange corns, a concentration of carotenoids is observed, which have activities of great importance to human health, becoming part of Provitamin

A from the carotenes, and an excellent option for the biofortification of crops. Fifty-nine native corn breeds are distributed In Mexico. In 2011, these breeds had reported 9136 records for native corns (also named "Exotic"), of which 7189 are sheltered in different national institutions or stored in cold rooms. Mexico is considered the center of origin of many corn breeds, which include Chapalote, Palomero Toluqueño, Arrocillo Amarillo, among others. These breeds come directly from the Teocintle, considered the wild ancestor of maize, dating from 8700 years ago. Because of the mutations that maize has undergone over the decades; its different breeds have developed methods by which they adapt and defend themselves from the environment surrounding them, not only through the use primary metabolism, but also through the secondary metabolism, which is present in conditions by both biotic and abiotic factors. Climate projections indicate that there will be greater variation in precipitation as well as high temperatures, although drought is considered to affect more than the effect of high temperatures on the metabolites found in the leaf, the combined effect of drought stress and high temperatures can damage the growth and productivity of plants, compared with the individual effect. According to data from the Agri-Food and Fisheries Information Service, nearly 77% of the area planted in Mexico is managed by rainfed systems, thus concluding that climate change could have a highly significant impact on food security for the country. The overall planet is in a critical situation for human development, from the ecological side, as well as the alimentary side. As noted above, climate change can bring a decrease in the yields of some species of high economic and social interest, scientists should be prepared to get face of the climate crisis. The genetic improvement of crops should take a turn to make a more efficient way of energy consumption, for different agricultural activities, as well as based on the selection of genotypes with mechanisms to defend against drought stress and high temperatures, among them, the formation of antioxidant enzymes, plant architecture, photosystem efficiency, the use of secondary metabolites, such as the previously mentioned anthocyanins, phenolic acids, and carotenoids could make easier the process of analysis to select genotypes with better adaptation to abiotic stresses. It is not too late.

A bioactive compound is defined as a substance with the ability to interact with one or more molecules, components or compounds of a biological system, found in animals, bacteria, or plants, being referred to, in the latter, as part of secondary metabolism (Guaadaoui et al., 2014). The secondary metabolism in plants contributes to proper functioning against insects, pests, as well as nonbiological conditions such as high radiation or tempera-ture. This part of the metabolism is controlled by the expression of genes associated with external signals that affect the development of the plant, generating secondary metabolites, which have been of interest due to their beneficial properties for human health caused by their antioxidant activity (Armendáriz-Fernández et al., 2019; Van Der Fits and Memelink, 2000).

There are different types of secondary metabolites, such as phenols, terpenoids, alkaloids, flavonoids, carotenoids, tannins, glycosides, saponins, and essential oils, which have different functions depending on the plant, phenological stage, or biotic/abiotic stress during which they occur (Kabera et al., 2014). Table 14.1 presents some of the compounds of interest to future genetic improvement programs against biotic and abiotic stress.

14.1 SECONDARY METABOLISM IN CROPS

Different authors have linked secondary metabolites, such as anthocyanins, flavonoids, phenolic acids, and carotenoids with biotic and abiotic stress (Garzon, 2008; Gould, 2004; Isah, 2019; Kaur et al., 2019; Pietrini et al., 2002; Sepúlveda-Jiménez, 2003; Rabêlo et al., 2019), but in a way larger studies, with the attributes of importance for human health (Guillén-Sánchez et al., 2014; Rosales et al., 2016; Scrob et al., 2014; Suwarno et al., 2019; Troncoso-rojas and Zamora-bustillo, 2015; Zhang et al., 2019), demonstrating the importance of the different compounds obtained in plants for the development of biotechnology, regardless of the trendiest perspective.

14.1.1 PHENOLIC COMPOUNDS

Phenolic compounds are the secondary metabolites with the greatest abundance in plants, which have different aromatic rings with one or more hydroxyl groups. These have gained attention due to their antioxidant potential, as well as their protective functions to UV rays, diseases and pests, as well as a characteristic and attractive coloration in plants (Dai and Mumper, 2010; Ozmianski et al., 2015).

To counteract the stress caused by the hypersensitivity response to abiotic conditions, it is believed that plants began to synthesize a large number of phenolic compounds, as they do not only help against reactive oxygen species (ROS) but also against stress caused by predators, as well as UV radiation. Phenolic compounds can absorb light radiation from 40 to 320 nm, which are the most harmful radiation against proteins, as well as nucleic acids (Chalker-scott, 1989).

TABLE 14.1 Bioactive Compounds of Secondary Metabolism of Interest for Genetic Improvement against Biotic and Abiotic Stress.

Secondary metabolite	Features	Graphic representation	Sources
Phenolic acids	Higher abundance of metabolites in plants with functional potential in processes, such as photosynthesis, protein synthesis, enzymatic activity and allelopathy	Representation of benzoic phenolic acids	Bravo et al. (2013)
Anthocyanins	Main water-soluble pigments among those visible to the human eye. The color of this pigment is based on the positions of the glycoside in carbon 3 and/or 5, being a mono, di or trisaccharide	Representation of anthocyanins	Guillén-Sánchez et al. (2014)
Carotenoids	They are pigments involved in photosynthetic metabolism and help prevent oxidative stress damage by their antioxidant capacity	Representation of a zeaxanthin molecule	Luo et al. (2020)

Phenolic compounds include phenolic acids, which are simple molecules of phenolic compounds, divided into benzoic acid derivatives and cinnamic acid derivatives. They are found in high concentrations and have been associated with functions, such as photosynthesis, protein synthesis, enzymatic activity and allelopathy (Dai and Mumper 2010; Hura et al., 2008). Phenolic acids are produced by the phenylpropanoid route through shikimic acid, during the monolignol route and are obtained by breaking polymers from the cell wall, such as lignin (Kumar and Goel, 2019). According to Mesarovic (Mesarović et al., 2017), the readings for protocatechuic acids and cumaric acids are set at 300 nm, ferulic and caffeic acids are read at 290 nm, and gallic acid is read at 278 nm.

The great antioxidant capacity of phenolic compounds also helps in the neutralization of the oxidative effects of ROS by donating hydrogen atoms for stabilization of these, although several direct and indirect mechanisms for antioxidant activity are known (Kumar and Goel, 2019), and these compounds have been shown to have chelation potential in heavy metal ions, as well as antibiotic and antifungal capabilities (Kulbat, 2016). Different authors have presented results regarding the concentration of total phenolic compounds in maize leaf, such as those presented by Rabelo (Rabêlo et al., 2019), who mentioned in some compounds that the concentration values of 1800 g/g of equivalent slug acid vary according to the treatment of added drought. Vazquez-Olivo (Vazquez-Olivo et al., 2019) presented concentrations of up to 578.25 mg·100 g^{-1} of gallic acid equivalents, these reported as total polyphenols, referring to the sum of bound and free compounds.

14.1.2 FLAVONOIDS

Flavonoids are organic compounds made up of phenylalanine and malonyl-CoA found on plants and protect against the damage caused by oxidizing agents, ultraviolet rays, and even environmental pollution. These have phenolic hydroxyl groups and have chelation properties of iron and other transition metals, giving them a great antioxidant capacity (Martínez-Flórez et al., 2002).

Their optimal conditions occur with low temperatures, UV stimulation, and microbial infection, phytoalexins, which help prevent the proliferation of pathogens. Coumarins that have toxic properties against herbivorous organisms, and tannins that are separated into hydrolysable (by polymerization of some phenolic acid and sugar) or condensates (formed from multiple flavones), and have insect repellent properties (Kulbat, 2016).

Flavonoids are considered part of the main antioxidant-capable compounds in plants, they are synthesized primarily in chloroplasts or cytoplasm through the route of phenylpropanoids and have a wide range of functions in plants, such as biotic stress defense, protection of oxidative agents for DNA, and reduction in damage of photosystems (Salama et al., 2015; Zhang et al., 2018). The total flavonoid analysis in leaf shows concentrations ranging from 1.21 to 1.50 mg/g of fresh matter (Salama et al., 2015).

14.1.2.1 ANTHOCYANINS

One of the coloring pigments found in colored corns is anthocyanin, which are glycosides responsible for the red, violet, blue, and purple pigmentation seen in flowers, fruits, leaves, and other plant tissues, but when talking about maize, in the right genotype, it is easy to detect in the grain due to the pigmentation found in the pericarp as well as in the aleurone layer (Soto Mooner et al., 2013). Between the purple maize genotypes, scientists have found various anthocyanins in greater quantity like cyanidin-3-glucoside in the grain, which can give a purple-blue hue. However, peonidin-3-glucoside can also be found, which results in purple-red tonality, as well as cyanide-6-malonyl glycoside in smaller amounts (Guillén-Sánchez et al., 2014).

Anthocyanins are glycosylated salts with a flavylium cation (without glycoside is called anthocyanidin). Anthocyanidins consist of two aromatic groups, a benzopyrilium and a phenolic group. In nature, 90% of anthocyanins are based on six basic structures, such as cyanidin, delfinidin, pelargonidin, peonidin, malvidin, and petunidin (Aguilera-Ortíz et al., 2011; Mangalvedhe et al., 2015).

Anthocyanins are the most important group of water-soluble pigments within the visible region. They have a wide range of functions in the plant, such as the attraction of pollinators, protection against UV radiation effects, viral and microbial infections, among others, thus gaining importance thanks to their antioxidant effects. In addition, they can confer added value on products for human consumption (Garzon, 2008). Anthocyanins have also been found in the leaves of various plants, presenting beneficial characteristics, such as the protection of photosensitive compounds and photosynthetic systems, being compounds of high interest for their diversity of uses (Gould, 2004).

A relationship has been found between the synthesis of anthocyanins in grain and drought in corn seedlings, obtaining results from 0.028 to 0.1 mg/g (Efeoğlu et al., 2009) and also, the corncob, or as called in Mexico, the Olote, and grain analyses have been performed by Salinas-Moreno (Salinas-Moreno et al., 2013). However, there are few works involving the concentration of anthocyanins in tissues, such as lamina, ligule, and corn sheath, as the one shown in Figure 14.1.

The contents of anthocyanins may vary, as shown by Pietrini et al. (2002), with a value of 8.1 g/cm in maize leaves, while Gu et al. (2018) found values from 0.09 to 44.3 mg/g. Even though, the genotypes used in this experiment had a mostly purple tonality. There is a great variation in the shades of colors of native corns in Mexico that must be taken advantage of. Currently, the team of the Faculty of Agronomy of the UANL makes determinations of the content of anthocyanins of native varieties of the state of Nuevo León, most located in high and southern areas of the state. However, there are several regions of the country waiting to be explored.

FIGURE 14.1   Corn sheath with the presence of high concentrations of anthocyanins (Own authorship).

14.1.3   *CAROTENOIDS*

In yellow or orange corns, a concentration of carotenoids is observed, which have activities of great importance to human health, becoming a great source of Provitamin A from the carotenes, and an excellent option for the biofortification of crops (Khamkoh et al., 2019). Carotenoids are pigments that support light collection during the photosynthetic process, as well as protecting the plant due to its photoprotective capability, since these inhibit the spread of ROS species and other free radicals, thus preventing the harmful action that these can cause at the cellular level (Meléndez-Martínez et al., 2004; Mínguez Mosquera et al., 2005).

These compounds are synthesized in the inner membranes of chloro-plasts, which are formed by the independent mevalonate route. In Corn, concentrations of compounds, such as carotene have been reported, as well as Lutein (Silva-Pérez et al., 2012). Zhao et al. (2003) mentioned that the concentration of total carotenoids in maize leaf is between 40 and 60 mg/m², indicating its function as antenna compounds for plant cell photosystems. These pigments are of great importance for diet and human health since their accumulation in the retina can reduce the incidence of cataracts and eye degeneration, being substances that do not form within the human body as well as that have antioxidant functions and support the immune system, even reducing the incidence of cancer in animal cells (Luo et al., 2020).

14.2 CORN AND SECONDARY METABOLITES

In Latin America, more than 200 races of corn have been identified, which were developed through sociocultural processes, which have led to an incredible biodiversity of the species. However, in today's times, these breeds are being threatened by habitat loss, as well as the abiotic stress to which they have been grown (Aguirre-Liguori et al., 2019). In Mexico, 59 native corn races are distributed around the country as presented in Figure 14.2. In 2011, these breeds had reported 9136 records for native (also named "Exotic") corns, of which 7189 are sheltered in different national institutions or stored in cold rooms (CONABIO, 2011).

FIGURE 14.2 Distribution of native corns around the country.

Source: Adapted from CONABIO (2011), https://www.biodiversidad.gob.mx/genes/proyecto Maices.html.

Mexico is considered the center of origin of many corn breeds, which include Chapalote, Palomero Toluqueño, Arrocillo Amarillo, among others. These breeds come directly from the Teocintle, considered the wild ancestor of maize, dating from 8700 years ago in the region of Iguala, Guerrero. Teocintle is currently endangered due to its market application being too low (Bedoya and Chávez Tovar, 2013; Vázquez-Carrillo et al., 2019). These breeds require greater appreciation as genetic resources, so the conservation and characterization strategies of the diversity of Mexico and the world must be implemented, in order to establish germplasm banks, obtaining living units, represented by samples containing the genetic makeup of a population and having the ability to reproduce (Vidal et al., 2020). Imagine the mysteries that have been forgotten because of ignoring the wild ancestors, or even more alarming, the mysteries that are about to be forgotten.

One of the main factors for the diversification of maize in Mexico is associated with the variation of orographic, climatological, as well as edaphological, presenting a great diversity in genotypes by the wide range of environments that can be found in the country (Ávila-Bello et al., 2016). This variety of environments has generated genetic diversification, as maize crops from 0 to 3400 m above sea level have been found. In Figure 14.3, it can be seen how the different main breeds of maize adapt to different conditions of precipitation as well as altitude. An example of this adaptation is the conical type maize, which has resistance to low temperature, as well as that some genotypes may come to present pure hues as a defense mechanism against ultraviolet rays (Boege, 2008).

Because of the mutations that maize has undergone over the decades, its different breeds have developed methods by which they adapt and defend themselves from the environment surrounding them, not only through the use primary metabolism, but also through the secondary metabolism, which is present in conditions by both biotic and abiotic factors, and the substances produced by the secondary metabolism can help to overcome the affections caused by the stress conditions (Isah, 2019; Sepúlveda-Jiménez, 2003).

14.2.1 ENVIRONMENTAL CONDITIONS AND PLANTS

High temperatures generate anatomical, morphological, and functional changes in plants, some are similar to those produced by drought stress, such as reduced cell size, reduced stomatic conductance, stomatal closure, changes in membrane permeability, increases in stomata and trichome density, and larger xylem vessels. Cell wall and plasma membrane systems (chloroplasts,

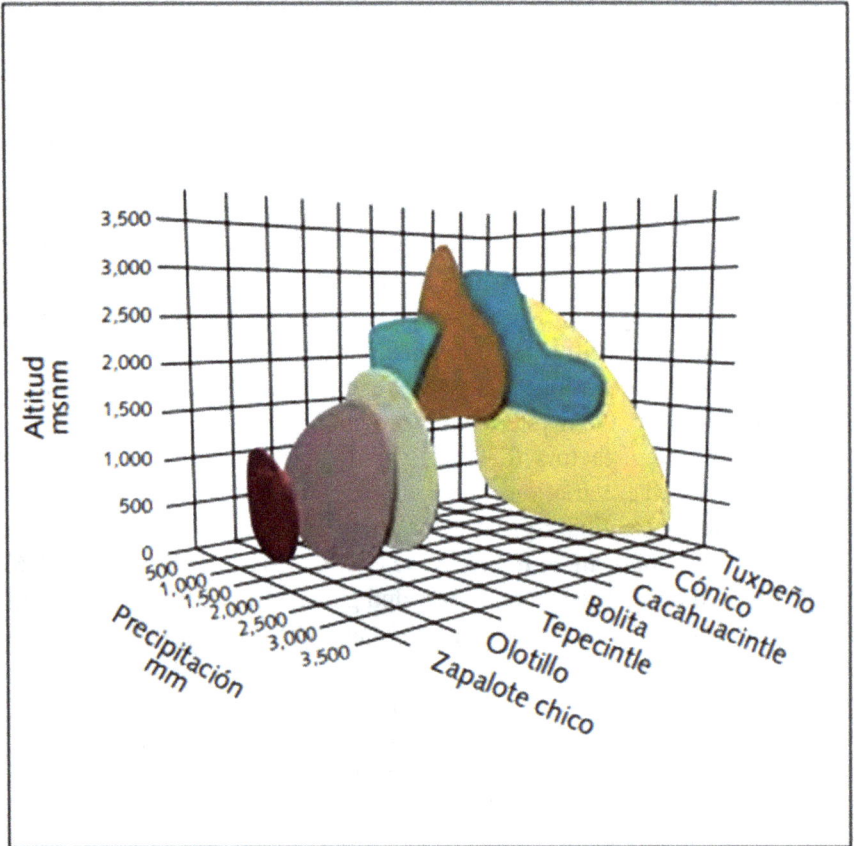

FIGURE 14.3 Graphical representation of the adaptation of different maize breeds to precipitation conditions represented in millimeters and altitude represented in meters above sea level.

Source: Adapted from CONABIO (2011), https://www.biodiversidad.gob.mx/genes/proyecto Maices.html.

mitochondria, Golgi apparatus, tonoplast, and endoplasmic reticle) are the first translators of stress in plants. The main consequence of the exposure of plants to high temperatures is the decrease in the carbon balance as a result of the inactivation of ribulose-1,5-biphosphate carboxylase/oxygenase, the reduction of photosynthesis and the imbalance between photosynthesis and respiration. The latter mentioned occurs at temperatures close to 50°C, where dark respiration and photorespiration increase as photosynthesis drops considerably. Larcher (1980) shows how the changes occur in the rate of photosynthesis and respiration according to temperature, explaining how from having low temperatures stress, the photosynthetic rate is reduced, being null at $-10°C$, and by exceeding $50°C$, the photosynthetic rate begins to decline. This situation is attenuated in CO_2-rich environments (Wahid, 2006). Even with that said, it cannot be ensured that water deficit affects plant photosynthesis above stomatic limitation or by metabolic alterations (Signarbieux and Feller, 2011). Drought conditions can inhibit the rate of electron transport, increasing oxidative stress (Masojidek et al., 1991), which can seriously affect the leaf photosynthetic machinery (Ort, 2001). The components of photosynthetic devices, and in particular, the photosystem II (PSII) located in the thylakoids of mesophyll cells, are considered the most sensitive components to drought stress and the main source of fluorescence signals (Havaux, 1992). Stress can cause the synthesis and accumulation of metabolites, such as ethylene, abscisic acid, salicylic, and jasmonic acids, compatible solutes, such as proline, secondary metabolites (alkaloids, terpenoids, phenols, phytoalexins, cyanogenic compounds), and stress-related proteins of various types, such as dehydrins and heat-shock stress proteins (Chávez-Barrantes and Gutiérrez-Soto, 2017).

Secondary metabolism in maize is of great importance due to the production of compounds, such as phenolic acids, which, as mentioned before, neutralize the radiation by absorbing and transforming radiation, filtering and limiting the overexcitation of chlorophylls during unfavorable conditions in photosystems (Hura et al., 2008). Taking this information into account, and in conjunction with agroclimatological, physiological, and agronomic data, it is possible to implement the way in which the genetic improvement of the different species of plants of agronomical interest is attended in order to fight against climate change. It is interesting to note how bioactive compound analysis usually emphasizes the products obtained by the crop, whether fruits, grains or vegetables. However, it is important to observe the concentration patterns in different parts of the plant, such as in the vanilla leaf as an example, which has presented a variety of concentrations of phenolic acids, depending on the altitude where it develops (Andrade-Andrade et al., 2018). In the Santacoloma (Santacoloma and Enrique, 2012) experiments, it was observed that at higher altitude, *Gliricidia sepium* does not produce any more total polyphenols in the leaf tissue, as confirmed by Arteaga (2016) in *Justicia pectoralis,* which mentions that altitude level does not affect functional properties as well as polyphenol content, although, the concentrations of bioactive compounds in the grapefruit in Brazil, presented by de Oliveira (de Oliveira et al.,

2019) show an opposite case, in which it is observed that grapefruit at higher altitudes are characterized by the presence of greater amounts of total polyphenols, total flavonoids, and total anthocyanins. Having this in mind, a viable conclusion is that the altitude affection depends on the crop, but still, there are more and more studies needed to confirm that height, or any other environmental factor, determines the condition for the biosynthesis of bioactive compounds. No studies were found on corn leaf that talked about altitude allocated in corn metabolome.

Climate projections indicate that there will be greater variation in precipitation as well as high temperatures, although drought is considered to affect more than the effect of high temperatures on the metabolites found in the leaf (Orians et al., 2019), the combined effect of drought stress and high temperatures can damage the growth and productivity of plants, compared with the individual effect, considering stress by temperature when exceeds 35°C, affects the different stages of maize from vegetative development to grain filling (Hussain et al., 2019). Drought comes to affect plants in different ways, such as leaf rolling, root increase, decrease in the aerial part, reduced protein synthesis, increased wax in the leaves, changes in respiration, photosynthesis, and distribution of nutrients in order to survive in the inhospitable environment (Robles, 2007). In response to drought, the stomata of the plant get closed, causing the amount of water lost through respiration to decrease, but generating an excess of energy to accumulate that must be handled by other pathways, such as the increase in the biosynthesis of secondary metabolites to reduce the energy reception (Selmar and Kleinwächter, 2013). It has been observed that maize shows a noticeable change in amino acids, sugars, and organic acids concentrations (Michaletti et al., 2018), which are metabolites used by the plant to try to overcome the osmotic stress caused by the lack of water (Chávez-Barrantes and Gutiérrez-Soto, 2017), but many more studies are needed to associate the metabolome response to certain physiological effects.

Different authors have shown how does the drought stress affect the concentration of phenolic compounds positively in foliar maize tissue, as well as negatively in foliar grape tissue (Król et al., 2014) Although Gharibi et al. (2016) have shown in their experiment with foliar tissue of *Achillea* species that the amount of phenolic compounds can be altered depending on the type of crop used. It has been shown that drought and high temperature stress can cause oxidative damage from an overproduction of ROS, as well as an increase in malondialdehydes, leading to a reduction of photosynthetic components (Hussain et al., 2019).

14.3   PERSPECTIVE TO THE FUTURE

The global climatic change has been the main reason on yield decline in basic crops, mainly due to extreme temperatures, inconsistent rainfall patterns, as well as biotic stress (Abberton et al., 2016) mainly due to the increase in greenhouse gas emissions. However, in certain areas, these effects could be beneficial as those for lack of heat, although they would be affected by high concentrations of CO_2, which can cause problems in stomatal conductance and excessive absorption of water (Chen et al., 2018).

Using prediction analyses, it has been observed that seasonal weather factors are affected by anomalies at approximately 20–49%, of which, a significant percentage is part of the extreme climate changes expected, although they affect differently in each type of crop (Vogel et al., 2019). One example is maize, which expects a 14% reduction in yield for dryland farming in the Malawi area (Msowoya et al., 2016). In Mexico, something similar is expected, where a negative impact on the yields of the temporary crops, being mainly affected by high temperature, while the crops under irrigation will remain constant (Ureta et al., 2020). According to data from the Agri-Food and Fisheries Information Service (SIAP, 2019), nearly 77% of the area planted in Mexico is managed by rainfed systems, thus concluding that climate change could have a highly significant impact on food security for the country.

The plant breeders require access to genetic material that can withstand the agroclimatic alterations faced by the planet, so the search for resistance qualities should be made in native varieties, which present a genetic base with great variability and complexity (Brozynska et al., 2016) due to the patterns of coincidence between genetic variation and environmental variation (Cerda-Hurtado et al., 2018).

The planet is in a critical situation for human development, from the ecological side, as well as the alimentary side. As noted above, climate change can bring a decrease in the yields of some species of high agricultural interest, although it may be beneficial for the planting of controlled irrigation in some parts of the planet. Either way, scientists should be prepared to get past the climate crisis. The genetic improvement of crops should take a turn to make a more efficient way of energy consumption, opting for different agricultural activities, as well as based on the selection of genotypes with mechanisms to defend against drought stress and high temperatures, among other factors such as the formation of antioxidant enzymes, plant architecture, photosystem efficiency, the use of secondary metabolites, such as the previously mentioned anthocyanins, phenolic acids, and carotenoids could make easier the process of analysis to select genotypes with better adaptation to abiotic stresses. It is not too late.

KEYWORDS

- **secondary metabolite**
- **bioactive compounds**
- **maize**
- **abiotic stress**
- **climate change**

REFERENCES

Abberton, M.; Batley, J.; Bentley, A.; Bryant, J.; Cai, H.; Cockram, J.; Costa de Oliveira, A.; Cseke, L. J.; Dempewolf, H.; De Pace, C.; et al. Global Agricultural Intensification during Climate Change: A Role for Genomics. *Plant Biotechnol. J.* **2016,** *14* (4), 1095–1098. https://doi.org/10.1111/pbi.12467.

Aguilera-Ortíz, M.; Reza-Vargas, M. del C.; Chew-Madinaveita, R. G.; Meza-Velázquez, J. A. Propiedades Funcionales De Las Antocianinas. *Biotecnia* **2011,** *13* (2), 16. https://doi.org/10.18633/bt.v13i2.81.

Aguirre-Liguori, J. A.; Ramírez-Barahona, S.; Tiffin, P.; Eguiarte, L. E. Climate Change Is Predicted to Disrupt Patterns of Local Adaptation in Wild and Cultivated Maize. *Proc. R. Soc. B Biol. Sci.* **2019,** *286* (1906), 20190486. https://doi.org/10.1098/rspb.2019.0486.

Andrade-Andrade, G.; Delgado-Alvarado, A.; Herrera-Cabrera, B. E.; Arévalo-Galarza, L.; Caso-Barrera, L. Variación de Compuestos Fenólicos Totales, Flavonoides y Taninos En Vanilla Planifolia Jacks. Ex Andrews de La Huasteca Hidalguense, México. *Agrociencia* **2018,** *52* (1), 55–66.

Armendáriz-Fernández, K. V.; Herrera-Hernández, I. M.; Muñoz-Márquez, E.; Sánchez, E. Characterization of Bioactive Compounds, Mineral Content, and Antioxidant Activity in Bean Varieties Grown with Traditional Methods in Oaxaca, Mexico. *Antioxidants* **2019,** *8* (1), 1–17. https://doi.org/10.3390/antiox8010026.

Arteaga, M. C. Genomics Data Genomic Variation in Recently Collected Maize Landraces from Mexico. *Genomic data* **2016,** *7*, 38–45. https://doi.org/10.1016/j.gdata.2015.11.002.

Ávila-Bello, C. H.; Morales-Zamora, J. A.; Ortega-Paczka, R. *Los Maíces Nativos de La Sierra de Santa Marta*, 1a Edición.; Quehacer científico y tecnológico: Xalapa, Veracruz, 2016.

Bedoya, C.; Chávez Tovar, V. H. *Teocintle: El Ancestro Del Maíz*; Ciudad de México, 2013. https://doi.org/10.1007/s11250-012-0216-z.

Boege, E. *El Patrimonio Biocultural de Los Pueblos Indígenas de México*, 1a Edición.; Instituto Nacional de Antrpología e Historia: Comisión Nacional para el Desarrollo de los Pueblos Indigenas: México, 2008.

Bravo, H. R.; Copaja, S. V.; Lamborot, M. Phytotoxicity of Phenolic Acids From Cereals. In *Herbicides-Advances in Research*; Price, A., Keltn, J., Ed.; IntechOpen, 2013; pp 37–49. https://doi.org/http://dx.doi.org/10.5772/55942.

Brozynska, M.; Furtado, A.; Henry, R. J. Genomics of Crop Wild Relatives: Expanding the Gene Pool for Crop Improvement. *Plant Biotechnol. J.* **2016,** *14* (4), 1070–1085. https://doi.org/10.1111/pbi.12454.

Cerda-Hurtado, I. M.; Mayek-Pérez, N.; Hernández-Delgado, S.; Muruaga-Martínez, J. S.; Reyes-Lara, M. A.; Reyes-Valdés, M. H.; González-Prieto, J. M. Climatic Adaptation and Ecological Descriptors of Wild Beans from Mexico. *Ecol. Evol.* **2018,** *8* (13), 6492–6504. https://doi.org/10.1002/ece3.4106.

Chalker-scott, L. Low Temperature Stress Physiology. In *Low Temperature Stress Physiology in Crops*; CRC Press: Boca Raton, 1989; pp 67–76.

Chávez-Barrantes, N. F.; Gutiérrez-Soto, M. V. Crop Physiological Responses to High Temperature Stress. I. Molecular, Biochemical and Physiological Aspects. *Agron. Mesoam.* **2017,** *28* (1), 237–253.

Chen, Y.; Zhang, Z.; Tao, F. Impacts of Climate Change and Climate Extremes on Major Crops Productivity in China at a Global Warming of 1.5 and 2.0°C. *Earth Syst. Dyn.* **2018,** *9* (2), 543–562. https://doi.org/10.5194/esd-9-543-2018.

CONABIO. *Proyecto Global de Maíces Nativos: Informe de Gestión.*; Ciudad de Mexico, 2011.

Dai, J.; Mumper, R. J. Plant Phenolics: Extraction, Analysis and Their Antioxidant and Anticancer Properties. *Molecules* **2010,** *15*, 7313–7352. https://doi.org/10.3390/molecules 15107313.

Dai, J.; Mumper, R. J. Plant Phenolics: Extraction, Analysis and Their Antioxidant and Anticancer Properties. *Molecules* **2010,** *15*, 7313–7352. https://doi.org/10.3390/molecules 15107313.

de Oliveira, J. B.; Egipto, R.; Laureano, O.; de Castro, R.; Pereira, G. E.; Ricardo-da-Silva, J. M. Climate Effects on Physicochemical Composition of Syrah Grapes at Low and High Altitude Sites from Tropical Grown Regions of Brazil. *Food Res. Int.* **2019,** *121* (December 2018), 870–879. https://doi.org/10.1016/j.foodres.2019.01.011.

Efeoğlu, B.; Ekmekçi, Y.; Çiçek, N. Physiological Responses of Three Maize Cultivars to Drought Stress and Recovery. *South African J. Bot.* **2009,** *75* (1), 34–42. https://doi.org/10.1016/j.sajb.2008.06.005.

Garzon, G. Las Antocianinas Como Colorantes Naturales y Compuestos Bioactivos: Revisión. *Acta Biológica Colomb.* **2008,** *13* (3), 27–36.

Gharibi, S.; Tabatabaei, B. E. S.; Saeidi, G.; Goli, S. A. H. Effect of Drought Stress on Total Phenolic, Lipid Peroxidation, and Antioxidant Activity of Achillea Species. *Appl. Biochem. Biotechnol.* **2016,** *178* (4), 796–809. https://doi.org/10.1007/s12010-015-1909-3.

Gould, K. S. Nature's Swiss Army Knife: The Diverse Protective Roles of Anthocyanins in Leaves. *J. Biomed. Biotechnol.* **2004,** *5*, 314–320.

Gu, X.; Cai, W.; Fan, Y.; Ma, Y.; Zhao, X.; Zhang, C. Estimating Foliar Anthocyanin Content of Purple Corn via Hyperspectral Model. *Food Sci. Nutr.* **2018,** *6* (3), 572–578. https://doi.org/10.1002/fsn3.588.

Guaadaoui, A.; Benaicha, S.; Elmajdoub, N.; Bellaoui, M.; Hamal, A. What Is a Bioactive Compound? A Combined Definition for a Preliminary Consensus. *Int. J. Food Sci. Nutr.* **2014,** *3* (3), 17–179. https://doi.org/10.11648/j.ijnfs.20140303.16.

Guillén-Sánchez, J.; Mori-Arismendi, S.; Paucar-Menacho, L. M. Características y Propiedades Funcionales Del Maíz Morado (Zea Mays L.) Var. Subnigroviolaceo. *Sci. Agropecu.* **2014,** *5*, 211–217. https://doi.org/10.17268/sci.agropecu.2014.04.05.

Havaux, M. Stress Tolerance of Photosystem II in Vivo: Antagonistic Effects of Water, Heat, and Photoinhibition Stresses. *Plant Physiol.* **1992,** *100* (1), 104–112.

Hura, T.; Hura, K.; Grzesiak, S. Contents of Total Phenolics and Ferulic Acid, and PAL Activity during Water Potential Changes in Leaves of Maize Single-Cross Hybrids of Different Drought Tolerance. *J. Agron. Crop Sci.* **2008,** *194* (2), 104–112. https://doi.org/10.1111/j.1439-037X.2008.00297.x.

Hussain, H. A.; Men, S.; Hussain, S.; Chen, Y.; Ali, S.; Zhang, S.; Zhang, K.; Li, Y.; Xu, Q.; Liao, C.; et al. Interactive Effects of Drought and Heat Stresses on Morpho-Physiological Attributes, Yield, Nutrient Uptake and Oxidative Status in Maize Hybrids. *Sci. Rep.* **2019,** *9* (1), 1–12. https://doi.org/10.1038/s41598-019-40362-7.

Isah, T. Stress and Defense Responses in Plant Secondary Metabolites Production. *Biol. Res.* **2019,** *52* (1), 1–25. https://doi.org/10.1186/s40659-019-0246-3.

Kabera, J. N.; Semana, E.; Mussa, A. R.; He, X. Plant Secondary Metabolites: Biosynthesis, Classification, Function and Pharmacological Properties. *J. Pharm. Pharmacol.* **2014,** *2* (January), 377–392. https://doi.org/10.1016/0300-9084(96)82199-7.

Kaur, H.; Kaur, K.; Gill, G. K. Modulation of Sucrose and Starch Metabolism by Salicylic Acid Induces Thermotolerance in Spring Maize. *Russ. J. Plant Physiol.* **2019,** *66* (5), 771–777. https://doi.org/10.1134/S102144371905008X.

Khamkoh, W.; Ketthaisong, D.; Lomthaisong, K.; Lertrat, K.; Suriharn, B. Recurrent Selection Method for Improvement of Lutein and Zeaxanthin in Orange Waxy Corn Populations. *Aust. J. Crop Sci.* **2019,** *13* (4), 566–573. https://doi.org/10.21475/ajcs.19.13.04.p1507.

Król, A.; Amarowicz, R.; Weidner, S. Changes in the Composition of Phenolic Compounds and Antioxidant Properties of Grapevine Roots and Leaves (Vitis Viniferal.) under Continuous of Long-Term Drought Stress. *Acta Physiol. Plant.* **2014,** *36* (6), 1491–1499. https://doi.org/10.1007/s11738-014-1526-8.

Kulbat, K. The Role of Phenolic Compounds in Plant Resistance. *Biotecnol. Food Sci.* **2016,** *80* (2), 97–108.

Kumar, N.; Goel, N. Phenolic Acids: Natural Versatile Molecules with Promising Therapeutic Applications. *Biotechnol. Reports* **2019,** *24*, e00370. https://doi.org/10.1016/j.btre.2019.e00370.

Larcher, W. *Physiological Plant Ecology*, 2nd ed.; Springer-Verlag: Munich, 1980.

Luo, H.; He, W.; Li, D.; Bao, Y.; Riaz, A.; Xiao, Y.; Song, J.; Liu, C. Effect of Methyl Jasmonate on Carotenoids Biosynthesis in Germinated Maize Kernels. *Food Chem.* **2020,** *307* (May 2019), 125525. https://doi.org/10.1016/j.foodchem.2019.125525.

Mangalvedhe, A.; Danao, M. G.; Paulsmeyer, M.;, Rausch, K.; Singh, V.; Juvik, J. Anthocyanin Determination in Different Corn Hybrids Using near Infrared Spectroscopy. In *ASABE Meeting*; ASABE: New Orleans, 2015; Vol. **7004,** pp 3–14.

Martínez-Flórez, S.; González-Gallego, J.; Culebras, J. M.; Tuñón, M. J. Los Flavonoides: Propiedades y Acciones Antioxidantes. *Nutr. Hosp.* **2002,** *17* (6), 271–278. https://doi.org/10.3305/nutr hosp.v17in06.3338.

Masojidek, J.; Trivedi, S.; Halshaw, L.; Alexiou, A.; Hall, D. O. The Synergistic Effect of Drought and Light Stresses in Sorghum and Pearl Millet. *Plant Physiol.* **1991,** *96* (1), 198–207.

Meléndez-Martínez, A.; Vicario, I.; Francisco J. H. Importancia Nutricional de Los Pigmentos Carotenoides. *Arch. Latinoam. Nutr.* **2004,** *54* (2), 149–155.

Mesarović, J. Z.; Dragičević, V. D.; Mladenović Drinić, S. D.; Ristić, D. S.; Kravić, N. B. Determination of Free Phenolic Acids from Leaves within Different Colored Maize. *J. Serbian Chem. Soc.* **2017,** *82* (1), 63–72. https://doi.org/10.2298/JSC160512104M.

Michaletti, A.; Naghavi, M. R.; Toorchi, M.; Zolla, L.; Rinalducci, S. Metabolomics and Proteomics Reveal Drought-Stress Responses of Leaf Tissues from Spring-Wheat. *Sci. Rep.* **2018,** *8* (1). https://doi.org/doi:10.1038/s41598-018-24012-y.

Mínguez Mosquera, M. I.; Pérez Gálvez, A.; Hornero-Méndez, D. *Pigmentos Carotenoides En Frutas y Vegetales: Mucho Más Que Simples "Colorantes" Naturales*; Madrid, 2005.

Msowoya, K.; Madani, K.; Davtalab, R.; Mirchi, A.; Lund, J. R. Climate Change Impacts on Maize Production in the Warm Heart of Africa. *Water Resour. Manag.* **2016,** *30* (14), 5299–5312. https://doi.org/10.1007/s11269-016-1487-3.

Orians, C. M.; Schweiger, R.; Dukes, J. S.; Scott, E. R.; Müller, C. Combined Impacts of Prolonged Drought and Warming on Plant Size and Foliar Chemistry. *Ann. Bot.* **2019,** *124* (1), 41–52. https://doi.org/10.1093/aob/mcz004.

Ort, R. D. When There Is Too Much Light. *Plant Physiol.* **2001,** *125*, 29–32.

Ozmianski, J.; Kolniak-Ostek, J.; Biernat, A. The Content of Phenolic Compounds in Leaf Tissues Of. *Molecules* **2015,** *20*, 2176–2189. https://doi.org/10.3390/molecules20022176.

Pietrini, F.; Iannelli, M. A.; Massacci, A. Anthocyanin Accumulation in the Illuminated Surface of Maize Leaves Enhances Protection from Photo-Inhibitory Risks at Low Temperature, without Further Limitation to Photosynthesis. *Plant, Cell Environ.* **2002,** *25* (10), 1251–1259. https://doi.org/10.1046/j.1365-3040.2002.00917.x.

Rabêlo, V. M.; Magalhães, P. C.; Bressanin, L. A.; Carvalho, D. T.; Reis, C. O. dos; Karam, D.; Doriguetto, A. C.; Santos, M. H. dos; Santos Filho, P. R. dos S.; Souza, T. C. de. The Foliar Application of a Mixture of Semisynthetic Chitosan Derivatives Induces Tolerance to Water Deficit in Maize, Improving the Antioxidant System and Increasing Photosynthesis and Grain Yield. *Sci. Rep.* **2019,** *9* (1), 1–13. https://doi.org/10.1038/s41598-019-44649-7.

Robles, A. A. C. Sobrevivir Al Estrés: Cómo Responden Las Plantas a La Falta de Agua. In *Biotecnologia*; 2007; Vol. 14, pp 253–262.

Rosales, A.; Agama-Acevedo, E.; Arturo Bello-Pérez, L.; Gutiérrez-Dorado, R.; Palacios-Rojas, N. Effect of Traditional and Extrusion Nixtamalization on Carotenoid Retention in Tortillas Made from Provitamin A Biofortified Maize (Zea Mays L.). *J. Agric. Food Chem.* **2016,** *64* (44), 8289–8295. https://doi.org/10.1021/acs.jafc.6b02951.

Salama, Z. A.; Gaafar, A. A.; Fouly, M. M. El. Genotypic Variations in Phenolic, Flavonoids and Their Antioxidant Activities in Maize Plants Treated with Zn (II) HEDTA Grown in Salinized Media. *Agric. Sci.* **2015,** *06* (03), 397–405. https://doi.org/10.4236/as.2015.63039.

Salinas-Moreno, Y.; Garcia-Salinas, C.; Coutiño-Estrada, B.; Vidal-Martinez, V. A. Variabilidad En Contenido y Tipos de Antocianinas En Granos de Color Azul/Morado de Poblaciones Mexicanas de Maíz. *Rev. Fitotec. Mex.* **2013,** *36* (3-A), 285–294.

Santacoloma, V. L. E.; Enrique, G. J. Interrelación Entre El Contenido de Metabolitos Secundarios de Las Especies Gliricidia Sepium y Tithonia Diversifolia y Algunas Propiedades Físicoquímicas Del Suelo Interrelation between the Content of Secondary Metabolites of Species Gliricidia Sepium A. *Rev. Investig. Agrar. y Ambient.* **2012,** *3* (2), 53–62.

Scrob, S.; Muste, S.; Has, I.; Muresan, C.; Socaci, S.; Farcas, A. Total Content of Carotenoids in Corn Landraces and Their Potential Health Applications. *Bull. UASVM Food Sci. Technol.* **2014,** *73* (2), 55–60. https://doi.org/10.15835/buasvmcn-fst.

Selmar, D.; Kleinwächter, M. Stress Enhances the Synthesis of Secondary Plant Products: The Impact of Stress-Related Over-Reduction on the Accumulation of Natural Products. *Plant Cell Physiol.* **2013**, *54* (6), 817–826. https://doi.org/doi:10.1093/pcp/pct054.

Sepúlveda-Jiménez, G. La Participación de Los Metabolitos Secundarios En La Defensa de Las Plantas. *Rev Mex Fitopatol* **2003**, *21* (3), 355–363.

SIAP. Servicio de Información Agroalimentaria y Pesquera http://www.siap.sagarpa.gob.mx (accessed May 12, 2019).

Signarbieux, C.; Feller, U. Non-Stomatal Limitations of Photosynthesis in Grassland Species under Artificial Drought in the Field. *Environ. Exp. Bot.* **2011**, *71* (2), 192–197.

Silva-Pérez, V.; Gómez-Merino, F. C.; García-Zavala, J. J.; Santacruz-Varela, A.; Burgueño-Ferreira, J.; Palacios-Rojas, N.; Tiessen, A. QTLs Associated to Carotene Content in Maize Leaves (Zea Mays L.). *Agrociencia* **2012**, *46* (4), 333–345.

Soto Mooner, A. L.; Ráez Guevara, L. R.; Robles Calderón, R. El Maíz Morado Como Materia Prima Industrial. *Ind. Data* **2013**, *16* (1), 85–91.

Suwarno, W. B.; Hannok, P.; Palacios-Rojas, N.; Windham, G.; Crossa, J.; Pixley, K. V. Provitamin A Carotenoids in Grain Reduce Aflatoxin Contamination of Maize While Combating Vitamin A Deficiency. *Front. Plant Sci.* **2019**, *10* (January), 1–12. https://doi.org/10.3389/fpls.2019.00030.

Troncoso-rojas, R.; Zamora-bustillo, R. Compuestos Fenólicos y Capacidad Antioxidante Presentes En Tres Variedades de Berenjena Cultivadas En El Valle de Mexicali, Baja California. *IDESIA* **2015**, *33* (3), 17–22.

Ureta, C.; González, E. J.; Espinosa, A.; Trueba, A.; Piñeyro-Nelson, A.; Álvarez-Buylla, E. R. Maize Yield in Mexico under Climate Change. *Agric. Syst.* **2020**, *177* (December 2018), 102697. https://doi.org/10.1016/j.agsy.2019.102697.

Van Der Fits, L.; Memelink, J. ORCA3, a Jasmonate-Responsive Transcriptional Regulator of Plant Primary and Secondary Metabolism. *Science (80-.)*. **2000**, *289* (5477), 295–297. https://doi.org/10.1126/science.289.5477.295.

Vázquez-Carrillo, M. G.; Santiago-Ramos, D.; Figueroa-Cárdenas, J. de D. Kernel Properties and Popping Potential of Chapalote, a Mexican Ancient Native Maize. *J. Cereal Sci.* **2019**, *86* (January), 69–76. https://doi.org/10.1016/j.jcs.2019.01.010.

Vazquez-Olivo, G.; López-Martínez, L. X.; Contreras-Angulo, L.; Heredia, J. B. Antioxidant Capacity of Lignin and Phenolic Compounds from Corn Stover. *Waste and Biomass Valorization* **2019**, *10* (1), 95–102. https://doi.org/10.1007/s12649-017-0028-5.

Vidal, R.; Silva, N. C. de A.; Ogliari, J. B. Old Tools as New Support for on Farm Conservation of Different Types of Maize. *Sci. Agric.* **2020**, *77* (1), 1–9. https://doi.org/10.1590/1678-992x-2018-0091.

Vogel, E.; Donat, M. G.; Alexander, L. V.; Meinshausen, M.; Ray, D. K.; Karoly, D.; Meinshausen, N.; Frieler, K. The Effects of Climate Extremes on Global Agricultural Yields. *Environ. Res. Lett.* **2019**, *14* (5). https://doi.org/10.1088/1748-9326/ab154b.

Wahid, A. Physiological Implications of Metabolite Biosynthesis for Net Assimilation and Heat-Stress Tolerance of Sugarcane (Saccharum Officinarum) Sprouts. *J. Plant Res.* **2006**, *120* (2), 219–228.

Zhang, Q.; Gonzalez de Mejia, E.; Luna-Vital, D.; Tao, T.; Chandrasekaran, S.; Chatham, L.; Juvik, J.; Singh, V.; Kumar, D. Relationship of Phenolic Composition of Selected Purple Maize (Zea Mays L.) Genotypes with Their Anti-Inflammatory, Anti-Adipogenic and Anti-Diabetic Potential. *Food Chem.* **2019**, *289* (November 2018), 739–750. https://doi.org/10.1016/j.foodchem.2019.03.116.

Zhang, T. J.; Zheng, J.; Yu, Z. C.; Huang, X. D.; Zhang, Q. L.; Tian, X. S.; Peng, C. L. Functional Characteristics of Phenolic Compounds Accumulated in Young Leaves of Two Subtropical Forest Tree Species of Different Successional Stages. *Tree Physiol.* **2018,** *38* (10), 1486–1501. https://doi.org/10.1093/treephys/tpy030.

Zhao, D.; Raja Reddy, K.; Kakani, V. G.; Read, J. J.; Carter, G. A. Corn (Zea Mays L.) Growth, Leaf Pigment Concentration, Photosynthesis and Leaf Hyperspectral Reflectance Properties as Affected by Nitrogen Supply. *Plant Soil* **2003,** *257*, 205–217. https://doi.org/10.1023/A.

Index

For Product Safety Concerns and Information please contact our EU
representative GPSR@taylorandfrancis.com
Taylor & Francis Verlag GmbH, Kaufingerstraße 24, 80331 München, Germany